I0822993

## Praise for *The Book of Cannabis*

"Every so often, a book comes along that is so intelligently written, lucidly organized, and comprehensively fascinating that it becomes *the* signature readable reference work for its field. *The Book of Cannabis* is such a book . . . If you have any interest whatsoever in cannabis, *The Book of Cannabis* by Jeremy Narby is a must-read—as entertaining as it is informative, and as fascinating as it is thorough."

—Jordan Gruber, JD, coauthor of *Microdosing for Health, Healing, and Enhanced Performance*

"Jeremy Narby has done it again—made a major contribution to our understanding of plant medicine. This time he presents a comprehensive history of cannabis. An objective and comprehensive approach to a psychoactive plant that is currently spreading its roots throughout our culture."

—Rachel Harris, PhD, author of *Listening to Ayahuasca* and *Swimming in the Sacred*

"Entertaining and profound, Narby gives us a new standard work that deepens humanity's long love affair with cannabis—indispensable for anyone who thinks they already know this plant."

—Thomas Grube, filmmaker

"A complete and thorough book that's a joy to read. If you're new to cannabis, it serves as an excellent introduction. And if you consider yourself well-versed in the subject—as I did—you'll be surprised by how much there is still to learn about this remarkable plant that has been part of human history for millennia."

—Jerónimo Mazarrasa, program director at the International Center for Ethnobotanical Education, Research, and Service

"You couldn't ask for a better guide to the intriguing history of humanity's relationship with the powerful plant. Trained as an anthropologist at Stanford University, Jeremy Narby has gone on to add the skills of investigative journalism, science writing, and deep historical research to his tool kit as a storyteller at the frontiers of science, Indigenous knowledge, and the agency and intelligence of plants. You won't find a more compelling, nonfiction, fact-based narrative of cannabis's mind-bending and world-altering story anywhere."

—Jon Christensen, director of the Laboratory for Environmental Narrative Strategies at the Institute of the Environment and Sustainability, UCLA

# THE BOOK OF CANNABIS

ALSO BY JEREMY NARBY

*The Cosmic Serpent*

*Shamans Through Time*
(coedited with Francis Huxley)

*Intelligence in Nature*

*Plant Teachers*
(cowritten with Rafael Chanchari Pizuri)

# THE BOOK OF CANNABIS

## The History and Future of the Plant and the Drug

JEREMY NARBY

ST. MARTIN'S ESSENTIALS
NEW YORK

The information in this book is not intended to replace the advice of the reader's own physician or other medical professional. You should consult a medical professional in matters relating to health, especially if you have existing medical conditions, and before starting, stopping, or changing the dose of any medication you are taking. Individual readers are solely responsible for their own health care decisions. The author and the publisher do not accept responsibility for any adverse effects individuals may claim to experience, whether directly or indirectly, from the information contained in this book.

First published in the United States by St. Martin's Essentials,
an imprint of St. Martin's Publishing Group

*EU Representative:* Macmillan Publishers Ireland Ltd, 1st Floor, The Liffey Trust Centre,
117–126 Sheriff Street Upper, Dublin 1, D01 YC43


Printed in the United States of America. For information, address
St. Martin's Publishing Group, 120 Broadway, New York, NY 10271.

www.stmartins.com

Library of Congress Cataloging-in-Publication Data

Names: Narby, Jeremy author
Title: The book of cannabis : the history and future of the plant and the drug / Jeremy Narby.
Description: First edition. | New York : St. Martin's Essentials, 2026. | Includes bibliographical references and index.
Identifiers: LCCN 2025045552 | ISBN 9781250436665 hardcover | ISBN 9781250436672 ebook
Subjects: LCSH: Cannabis—History—Popular works | Cannabis—Therapeutic use—Popular works | Marijuana—History—Popular works | Marijuana—Therapeutic use—Popular works
Classification: LCC SB295.C35 N37 2026
LC record available at https://lccn.loc.gov/2025045552



First Edition: 2026

10 9 8 7 6 5 4 3 2 1

*To Corinne*

# Contents

# Introduction

# Navigating the Narrative

Compared to other plants, *Cannabis* has experienced a turbulent journey through recent human history. It thrived for millions of years before our appearance on Earth, protecting, restoring, and detoxifying disturbed soils. When humans began domesticating plants around 12,000 years ago, *Cannabis* was among the first they cultivated. People valued its nutritious seeds, flexible and durable fibers, and psychoactive and medicinal flowers and resin. For millennia, *Cannabis* served as a Swiss Army knife plant—multipurpose and widely used.

More recently, some people began to decry the psychoactive substance, or drug, derived from the plant, which they associated with various harms, sometimes justifiably and other times less so. During the twentieth century, some authorities described *Cannabis* as "the weed that kills people" and proposed its extermination.[1]

In 1961, the United Nations adopted an international treaty prohibiting "the flowering or fruiting tops of the *Cannabis* plant (excluding the seeds and leaves when not accompanied by the tops) from which the resin has not been extracted." This treaty classified *Cannabis* among the world's most dangerous drugs, next to heroin and two other semisynthetic opioids created by chemists. No other plant species has had its reproductive organs targeted by international law.

After the worldwide implementation of this treaty, billions of individual plants were sprayed with herbicides or eradicated in large-scale military operations, while countless individuals spent years in prison for growing, using, or selling the plant and its products.

People have also celebrated *Cannabis*, giving it affectionate nicknames like "shrub of understanding" or commending it as an all-purpose remedy for those unable to afford other medications or intoxicants. Throughout the twentieth century, musicians, artists, and writers praised *Cannabis* for its capacity to inspire creativity and enhance the pursuit of knowledge. Rastafarianism, a religious movement born in Jamaica, regards the plant as a means of promoting spiritual awareness and connection to the divine.

During the 1970s, aficionados in many countries began to brandish the plant's leaf as a symbol of protest against its illegal status. *Cannabis* became "the world's most recognizable, notorious, and controversial plant." Meanwhile, global prohibition hindered scientific research, resulting in a knowledge gap. The "shrub of understanding" itself became poorly understood. But this did not stop people from expressing strong opinions about it, whether positive or negative.

Prohibition also led illicit cultivators to grow *Cannabis* indoors under electric lights. Gorged with chemical nutrients and drenched in artificial light, the plant responded with accelerated growth, adapting seamlessly to enclosed environments such as attics, basements, and closets.

*Cannabis* also lent itself to the work of plant breeders operating outside the law, whose main goal was to increase the levels of its primary psychoactive compound, THC, with each new generation. Some hybrid plants now have THC levels that are ten to twenty times higher than they would be if allowed to reproduce freely. The dried flowers of these hybrids can contain more than 30% THC—a concentration that no other plant species has matched for a single active substance.

* * *

*Cannabis* is known by many different names, adding to its complexity. Some of these names are used formally, such as *hemp* and *marijuana*, while others are slang terms, including *weed*, *grass*, *reefer*, *gage*, *pot*, *herb*, *ganja*, *skunk*, and *sinsemilla*. In this book, I use the term "cannabis" for simplicity. When referring to the plant, I capitalize and italicize the word, writing *Cannabis* in line with its Latin scientific designation, *Cannabis sativa* L. When discussing the drug or substance, I write "cannabis" in lowercase and without italics. However, the distinction between plant and substance can sometimes blur, as the unprocessed flowers and leaves of *Cannabis* can also be considered a drug. This creates a two-fold complication: The word "cannabis" refers to both the plant and the substance, and these two entities occasionally coincide. In material reality, *Cannabis*/cannabis can exist simultaneously as both a plant and a drug.[2]

In recent years, cannabis has been reinstated in the fields of medicine and scientific research and has even been re-legalized in some places for "nonmedical" or "recreational" use. Research has revealed, among other things, that the plant possesses remarkable abilities, such as producing an array of unique compounds, absorbing intense sunlight more efficiently than any other plant, and extracting toxins from its environment.

To tell the story of this singular plant, I have drawn upon botany, history, geography, chemistry, and biology, in addition to political, legal, and cultural studies. Chapters 1–9 discuss the past and proceed in chronological order, with some overlapping time frames. Chapters 10–15 focus on the present and explore what science has discovered in recent decades about the plant and the drug, along with significant questions that remain unanswered. Chapter 16 offers a glimpse into the future. Most paragraphs include an endnote that provides sources and supplementary information. Readers can navigate the text as they choose and use the index to get information about a specific topic or find out more about events in the plant's history. However, reading the chapters in numbered order shows how the plant's unusual fate unfolded and evolved.

In this book, I take the view that cannabis the drug is inseparable from *Cannabis* the plant. They share a name and physical reality, and people have long perceived them as one and the same. To tell the story of this controversial plant is also to tell the story of the drug derived from it. Accordingly, this account considers how THC moves through human bloodstreams, stimulates neurons, and lingers for days or even weeks in the body's fatty tissues. Several chapters focus on these effects because hundreds of millions of people have consumed cannabis to experience its psychoactive effects. I do not suggest that *Cannabis* can be reduced to a single compound; rather, I propose that the plant's story can only be fully understood in light of what happens when its main active component interacts with the human body.

* * *

When authors describe their personal experiences with a psychoactive plant, they sometimes share valuable insights into its effects. In this book, however, I have mostly kept myself out of the narrative, concentrating on the plant itself. Still, readers may want to know a little about my relationship with *Cannabis* before considering my account of its story.

Nearly forty years ago, I started growing *Cannabis* in my garden to see if consuming the plant could enhance my professional writing. I discovered that smoking small amounts of cannabis allowed me to reread my own words with detachment, as if someone else had written them. This proved valuable because I tended to be overly attached to my words. Inspired by cannabis, I could identify words that didn't feel right or lacked clarity. I could also recognize missing elements in the text due to my limited knowledge. For me, cannabis worked as a "plant editor."

At the time, cannabis was illegal almost everywhere, so I kept quiet about my work method. Discussing it would have meant confessing to a crime and could also have been understood as encouraging others to use cannabis, which was the last thing I wanted to do. The substance worked for me, but not for several people close to me. It was already clear that cannabis isn't everybody's cup of tea.

Thirty years passed, and I continued using the same method. I wrote about other psychoactive plants, such as ayahuasca and tobacco, without once mentioning the one I used to enhance my writing. Then, a few years ago, the idea of writing about *Cannabis* popped into my mind. Perhaps its recent re-legalization in certain countries reduced the stigma preventing open discussions about it. Or perhaps I had reached an age where I cared less about my reputation. In any case, I rolled up my sleeves and started reading about the plant and the drug.

The subject proved much more complex than anticipated. While exploring it as thoroughly as possible, I tried to keep the text easy to read.

I have written this book not as an aficionado seeking to promote or rehabilitate *Cannabis* but as a friend wishing to tell its unvarnished story—the good, the bad, and the ugly adventures of the world's most controversial plant. I hope you enjoy reading it as much as I enjoyed writing it.

# 1

# Useful Weed, Enigmatic Plant

A *Cannabis* plant can sprout from a seed smaller than a pea and grow into a large bush in just a few months. A mere herb, it grows so quickly that it produces biomass faster than any tree.[1] *Cannabis* plants also die rapidly. Male plants perish shortly after releasing their pollen, while female plants die several months later once their seeds have matured. Under natural conditions, most *Cannabis* plants will decompose into compost within a year, enriching the soil for future vegetation.[2]

In the wild, *Cannabis* grows in recently disturbed soils lacking plant cover, such as alluvial deposits in river valleys or scars on the vegetation caused by landslides, floods, or grazing animals. It thrives in nitrogen-rich soils, particularly where large animals have deposited excrement. While all plants require nitrogen to grow, most cannot tolerate soils overly rich in nitrogen. *Cannabis*, on the other hand, can absorb large amounts of this nutrient that would otherwise leach away.[3]

*Cannabis* belongs to the "pioneer plants," the first to grow back on soils exposed by naturally occurring disturbances. These plants help protect bare soil from erosion and enrich it with organic matter from their decomposition. By growing and dying quickly, they jump-start vegetal restoration.[4]

Like many pioneer species, *Cannabis* is a hardy plant that tolerates poor soils, grows rapidly, and produces abundant seeds. However, unlike some of these plants, which *contribute* nitrogen to the soil through certain bacteria in their roots, *Cannabis extracts* large amounts of nitrogen from the ground and returns it several months later when it dies and decomposes. *Cannabis* also detoxifies the disturbed soils where it grows because its long and deep roots allow it to absorb large quantities of naturally occurring heavy metals like copper, nickel, lead, cadmium, and zinc.[5]

The habitats that *Cannabis* has evolved to protect and repair are relatively rare, but this has not stopped the plant from thriving for a long time. Scientists have discovered fossil *Cannabis* pollen dating back approximately nineteen million years in Central Asia—the region they identify as the plant's likely place of origin—and about six million years ago in Europe. This suggests that *Cannabis* spread naturally across Eurasia, where it has been protecting, enriching, and detoxifying disturbed soils since long before humans appeared. In comparison, anatomically modern humans have been around for approximately 300,000 years.[6]

When early modern humans began wandering around Eurasia, they helped *Cannabis* spread to new territories. As nomads who hunted animals, gathered food plants, and moved on, they tended to leave behind campsites with disturbed soils and heaps of organic waste. *Cannabis* flourished in such circumstances, leading botanists to refer to it as a "camp-following plant."[7]

When humans began practicing agriculture about 12,000 years ago, they became large-scale soil disruptors. Most agriculture involves clearing forests and vegetation, plowing the soil, and fertilizing it, thereby setting the conditions that trigger the appearance of pioneer plants, such as *Cannabis*.

Pioneer plants have long provided essential ecosystem services, yet people, including scientists, often refer to them as weeds. The word *weed* means "a wild plant growing where it is not wanted and competing with cultivated plants." When left to its own devices, *Cannabis* seems to be just that—a plant that grows spontaneously and unwanted on vacant grounds and construction sites and in ditches and disturbed soils.[8]

Pioneer plants, or weeds, are known to grow faster than cultivated plants. They absorb as much sunlight and nutrients as possible and produce large quantities of fertile seeds. Agriculturalists understandably view them as unwelcome competition for their crops. But pioneer plants are just doing their job, jump-starting the vegetal restoration of bare soils.

*Cannabis*, like all pioneer plants, is known for its plasticity: Individual plants within the species can take on radically different forms and adapt to new environments with unusual flexibility. In inhospitable conditions, a *Cannabis* plant can survive as a dwarf shrub that produces just a few seeds, while in more favorable circumstances, it can grow into a large bush and yield tens of thousands of seeds. *Cannabis* has adapted to tropical, subtropical, temperate, and subarctic conditions. It can thrive from sea level to altitudes above three thousand meters. It can keep its roots shallow if the subsoil is waterlogged or send its taproot deep into dry soils in search of water. It can survive in the shade, although it prefers full sunlight. During their short life spans, *Cannabis* plants can change their sex from female to male or vice versa, and some plants can manifest both sexes simultaneously to self-pollinate. Some *Cannabis* plants produce large amounts of psychoactive resin, while others produce hardly any at all. And yet, all living *Cannabis* plants can interbreed and yield viable offspring.[9]

The malleability and vigor of pioneer plants make them particularly well-suited for human cultivation and selection. Early agriculture often involved the domestication of these fast-growing herbs. Most major crops—wheat, rice, barley, millet, sorghum, sunflower, and sugarcane—are closely related to wild pioneer herbs. This is also true for *Cannabis*, one of the world's oldest cultivated plants. As a fast-growing species capable of producing numerous easily sown seeds, *Cannabis* was preadapted to cultivation.[10]

The origin of crop domestication remains subject to debate. In the case of *Cannabis*, recent genetic studies indicate that its cultivation began approximately 12,000 years ago in what is now modern-day China. *Cannabis* provides seeds rich in digestible proteins and nutritious oils, strong and flexible fibers suitable for cordage and cloth, and flowers, leaves, and roots with medicinal virtues—food, oil, fiber, and medicine, all in one—a multipurpose plant.[11]

No one knows when or where people began cultivating *Cannabis* for its psychoactive properties. The plant seems to have a natural tendency to exude psychoactive substances in sunny and warm subtropical locations. Historically, this phenomenon has tended to occur in regions south of 30 degrees north latitude, although the reasons for this are debated. Research shows that *Cannabis* increases its production of psychoactive cannabinoids based on light intensity, and these substances can absorb harmful UV radiation to some extent.[12]

Early agriculturalists who cultivated *Cannabis* in subtropical locations were bound to have noticed the psychoactivity of mature female plants. When female *Cannabis* flowers are ripe, they glisten and reflect sunlight, and their resin is fragrant and sticky. Birds manifest a strong interest in the nutritious seeds contained in these flowers. When humans harvest psychoactive *Cannabis* flowers, they may experience light-headedness, possibly caused by the absorption of cannabinoids through the skin of the fingers.[13]

The true psychoactivity of *Cannabis* is only revealed when the flowers are heated, either by cooking them into a drink or food or by burning them and inhaling the smoke. In its raw form, the plant contains THCA, an acidic and nonpsychoactive compound that converts into THC under heat and other forms of energy. THC is the primary psychoactive substance of *Cannabis*. Though no one knows where or when people began heating and consuming cannabis, genetic studies suggest that deliberate selection and cultivation of psychoactive varieties began in South Asia about 4,000 years ago.[14]

The archeological evidence for "ritual cannabis smoking" is limited. The earliest directly dated and scientifically confirmed evidence dates back 2,500 years. It comes from the tombs of an ancient cemetery in Western China, where archeologists found burnt *Cannabis* resin in braziers made of juniper wood. Chemical analysis of the residue revealed the presence of cannabis with "high levels of psychoactive compounds" and "higher THC than typically found in wild plants." This demonstrated that people burned psychoactive cannabis during mortuary ceremonies, but not that they consumed the smoke. Perhaps they used the plant's resin as incense to produce a solemn atmosphere—a possibility reinforced by juniper wood's propensity to release a pine-like aroma when burned.[15]

Researchers in northwestern China also found large quantities of psychoactive *Cannabis* in graves dating back 2,700 years. One grave contained the skeletal remains of a man buried with close to two pounds of potent female *Cannabis* flowers—suggesting that people understood the full range of the plant's properties. The deceased man's grave also contained bows and arrows, a mortar for pulverizing cannabis, and a harp. The researchers concluded that he was a "shaman" and speculated that the gift of cannabis "may have been to enable him to continue his profession in the afterlife." However, getting buried with a large quantity of cannabis, some paraphernalia, and a musical instrument does not make one a shaman. In fact, this archaeological discovery did not yield conclusive evidence about how people used the plant outside of a funerary context.[16]

The earliest written mentions of *Cannabis* are disputed. An ancient Egyptian text written in hieroglyphs about 4,350 years ago lists a plant that might be *Cannabis*, but the hieroglyphs could also refer to other plants. Likewise, a text from ancient India, the Atharva Veda, written perhaps 3,000 years ago, refers to *bhang* as one of five herbs used to "release us from anxiety." However, Sanskrit scholars argue that *bhang* could have referred to other plants thousands of years ago—whereas nowadays the term is used to refer specifically and exclusively to *Cannabis*. And Chinese tradition holds that the legendary emperor Shennong, who lived about 4,800 years ago, mentioned *Cannabis* in his medical compendium as a treatment for a wide range of ailments, noting that if used in excess, it produced hallucinations, and if used over a long period, it made one "communicate with the spirits." Unfortunately, the original version of the text did not survive, and the oldest surviving copy dates back approximately 1,900 years.[17]

The earliest uncontested references to *Cannabis* come from ancient Mesopotamia, where 2,600-year-old clay tablets bearing cuneiform inscriptions mention a plant called *qunubu* or *qunnabu*. The resin of this plant was used as an all-purpose medicine and as incense during rituals because its "aroma was pleasing to the Gods." Several Mesopotamian tablets contain recipes for pounding and straining *Cannabis* plants to extract their resin and make incense. The tablets also show local doctors treating patients by exposing them to the smoke of burning *Cannabis* resin.[18]

When considering prehistoric or early historical times, distinguishing between medical, ritual, and psychoactive cannabis use can be challenging. A text from about 2,400 years ago by the ancient Greek historian Herodotus is a case in point. Describing a plant that Scythian nomads from Central Asia called "cannabis," which they consumed in fumigation tents set up after funerals, Herodotus wrote: "*Cannabis*, a plant that grows in Scythia, is very similar to flax, only much coarser and taller. Some grows wild; some is produced by cultivation. . . . The Scythians take some of this *Cannabis* seed and, crawling under the felt coverings, throw it upon the red-hot stones, where it burns like incense and produces a vapor so thick that no Grecian steam bath would surpass it, and the Scythians, delighted with the vapor bath, howl like wolves. This vapor serves them instead of washing, for in fact, they do not wash their bodies at all in water."[19]

Some historians interpret this passage by Herodotus as a description of ritualized psychoactive cannabis use. Others disagree, arguing that *Cannabis* seeds are not psychoactive and that burning them does not produce thick smoke or vapor. Burning seeded *Cannabis* flowers would produce thick and potentially psychoactive smoke, but Herodotus said nothing about flowers. Perhaps he only mentioned seeds because he based his story on the accounts of others, and some details got lost in the telling.[20]

One thing is sure: Herodotus's story introduced the word "cannabis" to the Greek language. The term would later be adopted in Latin and other European languages. However, the origin of the Scythian word transcribed by Herodotus remains a topic of scholarly debate. Indeed, the Scythian language itself is essentially unknown.[21]

* * *

Many early texts that refer to *Cannabis* in some detail have a medical focus. Physicians in Greece, China, Iran, and Egypt have mentioned the plant for its anti-phlegmatic, antiemetic, anti-inflammatory, analgesic, and aphrodisiac effects, as well as for its use as an anesthetic, appetizer, digestive, and life-prolonging agent. For example, the first-century Greek physician Dioscorides wrote that the juice extracted from green *Cannabis* plants soothed earaches, and the second-century Chinese doctor Hua Tuo used cannabis mixed with datura and wine as an anesthetic to perform surgery. In the ninth century, Iranian physician Sabur ibn Sahl wrote in his book of medical preparations that cannabis treated migraine and aching pains, and his contemporary, Arab physician al-Kindi, stated that cannabis acted as a muscle relaxant, while also noting that it could produce "senseless talk."[22]

Early texts that unmistakably refer to psychoactive *Cannabis* include several poems in Arabic from the 1100s and 1200s that refer to it as the "shrub of understanding" and other affectionate nicknames. An Indian text from around 1450, *Rajanighantu*, lists "speech-giving" and "inspiring of mental power" among the plant's effects. Another Indian text, *Dhurtasamagama*, written around 1500, notes that cannabis "corrects derangements of humors and produces a healthy appetite, sharpens the wit, and acts as an aphrodisiac."[23]

There are clear indications that psychoactive cannabis reached Swahili-speaking people on the eastern coast of Africa by the thirteenth century, brought by sea traders from the Indian subcontinent via Arabia. The Swahili word for cannabis, *bangi*, traces back to the Hindi *bhang*. Archaeologists have unearthed fourteenth-century water pipes in Ethiopia containing THC residues, demonstrating that they served to smoke psychoactive cannabis.[24]

There is little doubt that East Africans initiated the practice of smoking cannabis in a pipe. Before this innovation, people either cooked cannabis and ingested it or burned it and inhaled the fumes, like breathing incense—but they probably did not use smoking devices. Smoking cannabis in a pipe was an important innovation because it gave consumers greater control over the drug's psychoactive effects.[25]

People have cultivated psychoactive *Cannabis* in Egypt since at least the 1200s. Over the following centuries, African traders transported the plant down the east coast of Africa and across the southern half of the continent to the west coast. European observers documented psychoactive cannabis use in southern Africa in 1713 and western Africa in 1803, although its presence likely predated these observations.[26]

Once psychoactive *Cannabis* reached the west coast of Africa, it traveled across the Atlantic with enslaved African people taken by force on European ships to Brazil and Central America. Biogeographer Chris Duvall comments: "Since 1500, drug *Cannabis* has dispersed alongside labor underclasses because the drug enhances the ability of workers to endure lives of physically demanding but mentally dulling tasks, constant occupational hazards, poor nutrition, and exposure to infectious diseases. At least the *idea* that marijuana can be good for hard laborers has sustained the plant's migrations."[27]

Around the time that psychoactive *Cannabis* headed west across the Atlantic from Africa, it also traveled in the opposite direction, east from India, reaching Malaysia and Indonesia. By the late 1700s, people all across South Asia, the Middle East, and Africa, and some people of African descent in the Americas, were consuming psychoactive cannabis, even though most people in Europe and North America had not heard of it.[28]

* * *

Since the Bronze Age, people in Europe cultivated *Cannabis* for its fibers and seeds. In Roman times, it was understood that the plant possessed medicinal virtues, particularly in its roots. But when Europeans spread out around the world after 1500 and came upon *Cannabis* in South Asia and Africa, they were surprised to find that local people cultivated it not so much for its fibers as for its resin, which they consumed as a psychoactive substance. For instance, in 1578, Portuguese doctor and natural historian Cristóval Acosta described the "*bengue*" plant of India as similar in appearance to European *Cannabis*, except that Indians used it "to help in the sexual act and increase appetite, to forget their work and sleep without thoughts; or to enjoy sleeping with a variety of dreams and illusions; others to be drunk and act like funny jesters; others to charm women; officers and captains to forget their labors and to sleep without thoughts."[29]

To Europeans, *Cannabis* was a botanical enigma. How could the same plant produce esteemed fibers in temperate Europe while yielding no noticeable psychoactive substance, and do the opposite in tropical and subtropical lands, where it produced psychoactive resin and fibers of lesser quality?

European scientists grappled with this question during the second half of the eighteenth century. The species assignation of *Cannabis* was contentious from the start. In 1753, Swedish naturalist and taxonomist Carl Linnaeus included "*Cannabis sativa* L." as a single species in his encyclopedic work on plant species—with "sativa" meaning "cultivated" in Latin and "L." referring to Linnaeus himself, as the person who named the plant—even though German physician and botanist Leonhart Fuchs had named it "*Cannabis sativa*" more than two hundred years beforehand.

Further controversy arose in 1783 when French biologist Jean-Baptiste Lamarck suggested that there were, in fact, two species of *Cannabis*. One was *Cannabis sativa*, which Lamarck described as "an extremely interesting plant due to its uses and that one cultivates abundantly in Europe for the immense use we have for the filaments of its stalks and seeds. . . . Everybody knows that the most important product of this plant consists of the filaments that one takes from its bark, and with which, after appropriate preparations, one makes canvas and ropes, the utility of which is sufficiently well known." The other was a new species that Lamarck called *Cannabis indica* (*indica* meaning "from India") based on a sample plant that a friend had brought to him from India. The plant seemed quite different to Lamarck; it was shorter and had narrower leaves; its stalk "prevents it from providing filaments similar to those of the above-mentioned species from which we derive such great use. Its odor is strong and somewhat similar to the odor of tobacco. The main virtue of this plant consists in reaching the head, deranging the brain, bringing a kind of intoxication that makes one forget sadness, and gives a strong cheerfulness. . . ." Lamarck wanted to classify this intoxicating variety as a separate species. However, many scientists disagreed, arguing that *Cannabis* was a single species because all its varieties can interbreed and produce fertile seeds.[30]

Scholars still debate why "southern" *Cannabis* historically produced psychoactive resin and "northern" plants much less so. To this day, there is no consensus on the subject. One school of thought holds that human cultivation and selection have made the difference over the last 10,000 years. Another view proposes that the primary transformation occurred over geological time when the Indian subcontinent collided with the rest of Asia, causing the uplift of the Himalayas. *Cannabis* plants growing naturally in the area would have had to adapt to the high-elevation sunlight, which has one of the highest UV radiation levels on the planet. This gradual geological shift may have favored the evolution of *Cannabis* plants with an enhanced capacity to produce cannabinoids capable of absorbing UV radiation, including THC.[31]

The species assignation of *Cannabis* is another subject that remains debated. However, most contemporary researchers and textbooks have come to agree that *Cannabis* is a single species, not two or three.[32]

During the late twentieth century, human breeders brought further complexity to the question of *Cannabis* taxonomy by interbreeding different varieties to create hybrid plants. By combining "southern" psychoactivity and "northern" early maturation, these hybrid plants allowed people to grow psychoactive *Cannabis* that reached maturity at temperate latitudes. At present, most cannabis on the market comes from hybrid plants. In addition, common parlance has come to associate "sativa" with stimulating highs and "indica" with relaxing ones, and contemporary consumers commonly use these terms without any relation to their original botanical meaning.[33]

It now seems wise to avoid using "sativa" and "indica" to distinguish between *Cannabis* varieties. As cannabis expert John McPartland recently stated: "Categorizing cannabis as either 'Sativa' or 'Indica' has become an exercise in futility. Ubiquitous interbreeding and hybridization render their distinction meaningless."[34] This means that the external features of a contemporary *Cannabis* plant, such as its height, branching, or leaf shape, reveal nothing about its biochemical content. Human crossbreeding has made the psychoactive potential of *Cannabis* invisible to the naked eye.[35]

One might think that genetic studies would have established the identity of *Cannabis* as a species long ago. However, the plant's global prohibition in the twentieth century hindered such studies. Fortunately, this situation has begun to change. A recent study based on a worldwide sampling of plants found that *Cannabis* has an astonishingly variable genome that bears evident traces of human selection: Plants selected for their psychoactive resin have lost genes related to fiber production, while those chosen for their fibers have lost genes associated with the production of psychoactive substances. Yet all living *Cannabis* plants can interbreed.[36]

*Cannabis* is a versatile plant that has adapted to many different environments. Humans have long used it for a range of purposes. However, its genetic variability and plasticity continue to confound taxonomists and geneticists, making it a tricky plant to pin down.

# 2

# *Marijuana* and *Ganja* Go Traveling

The history of psychoactive *Cannabis* suffers from a general lack of documentation, as most people in historical times who produced written documents did not consume the plant or take much interest in those who did. As a result, the details of how *Cannabis* spread worldwide over the last few centuries remain shrouded in mystery.

Consider the plant's arrival in the Americas, where it was previously unknown. *Cannabis* seeds likely arrived on European ships sometime after 1500, but no one knows when or where this occurred or who transported them. Historians suggest that enslaved people taken from western Central Africa to Brazil may have been the first to grow psychoactive *Cannabis* in the Americas. However, no European observer interviewed enslaved Africans in Brazil or documented the living conditions on the plantations where they were forced to work. The evidence, in this case, is mainly linguistic: Several commonly used terms for cannabis in present-day Brazil—*maconha*, *maruamba*, *riamba*, *diamba*—are words for the plant in the Bantu languages spoken by the originally enslaved people.[1]

The earliest undisputed documentation of cannabis use in Brazil—and the Americas—dates to 1830, when the municipality of Rio de Janeiro published an edict banning its use and commerce, especially by enslaved Afro-Brazilians. This ban made Rio the first place in the Americas to prohibit cannabis.[2]

During the 1800s, people of African descent cultivated *Cannabis* in different parts of Brazil, usually smoking it in water pipes, as was the custom in much of Africa. However, no evidence exists that cannabis use spread from Brazil to other parts of South or Central America. Nor did the Indigenous peoples of Brazil take up cannabis in any significant way—perhaps because they already had an array of powerful psychoactive plants at their disposal.[3]

Beyond Brazil, enslaved people from western Central Africa likely introduced *Cannabis* to the Caribbean region. Here, too, the evidence is mainly linguistic, as variations of the Bantu word *mariamba*—meaning "some cannabis to smoke"—have persisted until now in several Spanish-speaking Caribbean countries. For example, *marimba* means cannabis in Colombian Spanish and Cuban Spanish, and *marihuana* is possibly a Mexican Spanish mispronunciation of the Bantu word.[4]

Scholars have long debated the origin of the word *marihuana*, which they have tended to consider as a "Mexican term of uncertain origin." It is true that *marihuana* first appeared in print in Mexico in two separate texts published in 1846. One was a compendium of Mexican medicinal plants listing "*mariguana*" as a plant with "narcotic leaves" identified botanically as "*Canavis indicus.*" The other was a newspaper article about a "rare illness" among Mexican soldiers who had smoked "an herb called *marihuana*," which caused them to become "weak, languid, and stunned." The word *marihuana* went on to become the common term for psychoactive cannabis in Mexico during the second half of the nineteenth century and beyond.[5]

Scholars have proposed various origins for this apparently Mexican word, including Chinese, Arabic, Aztec, Quechua, Portuguese, and Spanish. Most often, they have ignored Bantu as a possible origin, even though *marihuana* fits into "an Atlantic-wide pattern in which traceably African loanwords name the plant," as biogeographer Chris Duvall put it. During the height of the slave trade, between 1760 and 1860, historians calculate that some three million people were taken by force from western Central Africa and shipped to different locations in South and

Central America, including ports on the Atlantic side of Mexico, such as Veracruz and Campeche. As the last recorded slave ships arrived before 1800, enslaved Bantu speakers may have introduced *mariamba* smoking to Mexico sometime during the late 1700s. This Bantu word may then have made its way into local spoken Spanish in the mispronounced form of *mariguana* or *marihuana,* and when it ended up in print, it became a "Mexican" term.[6]

Scholars will undoubtedly continue debating how the word *marihuana* appeared seemingly out of the blue in mid-nineteenth-century Mexican Spanish. The term's possible Bantu origin will remain a hypothesis for some time. However, there is no question that the word *marihuana* passed from Mexican Spanish into American English during the early 1900s, where it became *marijuana* and subsequently entered many other languages worldwide.

* * *

Europeans did not document cannabis smoking by the African people they enslaved and transported to the Americas, but they paid considerable attention to tobacco smoking by Indigenous Americans. Europeans initially knew nothing about tobacco, a plant indigenous to the Americas. The first Europeans to witness tobacco consumption were the Spanish crew members of Christopher Columbus's initial expedition in 1492. After exploring the island of Cuba, these Spanish sailors reported that the natives rolled specific leaves into "cylinders they call *tobacos*," lighting one end with burning coals. On the other, "they suck the smoke or sip it, or take it in with their breath, and it numbs the flesh, and almost causes drunkenness, and with this, they say they do not feel tiredness." This initial Spanish-language report described inhaling smoke as "sucking" and "sipping" because no fifteenth-century European consumed smoke in this way nor used the verb "to smoke" in the sense of inhaling smoke.[7]

In most places where European explorers landed along the coasts of the Americas, they met Indigenous people who smoked tobacco. Europeans soon took to smoking the plant, and European sailors introduced tobacco smoking in Europe and wherever they traveled. The Portuguese, in particular, used American tobacco as a trade item along the coasts of Africa and Asia, where they had ports of call. Historians estimate that Portuguese sailors introduced tobacco to India, China, and Japan between 1590 and 1610. Meanwhile, the Spanish established a transpacific trade route between Mexico and the Philippines, where they also introduced the plant.[8]

People in Africa did not need to learn to smoke because they already used water pipes and dry pipes to consume cannabis and other plants. However, there is little evidence that people in Asia had experienced smoking before Europeans introduced tobacco from the Americas.[9]

In India, people at all levels of society took up smoking tobacco in the early 1600s, mainly using coconut-based water pipes called *hookahs*. The Portuguese may have brought these pipes from Africa when they introduced tobacco to India. Still, there is no direct evidence, and scholars continue to debate the geographic origin of water pipes.[10]

Sometime during the 1600s, people in India began adding cannabis to the tobacco they smoked. According to English mariner Thomas Bowrey, who witnessed people along the Bengalese coast using cannabis in the 1670s: "They study many ways to use it, but not one of them fails to intoxicate them to admiration. Sometimes, they mix it with their tobacco and smoke it, a very speedy way to be besotted; other times, they chew it, but the most pleasant way of taking it is [to drink it]."[11]

People in India had a long tradition of ingesting cannabis as a beverage called *bhang* or as female flowers called *ganja*, which they roasted and chewed and occasionally consumed through smoke inhalation. Ingestion of cannabis is slow-acting and challenging to dose, and inhaling the smoke of burning cannabis, as one might breathe in incense fumes, is inefficient because most of the smoke escapes into the air. In contrast, smoking cannabis in a pipe is an efficient, fast-acting, and easy-to-dose way of using the drug. Nevertheless, Indian cannabis users were slow to adopt this new form of consumption. Evidence suggests they did not commonly do so before the 1800s. Not only were Indians used to consuming cannabis in other ways, but the popularity of tobacco smoking at all levels of Indian society may have overshadowed the idea of smoking other plants.[12]

Indians who could not afford a hookah contented themselves with the top part of the apparatus, called *chillum*, which contains the burning tobacco. This funnel-shaped cone could be detached from the water pipe and used on its own, and it was also less cumbersome to carry around. During the 1800s, the chillum became the smoking device of choice for many poor Indians, including the wandering ascetics known as sadhus and fakirs, and cannabis mixed with tobacco became their preferred smoking material.[13]

* * *

India has a long tradition of militant ascetics who practice social detachment and spiritual and bodily discipline and who also consume cannabis. Historical documents suggest they used the plant for different reasons, such as enduring austerities and steeling their nerves before a battle. For example, a Persian study written in the 1650s, *Dabestan-e Mazaheb*, noted that sadhus and fakirs "wear their hair entangled. . . . They are always sitting at a fire; they drink a great deal of bang; and the most perfect among them go about without any dress, in severe cold, in Kabul, Kashmir, and such places." In a similar vein, Scottish sea captain Alexander Hamilton wrote about the Indian warriors who attacked his fleet near Bombay in the 1720s: "Before they engage in a fight, they drink bang, which is made of a seed like hemp-seed, that has an intoxicating quality, and, whilst it attacks the head, they are furious. They wear long hair, and when they let that hang loose, they'll give no quarter."[14]

In the 1760s, the sadhus and fakirs of Northern India clashed with the newly established British colonial authorities, who threatened to tax their pilgrimage routes and commercial interests. Armed groups of ascetics repeatedly attacked and killed British tax collectors and colonial representatives in a prolonged series of skirmishes that spanned several decades. Colonial authorities could not tolerate "recalcitrant sadhus wandering about the countryside armed, dangerous, often naked, and claiming to represent an alternate locus of authority," as historian William R. Pinch noted. The British banned armed ascetics but proclaimed tolerance for those "who quietly employ themselves in their religious function." They also used military force and firearms to kill those who resisted.[15]

During the early 1800s, most ascetic warriors in northern India laid down their arms and embraced peaceful resistance—such as jeering at colonial authorities. Around the same time, they also turned to smoking, rather than ingesting, cannabis, mixing ganja with tobacco in their chillum pipes. Contemporary Indian ascetics claim that smoking ganja inspires their religious sentiment and enhances their ability to meditate. Whether the sadhus and fakirs living in the early 1800s held similar views is difficult to know due to a lack of evidence. But there is little doubt that they tended to disregard colonial authorities, keep a vegetarian diet, wear long, matted hair, and smoke ganja in chillums. And courtesy of the British Empire, some of these nineteenth-century Indian ascetics would soon end up on the Caribbean island of Jamaica.[16]

* * *

No one knows when psychoactive *Cannabis* arrived in Jamaica. During the seventeenth and eighteenth centuries, the British transported hundreds of thousands of enslaved Africans to the island and forced them to work on plantations. Still, there is no evidence that these people brought *Cannabis* with them. In the 1830s, after abolishing slavery in most of their colonies, including Jamaica, the British found that many formerly enslaved people refused to continue working for them. To compensate for this labor shortage, they began importing poor and landless people from countries such as India in a new form of coerced labor known as indentured servitude. Between 1845 and 1917, the British imported some 36,000 Indians to Jamaica to work as indentured laborers. Most scholars think these people likely introduced psychoactive *Cannabis* to the island.[17]

The first written reference to cannabis in Jamaica comes from an 1862 letter by a local magistrate, who described how "Hindoo immigrants were carefully cultivating the Indian Hemp . . . for the purpose of preparing the powerful narcotic known as *gunjah*." The letter noted that the "Congo Africans" on the island also smoked cannabis "in preference to tobacco." This document does not resolve the question of the plant's initial provenance. However, it points to a difference between Indian and African cannabis smokers during the second half of the 1800s. The Indians smoked cannabis mixed with tobacco in chillum pipes, while the Africans smoked it on its own in water pipes.[18]

As time passed and these uprooted people from different continents lived and worked side by side, their traditions intermingled and fused into one of the most vibrant cannabis cultures of the modern era: Rastafari. The dreadlocked, religiously inspired, peacefully rebellious, vegetarian, chillum-using, ganja-smoking Rastas, made famous in the 1970s by Jamaican musicians Bob Marley and Peter Tosh, drew inspiration from Indian ascetics while simultaneously recognizing and celebrating their African heritage.[19]

Though the Hindi word *ganja* is a common term for cannabis in Jamaica, some Jamaicans also use African words to name the plant, such as *diamba* and *kaya*. *Kaya* means "smoked herbs" or "leaf" in several Bantu languages; it is also the title of an album by Bob Marley and the Wailers.[20]

Marley reportedly smoked cannabis without tobacco, African style. A minority of contemporary Rastas follow this lead, but most smoke a blend of cannabis and tobacco, Indian style.[21]

* * *

After 1850, thousands of Jamaican laborers migrated to Panama, Costa Rica, and Colombia to dig the Panama Canal and build railroads. They likely introduced psychoactive *Cannabis* to these countries but left no written testimonies of their acts. Likewise, Mexican and Caribbean migrant laborers brought marihuana to the United States sometime after 1895, but exactly when and where this occurred is unclear. The history of cannabis in the Americas lacks precise dates and identified locations and actors.[22]

Europeans contributed unintentionally to the plant's worldwide spread by shipping cannabis-using people across oceans and globalizing the practice of smoking, which made cannabis easier to use. However, before the middle of the nineteenth century, most Europeans showed little interest in psychoactive *Cannabis*, knew nothing of its effects, and often failed to notice when others used it.[23]

# 3

# From Grass of Fakirs to Assassins' Herb

Making a plant illegal takes special motivation because plants are the dominant life-form on Earth. Banning a fast-growing weed like *Cannabis* poses additional difficulties. Yet, people have spent considerable time and energy outlawing the plant.[1]

According to the historical record, the first people to consider prohibiting *Cannabis* were thirteenth-century Muslim jurists, most of whom lived in Egypt. They wrote at length about the plant commonly called *hashish,* meaning *grass* or *herb* in Arabic.[2]

The Quran, which forms the basis of Muslim law, does not mention the plant but recommends avoiding *khamr*, an Arabic word meaning both *wine* and *intoxication*. As cannabis use became widespread in certain medieval Muslim societies, governing authorities called for jurists to give their interpretation of the law regarding the plant's legal standing. Most of those who gave their opinion began by considering whether hashish caused "intoxication." As the word *khamr* comes from the verb *khamara*, to cover, they reasoned that any substance that "covers the mind" deserved banning. In their view, having a "covered mind" meant forgetting to pray and think about God and saying senseless things, and they mainly agreed that the *Cannabis* plant could have these effects on those who consumed it. They also raised concerns that people used cannabis for "enjoyment and pleasure" because substances that "generate joy" also "cover the mind." The Muslim jurists who concluded that cannabis did indeed cause intoxication usually recommended the same punishment for cannabis users as for wine drinkers, namely between forty and eighty lashes.[3]

Some medieval jurists denied that cannabis caused "intoxication," arguing that true intoxicants came in *liquid* form. In contrast, people consumed the *Cannabis* plant as a *solid* substance, which they cooked into a confectionery, baked into a paste, or toasted into a powder. According to this argument, it made no sense to consider *Cannabis* an "intoxicant" if people did not consume it as a drink.[4]

Other jurists noted that wine made people commit violent acts, but *Cannabis* did not, suggesting the plant was not a true "intoxicant." In the words of thirteenth-century Egyptian jurist Shihab al-Din Qarafi: "Wine is known to cause a strong tendency toward quarreling among drinkers. They go at each other with weapons and are ready to do frightful things they would not do when sober. . . . Nothing of the sort occurs when hashish eaters are together. In no way do they behave like wine drinkers. On the contrary, they are quiet and somnolent, as in a trance. If one were to take away their things, one would not encounter in them the strong violent reaction to be expected from wine drinkers in such a case. Hashish eaters are the closest thing to dumb beasts. Therefore, corpses of people who have died a violent death are frequently discovered among wine drinkers but not among hashish eaters."[5]

Another argument that favored *Cannabis* pointed out that Muslim law defined wine as "unclean," meaning that anyone who touched it needed to wash ritually or risk the "invalidation" of their prayers. In contrast, it did not classify any plant in this way, suggesting that *Cannabis* could not be considered "unclean."[6]

Yet, one unnamed Egyptian jurist quibbled, asserting that *Cannabis* was "clean" in its raw and vegetal form, but once cooked, it acquired "mind-destroying qualities" and became "unclean." However, this argument did not sway the overall debate. Muslim law did not lend itself to outlawing a plant by declaring it "unclean" and making it distinct from all other plants.[7]

Besides, another unnamed thirteenth-century jurist argued that banning *Cannabis* presented dangers of its own because "declaring forbidden what is not forbidden is forbidden." This argument had considerable force in medieval Muslim society and raised an additional obstacle to making the plant illegal.[8]

* * *

The writings of medieval Muslim jurists and scholars offer rich insights into how people consumed cannabis. For instance, in 1237, botanist and scholar Ibn al-Baytar described how individuals in Egypt used the hashish plant: "I have seen fakirs (*fuqara'*) use it in various ways. Some thoroughly bake the leaves, then rub them carefully by hand until they form a paste and roll them into pellets. Others dry the leaves slightly, toast them, husk them by hand, and mix them with a little husked sesame and sugar, put that dry into their mouth, and chew it for a long time, which gives them excitement and joy. It also intoxicates them and sometimes gives them fits of madness, or almost. These are the effects of this plant, which I witnessed."[9]

In Arabic, *faqir* means "poor person" (*fuqara'* is the plural form). Medieval Muslim scholars often mentioned that poor people used hashish, and some referred to the plant as the "grass of fakirs." There were practical reasons for this association. Anybody could harvest wild *Cannabis* or grow a few plants in their garden, meaning that cannabis was a cheap and accessible intoxicant. In contrast, wine was complicated to produce, expensive, and illegal. In this context, it made economic and legal sense to choose cannabis over wine.[10]

Some "poor" Muslims used cannabis by conviction. Historical sources suggest that mystical Muslims driven out of Persia by Mongol invasions introduced psychoactive *Cannabis* to Egypt in the early thirteenth century. Known as Sufis (from the Arabic word *sufi*, meaning *mystic*), they occasionally used the plant in their religious practices. They also believed in seeking God through direct personal experience rather than by following the recommendations of mainstream and traditionalist Islam. People generally referred to these ascetic Muslims as "the poor," *darvish* in Persian, *fuqara'* in Arabic, which gave rise to the

English words *dervish* and *fakir* (also spelled *faqir*). The Sufis' alternative religious beliefs and practices soon attracted followers among impoverished Egyptian peasants and laborers. However, the conservative Muslim elites of Egypt disapproved of these "heretical" interpretations of Islam and led the charge against cannabis.[11]

The objections to cannabis formulated by medieval Egyptian jurists make up the first known corpus of arguments in favor of banning the plant. Here is a summary of these objections as listed by fourteenth-century Egyptian jurist Az-Zarkashī: "It destroys the mind, cuts short the reproductive capacity, . . . attracts diseases, . . . makes the mouth smell foul, dries up the semen, causes the hair of the eyebrows to fall out, burns the blood, causes cavities in the teeth, brings forth the hidden disease [a euphemism for homosexuality], harms the intestines, makes the limbs inactive, causes a shortage of breath, generates strong illusions, diminishes the powers [of the soul], reduces modesty . . . It makes the well-spoken person dumb and the sound person stupid. It takes away every manly virtue and puts an end to youthful prowess. Furthermore, it destroys the mind, stunts all natural talent, and blunts the sharpness of the mental endowment."[12] The concern that cannabis modified the sexual orientation of consumers came up regularly in these discussions, which referred exclusively to men. Several medieval Muslim scholars claimed that cannabis made men "effeminate" or homosexual.[13]

Another recurring theme was that cannabis caused insanity. The thirteenth-century scholar Ibn al-Baytar first formulated this notion, writing: "Some people use it, and it alters their minds and leads them to insanity, and sometimes even to death."[14]

Medieval scholars also expressed concern that cannabis use caused permanent personality changes and made the mind absent or "remote from reality." In their view, having a mind was what distinguished humans from animals, leading to the claim that cannabis turned users into "dumb animals."[15]

Such statements served to counter cannabis's positive reputation among poets and ordinary people. In the 1200s, Syrian poet Ismail al-Is'irdi wrote that "the secret of hashish lifts the spirit in an ascent of disembodied thinking," and Arabic-speaking cannabis users commonly referred to the plant with endearing nicknames, such as "shrub of understanding," "the one that causes good appetite," or "morsel of thought and remembrance."[16]

The arguments against cannabis formulated by medieval jurists seem to have remained largely theoretical. Their texts did not address the application of Muslim law in everyday medieval life. Concerning the cultivation and consumption of cannabis, authorities intervened infrequently and with little success.[17]

For example, in 1245, the sultan of Egypt requested the governor of Cairo prohibit the cultivation of *Cannabis* in one of the town's parks, prompting the governor to cut down and burn the plants. In 1266, another sultan banned the sale and cultivation of cannabis in Egypt. In 1324, the governor of Cairo confiscated large quantities of cannabis and wine and had them destroyed. In 1378, Cairo authorities uprooted and burned the "accursed shrubs" and punished producers by extracting their molars. However, after each crackdown, the plant grew back, and people continued using it.[18]

Further bans on cannabis occurred in Egypt in 1516, 1632, and 1725, none of which succeeded. Up until 1800, these were the only documented attempts to make cannabis illegal anywhere in the world.[19]

* * *

Then, the French got involved. The French army, led by Napoleon Bonaparte, invaded Egypt in 1798 and declared it a French colony. Seeking to practice an "enlightened" form of conquest, Napoleon included members of the local elite in the new colony's government and designed policies that mixed Islamic law with the ideas of the French Revolution. Napoleon also encouraged his generals to marry the daughters of Egypt's ruling families.[20]

One general called Jacques-François Menou followed these recommendations to the letter. He converted to Islam, married a woman from an elite Egyptian family, and changed his middle name to Abdallah (meaning "servant of Allah"). In October 1800, three months after becoming the general-in-chief of the French army in Egypt, J. Abdallah

Menou banned the distribution and consumption of cannabis across the colony. Some scholars consider this "the first drug prohibition ban in the modern era."[21]

According to the first article of Menou's ban: "The use of the strong liquor made by a few Muslims with a certain herb called *hashish*, and the smoking of hemp seed are prohibited throughout Egypt. Those accustomed to drinking this liquor and smoking this seed lose their minds and fall into a violent delirium, which often leads them to commit excesses of all kinds."[22]

Claiming that cannabis caused users to lose their minds was nothing new in Egypt. However, the French ban of 1800 made the unprecedented claim that consuming the plant led people to commit violent acts. None of the previous arguments formulated against cannabis by Egyptian jurists had mentioned this; on the contrary, they had insisted that consuming cannabis made men quiet and sleepy, without reaction if one took away their belongings, and even "effeminate."[23]

History does not reveal where Menou got the idea that cannabis use induced violent behavior. At the time, only a few French-language texts mentioned cannabis in any detail, and the only recent study of Egypt's plants and animals sponsored by the French government and written in the 1770s reported, quite to the contrary: "In the absence of intoxicating liquors, the Arabs and Egyptians make various preparations with hemp from which they procure a kind of gentle intoxication, a state of reverie that provides happiness and pleasant dreams. This kind of annihilation of the faculty of thinking, this kind of sleep of the soul, has no relation to the intoxication occasioned by wine or strong liquors, and our language has no terms to express it. The Arabs call this voluptuous abandonment, this kind of delicious stupor, *kief*."[24]

However, this French study referred to a 1772 publication, *Travels Through Arabia and Other Countries in the East*, by German cartographer Carsten Niebuhr. It may have given French authorities in Egypt the idea that cannabis could make its users violent. Niebuhr mentioned "an herb called hashish" consumed by poor people in the cities of Arabia, telling the following story: "Those who love this plant assure that it gives them lots of courage. We have an example of this in one of our Arab servants in Loheia who had smoked hashish. He met four soldiers in the street and felt like herding them all in front of him. One of the soldiers gave him a good beating and brought him home. But one could not calm him, and he remained convinced that four soldiers would not be able to defend themselves against him."[25]

Niebuhr wrote the original account of this incident in German. But the French translation, published in 1774, added a twist to the story. The man no longer *herded* the soldiers before him but *chased them away*. This mistranslation of a German verb made the man's actions seem violent rather than disinhibited. (The 1792 English translation went further, stating that the man "attacked the whole party.")[26]

In 1800, when General-in-chief Menou ordered the cannabis ban, the scientists and scholars working for the French army in Egypt had most probably read the French translation of Niebuhr's book. Even without the mistranslation, the story about a plant that turned an "Arab servant" into a courageous person with no fear of soldiers could only have made colonial authorities uneasy. During their short time in Egypt, the French faced repeated uprisings and rebellions. Only months before the cannabis ban, a man disguised as a beggar knifed to death Menou's predecessor, General Jean-Baptiste Kléber. There was no evidence suggesting that the assassin used cannabis. However, French authorities did not want Egyptians consuming a plant that had the potential to make them feel courageous and disrespectful of soldiers.[27]

Besides their fears of rebellious Arabs, French authorities in Egypt had other reasons to ban cannabis. They had received reports of French soldiers consuming cannabis and alcohol and shocking local people with drunken behavior, and they wanted to put an end to this. They also sought to align the colony's laws with the values of Egypt's Muslim elites, who had long advocated making cannabis illegal. By ordering the ban, Menou aimed to achieve several goals at once.[28]

This first modern prohibition of cannabis did not last long. Less than a year after issuing the ban, Menou and the French army left Egypt, defeated by the Anglo-Ottoman army. They returned to France in late 1801 with few material riches but many stories to tell, including one about a plant that made people lose their minds and commit acts of violence.

* * *

Eight years later, in 1809, French linguist Silvestre de Sacy announced that he had resolved the long-standing mystery surrounding the origin of the word *assassin*. In a memoir presented to France's highest academic institution, de Sacy demonstrated that it derived from the Arabic *hashishin*, meaning "hashish eaters."[29]

The word *assassin* first appeared in European languages during the twelfth century, when European Crusaders used it in Syria to refer to Nizari Isma'ili Muslims. Known for eliminating their enemies through stealth, the "Assassins" became associated with targeted killings. Over time, the word evolved in Europe to mean a professional murderer. However, its origins remained unclear and became the subject of scholarly debate.[30]

De Sacy based his argument on a previously unknown Arabic manuscript he had unearthed at the National Library in Paris. Written by the thirteenth-century Syrian historian Abu Shama, it referred to Nizari Isma'ili killers as "*hashishin*." This demonstrated that the people whom the Crusaders called *Assassins* were sometimes referred to by this term by their Arabic-speaking contemporaries, thus establishing a strong connection between the two words.[31]

But de Sacy did not stop at this linguistic connection. He argued that hashish "can produce a violent mania." To support this claim, he cited the French ban on hashish in Egypt: "We have not forgotten that, during the stay of the French army in Egypt, the general-in-chief was obliged to prohibit the sale and use of these pernicious substances strictly." He also referenced Niebuhr's anecdote about the hashish-smoking servant and Marco Polo's legend of the Assassins, which described an old man in the mountains of Persia who turned his disciples into devoted killers by making them drink a mysterious potion. De Sacy proposed that this potion contained hashish, writing: "I tend to believe that among the Ismailis, only those specially trained for the function of assassin were called *Hashishi* or *Hashshash* and that they were conditioned for absolute obedience to their leader through the use of hashish."[32]

Most modern scholars agree that de Sacy was correct about the origin of the word *assassin* but was mistaken on most other points. He failed to notice that medieval Arabic sources that referred to the Nizari Isma'ilis as *hashishin* used the term as an insult, meaning roughly "low-class rabble." Despite the many Isma'ili texts discovered since de Sacy's time, there is no evidence that these groups consumed cannabis during the twelfth and thirteenth centuries or that they referred to themselves by this term. Only their enemies called them *hashishin*, along with other derogatory names. The European Crusaders called them "Assassins" because they mistook an Arabic insult for the Nizari Isma'ilis' proper name. While de Sacy was correct about the word's origin, the term itself was a misnomer.[33]

De Sacy's argument had further weaknesses. Neither the French cannabis ban in Egypt nor Niebuhr's anecdote about the intoxicated servant provided evidence that hashish consumption led people to commit acts of violence. As for Marco Polo's legend of the Assassins, it was a fictional amalgam of stories told by twelfth-century Europeans about Persian and Syrian Muslims, which did not mention hashish. And cannabis lacks the properties that would make it a helpful drug for anyone carrying out a dangerous assassination mission.[34]

A few contemporary authors think that the word *assassin* comes from another Arabic word, *assassiyun*, meaning *fundamentalists*, which the Nizari Isma'ilis applied to themselves, and which European Crusaders may have mistaken for their name. However, in either case, the term *assassin* derives from misunderstanding an Arabic word.[35]

Despite its inaccuracies, de Sacy's memoir went on to have a global impact by establishing as an academic fact that the term *assassin* comes from *hashish*. For subsequent generations of researchers and policymakers, this etymological connection framed cannabis as an intoxicant that induces madness and violent behavior—even though the original Assassins did not use the plant.[36]

# 4

# Shaky Entry into Western Culture

European people knew very little about psychoactive *Cannabis* in the early 1800s. Thanks to Silvestre de Sacy's work, some had heard stories about the oriental plant that had given its name to the infamous Assassins. Still, few had ever consumed it, and European researchers had yet to study its properties in detail.[1]

This absence of knowledge began to change in 1839 when Irish scientist and physician William Brooke O'Shaughnessy published a paper in a medical journal on the therapeutic potential of "Indian hemp." Many consider his study to be the foundational text of modern scientific knowledge about psychoactive *Cannabis*.[2]

O'Shaughnessy had spent several years during the 1830s teaching at the Medical College of Calcutta, India, while also traveling through Pakistan, Nepal, and Persia. After listening to native doctors who used cannabis as a remedy, he decided to explore the plant's potential. First, he made a liquid extract by heating and dissolving *Cannabis* resin in alcohol, which he tested on animals, chiefly dogs, to determine its safety and proper dosage. He observed that a medium dose made the dogs "stupid" and caused them to eat "with great delight," while a higher dose incapacitated and immobilized them. However, all the dogs continued breathing normally and recovered fully within four hours. This convinced O'Shaughnessy that he could safely administer the extract to human patients, particularly those suffering from severe pain and difficult-to-treat illnesses such as rheumatism, rabies, cholera, and tetanus. He found that while it did not cure these diseases, it provided comfort and alleviated suffering, making awful conditions more bearable by inducing a state of "harmless insensibility." O'Shaughnessy also noted that the drug's side effects were fewer and less problematic than those associated with "the over-indulgence in other powerful stimulants, or narcotics, namely, alcohol, opium, or tobacco."[3]

He described the experiences of several of his students who, on their own initiative, had tested the *Cannabis* extract: "The result of several trials was that in as small doses as the quarter of a grain, after an average interval of one hour, the pulse was increased in fulness and frequency; the surface of the body glowed; the appetite became extraordinary; vivid ideas crowded the mind; unusual loquacity occurred; and with scarcely any exception, great aphrodisia was experienced." He also noted that "no headache, sickness, or other unpleasant symptom followed the innocent excess."[4]

Unfortunately, O'Shaughnessy did not mention testing the extract on himself. Another regrettable point is that he gave credence to de Sacy's story of the Assassins. According to the version recounted by O'Shaughnessy, the leader of "a monastery of faqirs" supposedly discovered the virtues of the hemp plant on a hot summer day in the thirteenth century after noticing that the plant "danced in the heat as if with joy, while all the rest of the vegetable creation was torpid." The leading fakir then made a tincture from it—by cooking the plant in alcohol—and found that it imparted feelings of joy, gaiety, and excitement.

This story belonged to the realm of myth and had no place in a scientific publication. However, O'Shaughnessy's paper raised several important points about working with the plant, particularly regarding dosage. From Persian sources, he had learned that cannabis had "contrary qualities"—acting as a stimulant in small doses and as a narcotic in larger ones. He wrote: "My experience would lead me to prefer small doses of the remedy in order to excite rather than narcotize the patient." O'Shaughnessy warned against the use of high doses of cannabis, which could induce a "singular form of delirium"—a warning that would soon prove well-founded.[5]

* * *

A year after O'Shaughnessy's groundbreaking paper was published, French physician Louis Rémy Aubert-Roche documented his personal experience eating "hashish"—meaning, in this case, an edible paste made from boiled *Cannabis* flowers—during a stay in Egypt. His 1840 text marks the beginning of scientific accounts of subjective experiences with cannabis's psychoactive effects. Aubert-Roche wrote about his experience: "The most bizarre and diverse ideas raced through my mind with astonishing speed. Reason occasionally controlled the effects of the substance, but soon they took over. Apart from that, I felt a perfect sense of well-being—no pain, no awareness of the past, present, and future. There was only the present moment, which still managed to slip away from me. It was the most complete *dolce far niente* [pleasant idleness], with just enough self-awareness to appreciate the pleasure of it. Then, everything gradually calmed down, and I wished to sleep. The entire night felt like one long, pleasant dream." Aubert-Roche reported laughing a lot and experiencing "extravagances, either in speech or in action," as well as an "almost canine hunger." He also struggled to find the right words to describe the experience, writing: "Nothing in our customs or in what we know resembles what one feels under the influence of hashish."[6]

As a physician in Egypt, Aubert-Roche treated patients suffering from bubonic plague, a disease that he believed affected the nervous system. He first learned about the hashish plant through de Sacy's work. He became convinced that "if this plant had the property of inducing intoxication, it must also have some medical power over the nervous system." After experimenting on himself and several patients with hashish paste, he concluded that cannabis stimulated both the nervous system and the appetite and could potentially provide relief for such conditions.[7]

Soon, a colleague reinforced his claims by publishing several works on *Cannabis*'s therapeutic potential. Physician and psychiatrist Jacques-Joseph Moreau, who also practiced in Egypt, encountered the plant for the first time there and described it as follows: "Hashish is the name of the plant whose active principle forms the basis for a variety of intoxicating preparations used in Egypt, Syria, and, more generally, in almost all Oriental countries. This plant is common in India and South Asia, where it grows wild. It is a species of hemp similar to our European hemp." Moreau also described the substance made from it: "The most common hashish preparation, the one that is used as the basic ingredient in almost all other preparations, is a fatty extract. To make it is very simple: the plant's leaves and flowers are boiled in water to which fresh butter has been added."

Moreau also provided descriptions of his personal experience with hashish paste. After ingesting a high dose, he found himself having the strangest and most incoherent thoughts and experiencing extraordinary hallucinations while still managing to consider them with detachment and self-control. He noted that his body continued to function normally during the intoxication, which led him to compare the experience to madness.[8]

Like Aubert-Roche, Moreau saw cannabis as a potential medicine for patients with nervous system disorders. He administered hashish paste to "melancholic" patients and observed that they became cheerful and talkative for several hours before returning to their original state of silence and inner torment.[9]

To Moreau, occasional cannabis use could alleviate certain mental disorders, whereas excessive and prolonged use could lead to permanent insanity. He claimed to have observed individuals in the Middle East who had lost their sanity due to cannabis consumption. He described them as having expressionless, dejected faces, with "dull eyes rolling uncertainly in their sockets, or else staring with a fixed, automatic gaze, a gaping mouth, and slow, lethargic movements." However, he considered such cases to be "exceedingly rare" since the vast majority of regular cannabis users he encountered showed few long-term impairments. He wrote, "Wine and liquors are a thousand times more dangerous."[10]

For Moreau, the most intriguing aspect of cannabis was its capacity to induce a temporary form of madness in otherwise healthy individuals. He saw cannabis as a valuable research tool that could help psychiatrists explore insanity and unlock its mysteries. He wrote: "To understand the ravings of a madman, it is necessary to have raved oneself; but without having lost the awareness of one's delirium, without having lost the power to evaluate the psychic changes occurring in the mind. By its mode of action on the mental faculties, hashish gives to whoever submits to its strange influence the power of studying on themselves the moral disturbances of madness or at least the principal intellectual modifications from which all the different kinds of mental alienation originate."[11]

In the 1840s, doctors Moreau and Aubert-Roche collaborated to explore the potential of this unusual vegetal drug. In Paris, they consumed Egyptian hashish paste and introduced it to their colleagues and friends, paying little attention to dosage. One particular version of this paste, called *dawamesc* in Arabic, was an extremely potent edible resembling a bright green jam. A single teaspoonful was enough to propel users into the realm of hallucinations. To prepare *dawamesc*, *Cannabis* flowers were boiled with butter in water for several hours to produce the basic fatty hashish extract, then blended into a greenish puree before being incorporated into a heated mixture of honey, sugar, water, spices, and nuts, which was then cooked down into a concentrated paste. Once cooled, it became a potent psychoactive jam. Beginners beware! By some estimates, Moreau and Aubert-Roche were consuming and administering doses twenty to thirty times higher than the amounts recommended today for occasional recreational users of edible cannabis.[12]

There is little doubt that consuming such high doses could induce "a singular form of delirium," as O'Shaughnessy warned. On one occasion, during his early experiments, Moreau reported experiencing hallucinations for nine to ten hours. However, he did not fully appreciate the experience "due to the secret terrors that partially robbed me of my ability to judge freely." On another occasion, he accused his colleague Aubert-Roche of trying to kill him with an unusually high dose of hashish paste, shouting: "Aubert, you are an assassin; you have poisoned me!" After allowing himself to go to such extremes, Moreau acknowledged that he had reached "a high level of delirium." Yet, that was precisely his aim: to consume a high dose of hashish paste and experience madness for a few hours while observing its effects before returning to a lucid state of mind the following day.[13]

After ample self-experimentation, Moreau and Aubert-Roche enlisted psychiatrists, physicians, architects, artists, writers, and painters to see if they could help cast light on the temporary madness cannabis seemed to induce.

The first writer to swallow a spoonful of Moreau's paste was his friend Théophile Gautier, who published an account of the experience in a Parisian newspaper in 1843. Gautier explained that he had accepted the invitation of "several orientalist friends," including an unnamed doctor who had traveled in the East and wore Turkish garb. To introduce hashish, Gautier explained that it was a jam "which the Old Man of the Mountain fed to the executioners of the murders he ordered, and the word *Assassin* derives from *hashishin* (eater of hashish)." Gautier then described his experience. After washing down the green paste with several cups of coffee, he had dinner with the other participants and waited for the drug's first effects: "Some minutes later, I was pervaded by a general numbness. My body seemed to be dissolving and becoming transparent. I could see the hashish I had eaten within my chest, in the form of an emerald giving off millions of little sparks. My eyelashes were growing indefinitely long and winding themselves like gold threads onto small spinning wheels. Around me, multi-colored precious stones streamed and flowed, their patterns constantly renewed, which I can best compare to the games of a kaleidoscope; I could still see my comrades now and then, but they were disfigured, half men, half plants, with the pensive air of ibises standing on ostrich legs, beating their wings, so strange that I shook with laughter in my corner, and to participate in the buffoonery of the spectacle I began to toss my cushions in the air, catching and twirling them with the speed of an Indian juggler. . . . My hearing was fantastically sharpened. I could hear the sound of colors. Green, red, blue, or yellow sounds reached me in perfectly distinct waves. A spilled glass, the creaking of an armchair, a softly spoken word vibrated and echoed in me like rolling thunder; my voice seemed so loud that I did not dare speak for fear of shattering the walls or myself exploding like a bomb. . . . Two images of each object were reflected by my retina, producing a completely symmetrical pattern, but soon, the magical paste began acting with greater force on my brain because it was entirely digested, and I went completely mad for an hour."[14]

Moreau approved of Gautier's description of the experience, which confirmed that talented writers could use their skills of description and self-observation to provide information about the momentary madness induced by cannabis. This led doctors Moreau and Aubert-Roche to organize monthly cannabis soirées in a comfortable apartment on the Île St-Louis in the center of Paris. They convened renowned artists and writers such as Victor Hugo, Honoré de Balzac, and Charles Baudelaire. This loosely constituted group became known as the *Club des Hachichins*, the Club of Hashish-Eaters.[15]

* * *

Poet and essayist Charles Baudelaire attended several of the club's sessions, mainly as a spectator, because he preferred trying Dr. Moreau's paste in the comfort of his apartment. In 1851, he published an essay called *Du vin et du haschisch* (*Wine and Hashish*) that would cast a long shadow on cannabis.[16]

Baudelaire took a dim view of both cannabis and Moreau's scientific research. Yet he eloquently described the drug's effects, writing: "The senses develop an extraordinary degree of delicacy and acuteness. . . . The ear perceives the most ungraspable sounds amidst the most high-pitched noise. This is where the hallucinations start. External objects assume monstrous forms. . . . The most singular ambiguities, the most inexplicable transpositions of thought take place. Sounds take on colors, and colors contain music. . . . Other times, music recites infinite poems to you or places you within terrifying or magical dramas. . . . I have observed that water acquires a frightening charm for all artistically inclined minds under the influence of hashish. Running water, fountains, harmonious cascades, the blue immensity of the sea: all of it rolling, sleeping, singing deep in your mind."[17]

Contrary to Moreau, who claimed that cannabis could attenuate certain mental disorders, Baudelaire viewed it as "a kind of drug which certain dilettantes find delectable, whose effects are far more violent and powerful than wine. . . . There are temperaments in which this drug develops a noisy madness, a violent gaiety akin to vertigo—dancing, leaping, stamping, peals of laughter." Baudelaire had read de Sacy's theory about hashish and assassins and subscribed to the notion that cannabis could induce madness and violence.[18]

Seeking to "point out the disadvantages of hashish," Baudelaire contrasted it to France's favorite intoxicant, wine, which he praised as "the soul of our country." He claimed that humans and wine "play the parts of the Father and Son in the Trinity: they engender a Holy Spirit or a superior man who derives an equal measure from both." In contrast, he had no praise for cannabis, invoked no moral or religious values in its favor, and mentioned only in passing its use by "Orientals" from nations such as Egypt and India.[19]

In the penultimate paragraph of his essay, Baudelaire summed up his view: "Wine magnifies our will, hashish annihilates it. Wine supports us physically; hashish is a suicide weapon. Wine makes one kind and sociable. Hashish isolates. One is, so to speak, industrious, the other essentially lazy. What, in truth, is the use of working, plowing, writing, or manufacturing anything at all when one can reach paradise in one go? Lastly, wine is for those who work and deserve to drink it. Hashish belongs to the class of solitary pleasures; it is made for miserable idlers. Wine is useful and produces fruitful results. Hashish is useless and dangerous."[20]

Ten years after doctors Aubert-Roche and Moreau began introducing cannabis into French culture, Baudelaire shut the door with a single essay, arguing that cannabis was foreign and went against Western culture because it made people lazy and destroyed their will. "No rational government could ever survive with the use of hashish," he wrote. "It creates neither warriors nor citizens. . . . If a government interested in corrupting its citizens existed, it would only have to encourage the use of hashish."[21]

Several of these statements inverted Moreau's claims. Moreau affirmed that cannabis was safer than wine and could be a valuable tool for scientific research and a potential therapeutic agent, which allowed users to keep self-control. Baudelaire insisted it was more dangerous than wine, useless, and a killer of personal will. In the closing remarks of *Wine and Hashish*, he singled out Moreau and dismissed his research as pointless, writing: "It is only worth mentioning for memory's sake the recent attempt to apply hashish to the cure of mental illness. The madman who takes hashish contracts a madness which chases away the first, and when the intoxication has passed, true madness, which is the normal state of the madman, takes back its empire, as reason and health do with us. Someone has taken the trouble to write a book about this. The medical doctor who invented this ingenious system is in no sense a philosopher."[22]

With his essay, Baudelaire became Europe's most prominent, if not first, anti-cannabis voice. His patriotic and religious framing of the question and staunch defense of wine guaranteed the popularity of his opinion.[23]

Looking back at Baudelaire's statements, several things stand out. Science has since established that wine is nowhere near as useful, healthy, or productive as he claimed. Baudelaire also ignored the question of dosage in his comparison of the two intoxicants. His statement that cannabis was "far more violent and powerful than wine" referred to experiences based on ingesting about thirty times today's effective dose of edible cannabis. If one transposed this to wine, drinking thirty effective doses in quick succession, or thirty "standard" glasses of wine, would signify risking death. Dose per dose, wine is far more dangerous than cannabis. Regarding the relative safety of cannabis and alcohol, Baudelaire was wrong, and O'Shaughnessy and Moreau were right.[24]

To Baudelaire's credit, few of his European contemporaries, aside from O'Shaughnessy, paid much attention to cannabis dosage. Had these early European consumers smoked the drug instead of eating it, they would undoubtedly have had different experiences due to lower doses. When people smoke cannabis, they typically feel its effects within a minute or two, allowing them to stop if they feel overwhelmed. In contrast, those who have ingested edible cannabis have no comparable way to limit its effects.

* * *

Several participants in the Parisian soirées organized by doctors Moreau and Aubert-Roche turned away from using cannabis. Baudelaire only tried it on a handful of occasions, preferring alcohol and opium. Théophile Gautier renounced using it after about ten tries "not because it has done us any physical harm," he explained, "but because the true *littérateur* needs only natural dreams and does not like his thinking to be influenced by any outside agency."[25]

Still, Moreau and Aubert-Roche were precursors. They wanted to include a plant from another part of the world in the therapeutic arsenal of Western medicine; they emphasized self-experiments and self-reporting and worked with creative people from their own culture to try to harness the phenomenon. However, Moreau's insistence on connecting cannabis and madness distorted the initial scientific understanding of the drug, and his systematic use of high doses obscured its "contrary qualities"—O'Shaughnessy's term for its capacity to act as a stimulant at low doses and as a narcotic and source of delirium at higher doses.

Even though extremely high doses of cannabis are not known to cause death, they can be difficult to manage and comprehend. At such high doses, cannabis tends to become "useless," to borrow one of Baudelaire's descriptors.[26]

From O'Shaughnessy introducing the plant to modern medicine in 1839 to Baudelaire condemning it in 1851, *Cannabis* went from promising medicine to dangerous drug in just over a decade. Its reputation has yet to recover from this shaky entry into Western culture.

# 5

# Insanity

Shortly after Baudelaire denounced cannabis as a useless and dangerous drug "made for miserable idlers," scientists in Europe and North America raised alarms, describing it as an "abominable poison" that drove its users to "commit acts of violence" and "terminate their existence as lunatics."[1]

In 1858, British colonial authorities in India began arresting fakirs and mendicants and incarcerating them in "lunatic asylums." The British disliked these naked "loafers" who smoked ganja all day long and jeered at them as they passed in the streets. Locking them up to "treat" them for their "insanity" was a handy way of silencing them. During the 1860s and '70s, the British built "Native Lunatic Asylums" across India and filled them with thousands of fakirs and mendicants. "Treatment" consisted of forced labor, imposed purgatives, and blistering the penis of male inmates to dissuade them from masturbating.

The British also kept detailed records of their activities, filling out forms for each inmate stating the reason for their arrest and the cause of their "insanity." In many cases, "ganja smoking" was given as the primary cause.[2]

By 1871, the yearly reports of the asylums in British India prompted authorities to conduct a colony-wide inquiry into whether "the abuse of ganja produces insanity and other dangerous effects," including "the alleged influence of ganja and bhang in exciting to violent crime." After an investigation that lasted two years, the inquiry concluded: "It does not appear to be specifically proved that hemp incites to crime more than other drugs or than spirits. And there is evidence to show that, on rare occasions, this drug, usually so noxious, may be usefully taken. There can, however, be no doubt that its habitual use does tend to produce insanity."[3]

The inquiry by the British government in India used the statistics of its "lunatic asylums" to confirm that cannabis use caused "insanity." But the statistics were flawed. The number of cases was tiny in proportion to the total number of people in the population who consumed the plant. The concept of "insanity" that the British imposed on wandering mendicants was riddled with prejudice. And the information on which the statistics were based was deficient—as even some colonial authorities recognized at the time. For example, Surgeon Major Benjamin Simpson, the superintendent of the Patna asylum in Bengal, wrote in 1874 about the cases of "insanity" attributed to cannabis use: "It is by no means an easy matter to determine in all such cases the proximate cause of insanity. In many of those attributed to *ganjah,* there may have been, and probably was, some exciting cause at which, from the deficient details furnished by the [police's] descriptive roll, we are unable to arrive. In most of these, where the lunatic is addicted to *ganjah*, the latter, and perhaps with reason, gets the credit of being the cause, and it is put down as such. Judging from the style of answers furnished by the police in the descriptive rolls, it would appear that if the man be a *ganjah-smoker*, the drug is invariably put down by them as the cause of insanity."[4]

Indian mendicants routinely consumed cannabis, but no effort was made to determine if this had caused them to become troublesome and be arrested as "lunatics." Even if consuming cannabis disinhibited some of the individuals who taunted the British occupiers as they passed in the streets, this did not mean that it had made them "insane."[5]

Still, some of these cannabis-consuming Indian mendicants may not have been of entirely sound mind. Fakirs cultivated states of rapture, or modified consciousness, and devoted themselves to consuming cannabis. For instance, one man told a British doctor in the early 1850s that his full-time occupation was "eating and smoking gunjah." Several such individuals who spent their days consuming cannabis may well have experienced hallucinations, delusions, and thought disorders—symptoms that present-day science associates with "cannabis-induced psychotic episodes."[6]

Contemporary research has established that consuming cannabis can cause intoxications that mimic psychosis but dissipate once the effects wear off, usually in a matter of hours or days. Still, those who have consumed an overwhelming dose of cannabis know how difficult the experience can be. As early as 1854, American writer Bayard Taylor gave an account of his agonizing experience after ingesting the equivalent of "a sufficient portion for six men" of hashish paste in a hotel room in Damascus, Syria. At one point, Taylor touched his head with his hands, and it felt like the fleshless skull of a skeleton. He feared that he had gone permanently insane and contemplated jumping to his death from the balcony—before reconsidering. Finally, he fell into "a death-like stupor" that lasted thirty hours. When he awoke, he felt "prostrate and unstrung" and experienced "frequent involuntary fits of absence" for several more days.[7]

Depending on the dose and the individual, the length of a cannabis-induced "transient psychotic-like experience" can vary. The effects are short-term in the great majority of cases and dissipate if the person discontinues consuming cannabis. As distressing as such experiences may be, research indicates that they do not produce permanent mental illness.[8]

However, contemporary studies show that cannabis can accelerate the onset of schizophrenia in individuals with a preexisting vulnerability to the mental disorder. This does not imply that cannabis use causes schizophrenia but that it is a risk factor for those who are likely to develop the disorder regardless.[9]

Retrospectively, these contemporary findings suggest that certain individuals incarcerated in the 1860s and '70s in the "lunatic asylums" of British India may have suffered from psychotic disorders triggered by cannabis. So, there may have been a kernel of truth in some of the diagnoses of cannabis-related "insanity" made by British authorities.

Still, this does not mean that the statistics they produced had any value, nor that incarceration was an appropriate treatment.

* * *

In 1882, the British expanded their empire by invading Egypt and taking control of the Suez Canal, the maritime shortcut between Europe and Asia. During the next forty years, they ruled the country and managed its institutions, including its only insane asylum in Cairo.[10]

In 1895, the Cairo asylum's newly appointed superintendent, Dr. John Warnock, began introducing "order and organization" to the institution. On one occasion, he counted three thousand individual bugs, lice, and fleas in one inmate's bed to demonstrate the importance of quantifying problems. He also generated statistics like those produced by his colleagues in the asylums of India. After ten months on the job, Warnock claimed that the "abuse of hasheesh" had caused the insanity of 41% of the men and 7% of the women recently admitted to the asylum, leading him to conclude: "I have no doubt that in quite a considerable number of cases here, hasheesh is the chief, if not the only, cause of the mental disease."[11]

Warnock made such statements even though he worked without a translator and understood no Arabic and the asylum's patients and staff spoke no other language. Just how he made diagnoses in these conditions is unclear. Information gathered in such circumstances cannot produce reliable statistics. Yet, Warnock's work gained widespread attention because it suggested a striking conclusion: In countries as distinct as Egypt and India, where people routinely consumed cannabis, the drug's usage appeared to account for about one-third of all cases of insanity.[12]

Warnock did not hesitate to flesh out his statistics with sweeping generalizations. In 1903, he described the difference between women and men concerning cannabis and insanity in the following terms: "Very few female patients used hasheesh, and it is noteworthy that insanity is more than three times as common among the hasheesh-using sex as among women, who, comparatively, seldom use the drug. I think this difference in the insanity rate between the sexes is significant and goes a long way to prove the importance of hasheesh as a cause of insanity among Egyptian men. Let it also be remembered that in England, insanity is more frequent among women than among men." He added: "I find that persons insane from hasheesh have a proneness to commit crimes, especially those of violence, and I have a strong suspicion that much disorderly conduct results from hasheesh smoking, just as alcohol among Europeans leads to such misconduct. To sum up, the use of Cannabis Indica in Egypt seems to have graver mental and social results than in India and is responsible for a large amount of insanity and crime in this country."[13]

Remarkably, two decades later, Warnock reversed his position, writing that the "number of cases of hasheesh insanity admitted to hospital is much less than was formerly reported to be the case; and as a cause of insanity, it does not seem to be of great importance." Warnock gave the yearly admissions statistics of the Cairo asylum for 1922, which attributed less than 1% of all cases to "hasheesh insanity."[14]

Warnock did not explain this reversal, despite having single-handedly produced the Egyptian insanity statistics for over a quarter of a century. He had come to understand Arabic over the years, which doubtless helped him establish correct diagnoses. He had also taken to citing Baudelaire on hashish, and the French writer's work may have led him to a more nuanced understanding of the experiences of hashish users.[15]

At the end of his career, in 1924, Warnock minimized the problems associated with cannabis. However, that same year, Egyptian authorities paraphrased his early reports in their statements to the League of Nations as they tried to convince representatives from countries around the world to clamp down on the plant and drug.

* * *

In November 1924, the Second Opium Conference convened in Geneva, Switzerland, under the auspices of the League of Nations. Its goals were to establish the maximum amounts of morphine, heroin, and cocaine that countries could produce from opium and coca leaves and to amend the 1912 drug control treaty known as the International

Opium Convention. During this conference, the head of the Egyptian delegation, Dr. Mohamed Abdel Salam El Guindy, pulled off a diplomatic coup by forcing cannabis onto the agenda. Some legal experts consider El Guindy's performance "a crucial intervention in the history of drugs."[16]

In an initial address to the assembly, El Guindy declared that Egypt had only recently become a free country—having gained independence from the British Empire in 1922—but had taken measures against cannabis "a long time ago." He described cannabis as a "drug at least as harmful as opium, if not more so" and insisted on the necessity of including it in the list of narcotics that the conference proposed to regulate. Delegates from Turkey, Japan, Poland, and Greece supported this proposal, and the Turkish representative told the assembly that his country had already passed laws "to prevent the cultivation of, and trade in, hashish."[17]

Several weeks later, El Guindy delivered a half-hour speech to the assembly about the plant known as "*Cannabis indica* or *sativa*, also called by the name of hashish, in English: Indian hemp." He claimed that consuming large doses of hashish "predisposes to acts of violence." He gave a vivid description of the ravages caused by chronic use: "The facial appearance of the habituated user becomes gloomy, his eye is wild, and the expression of his face is stupid. The individual is silent, has no muscular power, and suffers from physical ailments, heart troubles, digestive troubles, etc.; his intellectual faculties gradually weaken, and the whole organism becomes anemic. The individual very frequently becomes neurasthenic, and this state ends in insanity. In general, the absorption of hashish produces hallucinations, an alteration of the notions of time and space, fits of trembling, and convulsions. A person under the influence of hashish presents symptoms very similar to those of hysteria."[18]

El Guindy described hashish as "a toxic product, a poison against which no effective antidote is known." He told the assembly that Egypt had forbidden the plant's cultivation in 1884. He borrowed several sentences from John Warnock's early writings, declaring: "The illicit use of hashish is the principal cause of most of the cases of insanity occurring in Egypt. In support of this contention, it may be observed that there are three times as many cases of mental alienation among men as among women, and it is an established fact that men are much more addicted to hashish than women. In Europe, on the contrary, it is significant that a greater proportion of cases of insanity occur among women than among men. Generally speaking, the proportion of cases of insanity caused by the use of hashish varies from 30 to 60 percent of the total number of cases occurring in Egypt."[19]

El Guindy concluded his speech with a warning. If the assembly's delegates omitted cannabis from their decisions about opium, cocaine, and their derivatives, it "would soon replace the other narcotics and become a terrible menace to the whole world." He pleaded with them to help prevent cannabis from wreaking havoc on the world's citizens by banning "this scourge, which reduces man to the level of the brute and deprives him of health and reason, self-control, and honor."[20]

Even though many of the conference's delegates knew very little about cannabis, they greeted El Guindy's speech with "prolonged applause," according to the official transcript. The first person to rise and comment was the conference's vice president and head of the Chinese delegation, Sao-Ke Sze, who declared: "I am greatly moved by the statement made by the honorable delegate of Egypt. While I know next to nothing about the subject, I wish, in view of the statement the Egyptian delegate has made about the danger which this drug is to humanity, to second his request that this Conference should make a study of the question and do everything possible to put an end to this dangerous form of drug named hashish." The following speaker was the delegate of the United States of America, Stephen Porter, who said: "My knowledge of hashish and its use is quite limited. The very carefully prepared statement of the delegate of Egypt, together with my own knowledge on the subject, have satisfied me that we are under an obligation in this Conference to do everything we can to assist the Egyptian and Turkish people to rid themselves of this vice." The US delegate proposed that the Conference's Sub-Committee F, appointed to deal with various scientific and technical issues, could immediately take up the question of cannabis. The British delegate for India, Harold Clayton, objected that the "government of India was not aware that the question of hashish would be raised at this conference, and consequently, the Indian delegation is entirely without instructions." The following speaker, Sir Malcolm Delevingne, representing the British Empire, reminded the assembly that the conference had not been summoned to discuss cannabis, meaning the matter was still "in an unprepared state." He argued that the assembly should name a separate committee to consider cannabis at a later date. The delegate for France, Gaston Bourgois, agreed with the British proposal. He declared that "from a medical point of view, there can be no doubt that hashish is very dangerous." He also pointed to the practical difficulties of controlling cannabis consumption in some instances, telling the assembly, "In Congo, for example, there are several tribes of savages and even cannibals among whom the habit is very prevalent."[21]

The Chinese and American delegates received applause for their comments; the British and French representatives did not. The assembly's Danish president, Herluf Zahle, asked El Guindy if he approved any of these suggestions. The Egyptian delegate replied that he accepted the American proposal. As no member of the assembly objected to this, the president declared that "the Egyptian proposal regarding hashish may be considered as unanimously referred to Sub-Committee F."[22]

Just like that, cannabis was inserted into the agenda of the Second Opium Conference.

* * *

During the following weeks, El Guindy participated in the meetings of Sub-Committee F and drafted the section on cannabis for its report. When the subcommittee's rapporteur, French pharmacologist Emile Perrot, introduced the report to the assembly, he declared: "As you are aware, Indian hemp is not a preparation, but is simply the upper part of the female hemp plant, which grows chiefly in India. This question was particularly complex because the hemp used for manufacturing cloth belongs to the same botanical species. Accordingly, it is extremely difficult to abolish its cultivation, unlike the opium poppy. . . . The question concerning narcotics may be stated as follows: This variety of Indian hemp, which grows at a somewhat higher altitude, yields a resin that is widely sold in Central Asia and is a dangerous narcotic akin to heroin. Thus, the harmful custom of smoking hemp and absorbing preparations composed chiefly of hemp resin is widespread throughout Central Asia and also in parts of Africa. It is necessary, therefore, to endeavor to suppress its use, particularly in Egypt and Northern Africa."[23]

The subcommittee's report defined "Indian hemp" and affirmed that "something could be done to protect the world" from its dangers. It recommended that the forms of the plant "not being at present utilized for medical purposes and only being susceptible of utilization for harmful purposes, in the same manner as other narcotics, may not be produced, sold, traded in, etc., under any circumstances whatever."[24]

The assembly's president responded that the subcommittee's report was by its nature "technical" and asked its rapporteur to form a new subcommittee to draft a text about cannabis with applicable principles for inclusion in the final convention. El Guindy participated in this new subcommittee but did not manage to convince his colleagues to ban the plant altogether. The British delegate for India stuck to the position that attempting to control cannabis would come up against "very considerable administrative difficulties" because "the plant from which the drug is prepared grows wild, and the drug is prepared by very simple processes."[25]

The final convention was a compromise that focused on restricting the international trade of cannabis but said nothing about banning the plant. Contracting parties merely committed themselves to exercise controls "to prevent the illicit international traffic in Indian hemp and especially in the resin." The text defined "Indian hemp" as "the dried flowering or fruiting tops of the pistillate [or female] plant *Cannabis sativa* L. from which the resin has not been extracted, under whatever name they may be designated in commerce."[26]

On February 19, 1925, the assembly of the Second Opium Conference approved the amended version of the International Opium Convention, which included the freshly drafted sections on cannabis. This Convention was enacted in September 1928, placing "Indian hemp" under international legal control for the first time. It also led people worldwide to view the *Cannabis* plant as a "dangerous drug" akin to cocaine and heroin.[27]

* * *

El Guindy and the Egyptian delegation used outdated and misleading statistics to argue that cannabis caused insanity and presented grave dangers to humanity. They did not explain their motivations, but their tactics had deep historical roots. The list of ravages that El Guindy ascribed to cannabis during his speech drew in both form and content from Az-Zarkashī's fourteenth-century diatribe against hashish (see chapter 3).[28]

Surprisingly, El Guindy's claims regarding the dangers of cannabis faced little contradiction during the conference. British authorities, in particular, could easily have mentioned Warnock's recent work to refute El Guindy's statistics. After all, the British had ruled Egypt until 1922, and Warnock, as a medical officer appointed by British authorities, had produced all the statistics concerning "hashish insanity" in Egypt. But the British delegates in Geneva remained silent on the matter. Nor did they mention the work of the Indian Hemp Drugs Commission, established in 1893 at the request of the British Parliament to study cannabis usage in British India. This commission's eight-volume report remains the most extensive and well-documented investigation of a cannabis-consuming society. It notably concluded that moderate cannabis use has no significant ill effects and that "even the excessive consumer of hemp drugs is ordinarily inoffensive." It stated that cannabis can sometimes produce insanity in excessive users with a hereditary predisposition to mental troubles, but "the effect of hemp drugs in this respect has hitherto been greatly exaggerated." It also discredited as "worthless" the asylum statistics used to justify the claim that cannabis caused insanity.[29]

By failing to mention the conclusions of the *Report of the Indian Hemp Drugs Commission*, the delegates representing the British Empire and British India effectively withheld evidence from the other delegations present at the 1925 conference in Geneva. Why they chose to do so remains subject to debate. Still, this much is clear: The assembly of the Second Opium Conference, organized under the auspices of the League of Nations, formally placed cannabis under international control without duly considering the relevant evidence.[30]

During the conference's closing courtesies, the Egyptian delegation expressed dissatisfaction with the leniency of the convention's measures. El Guindy told the assembly that the issue of regulating cannabis was "not entirely settled" and formed the wish to see international legislation mandating "the complete suppression of the use of Indian hemp as a narcotic, and even as a medicine."[31]

The delegation of the United States agreed with El Guindy about the convention's "lenient" approach to drug control, refused to sign the final document, and walked out of the conference before the end. The Americans had come to Geneva in the hope of establishing an international ban on opium, and they had little interest in, or knowledge of, cannabis. Still, they recognized that the Egyptians could serve as valuable allies in their attempts to impose international prohibitions on "dangerous drugs." As Stephen Porter, the head of the American delegation, remarked after El Guindy's second speech on cannabis, "Many countries of the world have their own problems; by helping each other, we can make the world much happier and much better."[32]

# 6

# The Prohibition Effect

The United States government was certainly not opposed to the principle of prohibition, having passed a federal law regulating opiates and cocaine in 1914 and prohibiting the production, transport, and sale of alcohol in 1919. However, federal authorities considered that cannabis restrictions were the responsibility of the country's various states.[1]

Indeed, between 1911 and 1918, ten American states passed laws regulating cannabis use: Massachusetts, California, Indiana, Maine, Wyoming, Vermont, Utah, Colorado, Nevada, and Rhode Island, in order of passage. Most had previously restricted the use of opiates and cocaine and had either prohibited alcohol or were in the process of doing so. They enacted cannabis laws for various motives, such as discouraging people from switching to cannabis as a substitute for recently outlawed drugs and alcohol, protecting people from a plant reputed to cause insanity and violence, and preventing "euphoric drugs" from corrupting "the morals of youth." Most of these early American cannabis laws were preventative measures taken by states with puritanical and prohibitionist values and with a notable absence of problems caused by cannabis consumption.[2]

Cannabis had long been available as a liquid alcohol extract in pharmacies and drugstores across the United States and other Western countries. In 1851, the *Pharmacopeia of the United States of America*, the official reference for the country's pharmacists, listed *Extractum Cannabis,* or extract of hemp, as a medicine. Made from the dried tops of *Cannabis* plants imported from India and boiled in alcohol, *Cannabis* extract served as a painkiller and a means to alleviate various conditions, including gout, rheumatism, tetanus, cholera, convulsions, hysteria, mental depression, insanity, and uterine hemorrhage.[3]

As a medicine, *Cannabis* extract had several drawbacks. Its potency depended on the strength of the plants used in its production and varied from bottle to bottle, making it impossible to estimate the appropriate dose. The liquid also tended to separate, with the cannabis residue sinking to the bottom. If the bottle was not shaken before use, the final doses could become so concentrated that they caused consumers to experience intense hallucinations. Doctors expressed concern about these "overdoses"—because similar misadventures with opium could be fatal—and referred to such cases as "cannabis poisoning," even though these episodes were not life-threatening. By the end of the 1800s, medical interest in *Cannabis* extract had declined in the United States and other Western countries, mainly because of the variability of its effects and its therapeutic unreliability. Nevertheless, American pharmacists and pharmaceutical companies continued to produce and commercialize *Cannabis* extract and include it as a minor ingredient in corn remedies and veterinary medicines.[4]

During the discussions leading up to the formulation of the Narcotics Act of 1914, American authorities considered regulating cannabis along with opiates and cocaine. However, the American pharmaceutical industry lobbied against this, arguing that a minor medicine such as cannabis had no place in federal anti-narcotics legislation. In the early 1910s, cannabis use in the United States was too infrequent and unproblematic to warrant a federal restriction.[5]

This changed entirely during the nationwide prohibition of alcohol that began in early 1920 and lasted until late 1933. According to historian Ronald Hamowy: "Marijuana's popularity as a recreational drug in the United States emerged largely as a result of passage of the Eighteenth Amendment, which raised the price of alcohol to prohibitively high levels for many poorer drinkers, leading them to substitute use of this far less expensive weed. . . . Before prohibition, marijuana use had been almost completely confined to immigrants from the West Indies and Mexico. . . . Despite the growing popularity of marijuana among non-Hispanics, however, the drug remained associated with Mexicans."[6]

* * *

During the 1920s, the port city of New Orleans, Louisiana, at the confluence of the Mississippi River and the Gulf of Mexico, was one of the most important entry points of cannabis into the United States. Merchandise shipped into New Orleans from abroad tended to make its way up the Mississippi Valley to supply the American heartland. In the case of cannabis, importers found an initial market in New Orleans. They only later supplied the drug to large Midwestern cities such as Chicago, Cincinnati, and Kansas City. Historian Adam Rathge considers New Orleans as "perhaps the best place in the United States to witness the emergence and consolidation of anti-marijuana sentiment" because events that transpired there "influenced and previewed" what emerged subsequently in other states and at the federal level.[7]

As early as 1921, the New Orleans newspaper, the *Times-Picayune*, wrote about "the narcotic preparation of a plant called 'marijuana,'" which it identified as "Indian hemp (Cannabis indica)." During the following decade, the newspaper published hundreds of articles referring to the plant by its Mexican name, *marihuana*, or some variant of the word, such as *mirauana*, *marajuana*, and *mariahuana*. Many of these newspaper articles also mention its American English slang names, including *muggles*, *reefers*, *moota*, and *Mary Warner* (the latter being a play on words based on pronouncing marihuana like the name of an American woman).[8]

In 1922, the *Times-Picayune* reported under the headline NEW DRUG HABIT RAPIDLY GROWING that Louisiana's Board of Health was seeking to pass "a drastic law to curb the constantly growing practice of selling and smoking marijuana, also known as muggles." According to the Board of Health's president, Dr. Oscar Dowling, "the high cost of liquor and 'dope' [morphine] has caused thousands of Orleanians of both sexes to acquire the habit of smoking marijuana" and "'muggles parties' are becoming frequent among those who crave an artificial stimulant." In the same article, the chief narcotics officer for Louisiana, G. W. Cunningham, deplored the absence of a law against the plant's sale and use, as well as its low price, noting that "twenty-five cents will purchase enough marijuana to roll at least two dozen cigarettes."[9]

At the time, a glass of bootleg whiskey cost between thirty and fifty cents in New Orleans. Alcoholic drinks were not hard to find during Prohibition; they were just expensive, costing roughly three times more than previously. The law did not punish those who consumed alcohol, only those who produced, transported, and sold it. One effect of this was to drive up the cost of liquor—to pay those who took risks providing the product.[10]

In 1923, the New Orleans newspaper reported that a "'Mary Warner' epidemic" had befallen the city, and "the Mexican dope 'Marijuana'" was "ravaging the brains of thousands of local addicts," including "newsboys and school children who haven't means to purchase a more expensive drug." The newspaper commented: "The tragedy of the situation is that this drug is striking at the very roots of society in attacking the children, making them slaves, not only to the drug, but to those unscrupulous boys and men who find it to their advantage to 'dope' the children, taking from their hard-earned pennies, gained by selling papers, shining shoes and so on, leaving children sleeping in the alleys, in gutters, and in the streets."[11]

The city council of New Orleans promptly prohibited the possession and sale of "'cannabis indica,' better known as 'Mari Juana' or the 'Mexican happy smoke,'" except by medical prescription, with violations punishable by a fine of up to $25 or thirty days of imprisonment. The local newspaper commented: "The weed, which grows wild in Mexico and parts of Texas, is usually indulged in the form of cigarettes, and evidently was not intended for the use of Anglo-Saxons for it usually has the same effects as knockout drops, with the user ending up in the hospital unconscious sans valuables, and unable to tell what has happened to him." It added that Dr. John Fletcher of Tulane University had analyzed a sample of "marihuana . . . which so many of the boys of the city are accustomed to smoke as 'muggles' or 'Mary Warner cigarettes'" and identified it as "the flowering top and the tender parts of cannabis indica, or Indian hemp." Fletcher vouched that the plant had been "in use for centuries as a narcotic stimulant," and claimed that its regular consumption led to "crime, insanity, and idiocy."[12]

The following year, 1924, Louisiana passed a statewide law prohibiting the possession, use, and sale of "Cannabis Indica, Cannabis America, or Marajuana," except when sold in pharmacies on a prescription by a licensed physician. The law forbade manufacturing or smoking cigarettes containing "the Mexican plant known as Marajuana." But why the insistence on Mexico? The most significant cannabis shipments did indeed come into Louisiana on steamboats and ships from Mexico—but a review of all the documented arrests of cannabis users and dealers in New Orleans between 1924 and 1929 shows that Mexicans made up less than 5% of offenders, the majority of whom were young white men living in the city.[13]

In 1926, a criminologist and "international narcotic expert," Dr. Carleton Simon, conducted a survey for the Louisiana Board of Health and found that "thousands of youths in this state are using marihuana, a form of hasheesh generally known as 'muggles.' This weed is literally threatening your young men and women by the thousands." Simon called the situation "appalling" and referred to the port city of New Orleans as "the hypodermic needle feeding the entire Middle West with illicit drugs." It was true that illegal products, such as cannabis, morphine, cocaine, and large quantities of whiskey and rum, were shipped into New Orleans from abroad and smuggled out of town by automobile, truck, or train to points up the Mississippi Valley, including Chicago.[14]

In 1929, the *Chicago Tribune* reported that smoking "marihuana" had "become widespread among American youths" in the Chicago area and other parts of Illinois. But the state had yet to pass a law prohibiting cannabis use—and would not do so before 1931. According to the newspaper, this legal void had a clear consequence: "There being no legal ban such as makes other drugs scarce, 'loco weed' is cheap. The rush of its popularity in Chicago and all over the country since the oncoming of Prohibition is partly explained by the price of [cannabis] cigarettes, 3 for 50 cents or at most 30 cents apiece."[15]

Similar reports came from other Midwest towns. In Cincinnati, where cannabis had "become quite extensive since 1928," one medical doctor declared: "The loco-weed can be purchased in most public places and on the streets with much more ease than other intoxicants." *The Cincinnati Tribune* described boys and girls of high school age sitting around "in these 'muggles trains,' or circles, passing muggles cigarettes from one mouth to the next, each taking a puff, and as the night wore on, becoming more and more dreamy, incoherent in speech and looking into space with an absent expression." In Kansas City, newspapers told of "marihuana parties held by young boys and girls," reporting that "habitual smoking is at present almost exclusively confined to young persons among the white people."[16]

Meanwhile, back in New Orleans, local authorities attributed several brutal murders to cannabis smokers. In April 1929, the *Times-Picayune* reported that one man "smoked half a dozen 'mooters'" and axed to death a woman and child; and another man, described as a "mugglehead" [cannabis user], shot to death a cashier for no apparent reason. The superintendent of police, Thomas Healy, commented that "most robberies and shootings lately have been perpetrated by marihuana addicts." According to the newspaper: "Gunmen smoke marihuana because it makes them feel happy and gives them an oversupply of false courage, police say."

New Orleans had become "the distribution center for the United States," and individuals with suitcases full of "Mexican loco weed" were traveling from the port city to locations as distant as New York to supply the underground market. According to the *Times-Picayune*: "Greenwich Village, New York's purlieus of personages who do not acknowledge the hinterlands, has taken up muggles in a big way. Only it is called hasheesh up there."[17]

By 1931, some Americans had started growing *Cannabis* in Louisiana rather than relying on others to smuggle it in from abroad. One man, George Bonoa, was convicted of "wilfully and unlawfully possessing 500 plants" at his home in the lower section of New Orleans and condemned to pay a fine of $500 and serve a six-month term in prison. According to the police, Bonoa had a "muggles farm in the yard of his house." Bonoa appealed the verdict to the Louisiana Supreme Court, where his lawyers contested the law's constitutionality. They declared that they had done "a careful search in dictionaries and could not find *marihuana* or *marihuana plant* listed in any of them," which led them to consider that "there is no such plant known as *marihuana*" and therefore that "the law was an attempt to prohibit something which is unknown." Bonoa's lawyers explained that the plants found in their client's possession were *Cannabis* plants, which could be used "for valuable purposes, such as the manufacture of hemp rope, twine, in the preparation of useful drugs, and for the production of seed which forms a large part of the ration of the millions of canary birds in the country, and that, only insofar as the plant is sold, used and possessed for deleterious purposes, there would be a violation." They claimed that by prohibiting "the possession of the plant or the growing of the plant," the law "went far beyond its constitutional power" and that "it cannot go so far as to prohibit the possession of harmless articles or plants, unless and until they are put to some unlawful use." If possession of the *Cannabis* plant may be prohibited simply because intoxicating resin may be extracted from its flowering tops, they argued, then the possession of corn, grapes, and poppies should also be banned since whiskey, wine, and opium may be extracted from them.[18]

District Attorney Eugene Stanley, representing the State of Louisiana, replied that "the Mexican plant is well known, both medically and pharmaceutically," and "its use results in a derangement of the central nervous system, causing fatal results both mentally and physically. Statistics show that many crimes have been committed by persons under its influence, their powers of resistance being lessened, and their criminal tendencies aggravated as a result of the use of this drug. That the suppression of this drug or plant is directly within the police power of the state is manifestly plain."[19]

The court's verdict rejected Bonoa's appeal. It upheld the validity of the state law, noting that "there is no doubt but that the plant contains a drug known as Cannabis, which is deleterious and of dangerous propensities" and that "to permit the plant to be possessed in Louisiana, even in its growing form, is virtually as unsafe as to permit its possession in the manufactured form of cigarettes and tobacco, so readily and easily may it be converted into those forms." The court condemned George Bonoa to twelve months in prison—six more than the original sentence because he could not afford to pay the fine.[20]

* * *

In April 1931, a medical doctor named Albert Emile Fossier gave a presentation called "The Mariahuana Menace" to the Louisiana State Medical Society in New Orleans, which would later have an international impact. Fossier started by referring to the Assassins, a "diabolical, fanatical, cruel, and murderous tribe" that derived their name from *hashish*, "a confection of hemp leaves, *Cannabis indica*," which they consumed before committing their heinous crimes. Fossier explained: "Mariahuana, vulgarly called 'muggles,' is *Cannabis sativa*, derived from the flowering tops of the female plant of hemp grown in semi-tropical and temperate America. It was once thought that only *Cannabis indica* grown in the far East was active, although the German hemp is inert, the American specimen, mariahuana, is equal in potency to the best weed of India."

Fossier quoted an authority from British India who considered cannabis a "cerebral poison" that caused insanity. He claimed that cannabis use removed the normal human inhibition of "committing willful theft, murder, or rape." He cited a recent survey conducted in the prison of New Orleans, according to which 27% of prisoners and 46% of murderers were "confirmed mariahuana addicts." Fossier noted that the problem was "twice more frequent in the whites than in the negroes" and that cannabis consumption was "not confined to the criminal class" since "school children of tender age have been detected smoking muggles." He also stated that the recent popularity of cannabis had "assumed formidable proportions since the advent of that 'noble experiment,' that fiasco, prohibition." In his conclusion, Fossier called on federal authorities to place cannabis under national control, pointing out that England had already done so.[21]

Several countries had passed laws restricting cannabis use in the preceding years, including Mexico in 1920, Canada in 1923, and England in 1925. At the time, in early 1931, twenty-seven American states and territories had laws restricting cannabis, but twenty-three others did not, and the resulting legal patchwork hampered effective enforcement. Several states, including Louisiana, had officially appealed to the federal government for help, but federal authorities continued to resist.[22]

Most experts in American law agreed that the states had the power to regulate the practice of medicine, including the distribution and possession of drugs, and the federal government only had the authority to intervene in questions of interstate drug commerce. This suggested that placing cannabis under direct federal control would be unconstitutional. As the legal advisor of the Bureau of Prohibition, Alfred L. Tennyson wrote in 1929 in reply to a senator from Colorado who had requested the inclusion of cannabis in the existing federal Narcotics Act: "It is thought that this evil may be more properly met by state and municipal legislation, for which there is more ample fundamental authority. I respectfully venture the suggestion that if the abuse of cannabis indica . . . exists to an appreciable extent in Colorado, the matter be referred to the state legislature for appropriate attention."[23]

Colorado prohibited the cultivation and sale of cannabis in 1917. It passed additional legislation in 1927 banning the plant's possession and tightening restrictions on its distribution, while increasing penalties for all relevant violations. However, this did not change the federal point of view that cannabis could not be included in existing federal drug-control legislation.[24]

The Narcotics Act of 1914 imposed national regulations on morphine and cocaine—two substances used in medicine and typically regulated by the country's various states. Since both drugs were produced abroad, their importation into the United States could legitimately fall under federal control. However, in the case of cannabis, this kind of sidestepping was not possible. Americans had cultivated the hemp plant for fiber since the seventeenth century. In the early 1900s, American scientists researching imported medicinal plants found that they could grow potent psychoactive *Cannabis* in Kentucky, which was "fully as active as the best imported Indian-grown *Cannabis Sativa.*" In 1915, the United States Department of Agriculture confirmed that *Cannabis* cultivated "for drug purposes" was "better suited to the warmer climates of the southern half of the United States." For federal authorities considering the issue in 1931, imposing federal controls on a medicinal plant that grew in many different parts of the country was not just a constitutional headache but could also become an enforcement nightmare.[25]

Besides, federal authorities already had their hands full enforcing the nationwide ban on alcohol. Even though large majorities had voted for Prohibition in 1917 and again in 1928, many Americans disliked its enforcement and found ways around the law or else chose to break it. A tension between puritanical, Protestant, and prohibitionist values on the one hand and the freewheeling use of intoxicants on the other runs through this period—and through much of American history.[26]

Prohibition did not stop Americans from drinking alcohol, but it taught them to violate the laws of their own country. It pushed up the price of alcoholic beverages and pushed down their quality; bootlegged alcohol could be dangerous to consume, and adulterated and unregulated liquor killed thousands of Americans during Prohibition. Crime increased and became "organized" because the enormous profits to be made from producing and selling illegal liquor attracted criminal organizations. Prisons across the country filled up with alcohol-related offenders. Corruption of federal agents became rampant. Prohibition also caused Americans to drink more potent forms of alcohol because almost all illegal production was distilled. Economist Mark Thornton explains: "Prohibition made it more difficult to supply weaker, bulkier products, such as beer, than stronger, compact products, such as whiskey because the largest cost of selling an illegal product is avoiding detection. Therefore, while all alcohol prices rose, the price of whiskey rose more slowly than that of beer." The price of beer increased by more than 700% during this period.[27]

Prohibition also led Americans to experiment with intoxicants they might otherwise have ignored, particularly cannabis. When offered the choice between a glass of bootleg whiskey or a couple of reefers for the same price, young people across the American heartland chose reefers.

In the context of a nationwide prohibition on alcohol, the sudden popularity of a smokable plant intoxicant from Mexico raised alarms. Its use as an intoxicant was cause enough to arouse disapproval; the plant itself had a reputation for causing crime, violence, and insanity; and its low price was a form of "unfair competition," given that Prohibition had made alcohol expensive rather than unobtainable. American authorities could only express hostility toward this new intoxicant, which they increasingly referred to by its Mexican Spanish name.

* * *

The word *mariguana* first appeared in an American English publication in 1874 when ethnographer Hubert Bancroft mentioned it as a "narcotic herb" used by "the wild tribes" of Central Mexico. During the late 1800s and early 1900s, American newspapers and journals occasionally mentioned marihuana/mariguana (or some other variant of the word), most often without identifying the plant as *Cannabis*. Conversely, American authorities who wrote about *Cannabis* during this period did not refer to it by its Mexican Spanish name. For example, when the United States Department of Agriculture published a booklet in 1915 advising American farmers on the cultivation of drug plants, the section on *Cannabis* mentioned the plant's other names, Indian hemp and *Cannabis sativa*, and even suggested eliminating male plants to enhance the resin production of female plants; however, it made no mention of marihuana.[28]

During the 1920s, Americans began smoking cannabis as a recreational intoxicant and a substitute for alcohol. American lawmakers, politicians, law enforcement officers, and journalists referred to it as *marihuana*, even though they were mostly familiar with the plant's names in English. They also increasingly spelled the Mexican Spanish word with a *j*. However, English-speaking Americans who smoked cannabis did not call it marihuana or by any other Mexican Spanish name; instead, they used American English slang, terms such as "muggles," "reefers," and "Mary Warner." Why, then, did American authorities insist on using an unfamiliar, Spanish-sounding word from Mexico, which they spelled and pronounced inconsistently and which no one besides themselves used in American English?[29]

Anti-Mexican sentiment has a long history in the United States, dating back at least to the Mexican-American War of the 1840s, when Americans viewed Mexicans as "alien," "savage," and "degenerate." In the 1920s and '30s, using a Mexican name was a way for Americans to signal disapproval to one another. Renaming cannabis to *marihuana* and emphasizing its Mexican origins made it seem unfamiliar and unappealing. *Cannabis* was a plant and medicine listed in the *Pharmacopoeia of the United States of America*, whereas marijuana was a Mexican drug that Americans had recently begun to smoke. In this context, "Mexican" meant *imported from Mexico* and implied *alien* and *bad.* When prohibitionist authorities introduced the term *marihuana* to American English, they intended it as a slur.[30]

The word *marijuana*—spelled with a *j* or *h*—subsequently entered English dictionaries. In 1972, the *Oxford English Dictionary* defined it as: "Dried leaves of Indian hemp, used to make narcotic cigarettes (called *reefers*)." In 1999, the same dictionary updated this definition to "Cannabis, especially as smoked in cigarettes." In contemporary English, marijuana has come to refer to both the plant and the smokable herbal substance derived from it.[31]

From American English, marijuana passed into many other languages, such as French, German, Japanese, and Yoruba. French dictionaries define *marijuana* as "a drug made from the dried and chopped leaves and stems of Indian hemp." In German, *das Marihuana* means "the dried, resinous flowers and small leaves close to the flowers of the female hemp plant (cannabis)." A *drug* in one language, *flowers* and *leaves* in another—words can take on different meanings as they travel.[32]

They can also change through use and the debates people have about them. *Marijuana* is currently the single most used term for cannabis in the United States, but some people have renounced it because it "carries prejudicial implications rooted in racial stereotypes." Others argue that marijuana no longer conveys xenophobic or prohibitionist meanings, and banning this mainstream word would create confusion and incomprehension. Both views have merit. Prohibitionist authorities did indeed drag the Mexican Spanish term into American English in the 1920s and '30s while appealing to the xenophobia of the time. And most people today use the word *marijuana* without any xenophobic or prohibitionist intentions. Does the dilemma have a solution?[33]

One place to start, I find, is to pronounce marijuana as it sounds in Spanish, with a rolled *r* and four crisp syllables, *ma-RRi-hua-na*, to express appreciation of its Mexican Spanish melodiousness and heritage. Beyond this small personal practice, I think that telling people to stop saying marijuana in English would be one more prohibitionist act. Instead, I encourage those who use the word to know its story.

# 7

# To Kill the Weed That Kills People

A few months after the repeal of Prohibition, in early 1934, a wave of crime allegedly fueled by cannabis swept across the United States—at least, according to a report from the secretariat of the League of Nations, published from its headquarters in Geneva. Tasked with identifying and analyzing issues that could lead to international unrest, the secretariat attributed this supposed rise in crime to the absence of a federal law prohibiting cannabis consumption. The report used Fossier's paper "The Mariahuana Menace" as its primary source. It also claimed that cannabis trafficking was expanding globally due to inadequate international drug laws.[1]

The United States was not a member of the League of Nations. However, American authorities maintained working relations with the organization and occasionally sent representatives to its meetings. In November 1934, the US government submitted a memorandum to the Advisory Committee on Traffic in Opium and Other Dangerous Drugs—the League's body responsible for international drug regulation—titled "The Abuse of Cannabis in the United States." Written by American representative Stuart J. Fuller, this document largely confirmed the conclusions of the secretariat's report. Fuller claimed that regular cannabis users tended to commit violent crimes and that prolonged consumption led to "mental deterioration and eventual insanity." He also cited the head of California's Narcotics Enforcement

Bureau, who claimed that "marihuana has a worse effect than heroin. It gives men the lust to kill, unreasonably, without motive—for the sheer sake of murder itself." Fuller also wrote that "marihuana addicts are becoming one of the major police problems" in the United States and "fifty percent of the violent crimes committed in districts occupied by Mexicans, Turks, Filipinos, Greeks, Spaniards, Latin-Americans and Negroes, may be traced to the abuse of marihuana."[2]

No one at the League of Nations objected to this xenophobic and racist language. On the contrary, when the advisory committee convened in Geneva in late November 1934 and addressed the subject of cannabis, Austrian chairman Bruno Schultz acknowledged "the United States representative's interesting memorandum." The first speaker, Egyptian representative Miralai Baker Bey, thanked the American representative for "eloquently" describing "the real moral and physical dangers of hashish," the use of which was "very often the prologue to crime." The Egyptian representative also insisted on the inadequacies of the International Opium Convention's control measures concerning cannabis and stated that "Egypt very vigorously recommends, and would most warmly welcome, any project tending towards a worldwide outlawing of the *Cannabis indica* plant."[3]

These official reports and statements called for a reaction from the Advisory Committee, but many questions concerning *Cannabis* seemed confusing—starting with the plant's name. The secretariat's report described Indian hemp as a single species called *Cannabis sativa* L., with numerous varieties, including *indica*. The American memorandum referred to one plant with several names, including *marihuana*. The Egyptian delegate spoke of *hashish*, which he defined as "the flat cake of compressed resinous powder extracted from the female plant of the variety *indica* of the species *Cannabis sativa*." Were they all talking about the same plant?[4]

The plant's legal definition posed an additional problem. "Indian hemp," as defined in the amended version of the International Opium Convention, referred exclusively to the dried flowering or fruiting tops of female plants. The experts who had come up with this definition believed that only the tops of female *Cannabis* plants produced resin, which "had no current medical uses and could only be used for harmful purposes." They also omitted the plant's seeds and stalks from the definition to protect the industries that used them. The result may have made sense on paper but created confusion in practice. How could a police officer or a judge determine whether a given sample of dried cannabis came from the flowering tops of a female plant and not from its leaves or a male plant?[5]

The advisory committee realized it needed a better grasp of "the whole problem of Indian hemp" before formulating recommendations to the League of Nations. So, it appointed a subcommittee with the power to consult "experts, doctors, and others who are duly qualified in the matter of Indian hemp" to clarify the unresolved questions concerning the plant. Their queries included: Was there only one species of *Cannabis*? Was a reliable chemical test capable of detecting the plant's "narcotic substance"? Did this substance only appear in the resin of flowering female plants? Was it desirable to include the plant's leaves in the International Opium Convention's definition of Indian hemp? Was *Cannabis* medically indispensable? Did the abuse of cannabis cause insanity, criminality, and drug addiction? Was it possible to establish a single term for "Indian hemp" that would include all aspects of *Cannabis sativa* L.—the plant itself, its flowering tops, its resin—as well as be precise, scientific, and adapted "for ordinary popular use and for the use of Custom authorities?"[6]

In early 1937, several international experts contacted by the League of Nations Sub-Committee on Cannabis sent their replies—some of which advanced the scientific understanding of the plant. For example, French pharmacist Jules Bouquet, the inspector of pharmacies in Tunis hospitals, reported examining mature *Cannabis* plants under an optical microscope and observing that they produced and stored an amber-color resin in the swollen tips of tiny glandular hairs. Bouquet noticed dense concentrations of these hairs on the tops of flowering female plants and sparser ones on the underside of their leaves and the flowers of male plants. These observations indicated that male *Cannabis* plants and the leaves of female plants also produce resin, which led Bouquet to recommend revising the International Opium Convention's definition of "Indian hemp."[7]

One month after the publication of Bouquet's report, scientists working for the United States Treasury Department reached a similar conclusion using different methods. They soaked the leaves of mature female *Cannabis* plants in alcohol to prepare an extract. They did the same with flowering male plants, then fed capsules of these extracts to dogs and observed the animals' consequent incoordination. "If the dog's legs get tangled up, the drug is potent," one scientist explained. At the time, the assistant surgeon general of the United States considered this to be "the only reliable method" for determining whether a given sample of cannabis was indeed psychoactive. One pharmacologist wrote that dogs were the experimental animals of choice because "the reaction of dogs to this drug closely resembles the reaction of human beings." These scientists believed that "marihuana destroys mental fabric" and "if you go on giving the dog marihuana, his brain is destroyed," leading them to a fatal conclusion: "After the dog is used for a period of time, the detrimental effect of the drug on the brain demonstrates itself. The dog must then be destroyed." None of these scientists reported conducting autopsies on the dead dogs to verify whether the repeated administration of potent cannabis had damaged their brains.[8]

Present-day research no longer inflicts the administration of incapacitating doses of cannabis on dogs. The animals used to test cannabis potency in the 1930s may well have been put to death for no valid physiological reason. Nevertheless, the research on these dogs convinced American scientists that male *Cannabis* plants and the leaves of female plants produce intoxicating resin, albeit in smaller quantities than the flowering tops of female plants. This led them to conclude that the International Opium Convention's definition of "Indian hemp" needed reformulating to include all parts of the plant.[9]

* * *

The scientists consulted by the League of Nations Sub-Committee on Cannabis in 1936 and 1937 were all European or North American men. They tended to frown on cannabis consumption, which they did not seem to have experienced, and some held overtly racist views—which was unfortunately not uncommon for scientists at the time.[10]

French pharmacist Jules Bouquet, for instance, claimed in his reports that "race" affected how people experienced the effects of cannabis. Among "North Africans," Bouquet wrote, the "brilliant display of amusing and witty ideas which occurs in experimentally intoxicated educated Europeans does not take place. There is only a flow of crazy talk, coarse jokes, and obscenities, often on homosexual topics." According to him, Europeans living in North Africa "who have had a fancy to try hemp do not appear to have been driven to commit violent anti-social acts whilst under the influence of the drug itself," whereas "the petty criminal classes" in North Africa "are ardent devotees of hashish, a fact which has doubtless contributed in great measure to their removal from the ranks of the law-abiding community." Bouquet attributed these differences to "race-receptiveness" and claimed that "laziness is the basic element in the character of Moslems" and their "dislike of effort" led them to use drugs, especially cannabis, which he considered "the narcotic best adapted to their mentality." Bouquet did not say how he had reached these conclusions. Still, he admitted having trouble communicating with local cannabis users, writing: "In North Africa, many difficulties are encountered in collecting evidence concerning hashish intoxication by questioning native addicts. They are suspicious and reticent. Most of them, too, are incapable of analyzing their experiences. As a rule, they simply say that they take hemp 'because it makes them happy.' That is as much as can be elicited from them."[11]

Bouquet considered "the different effects produced by the drug on different 'races'" to be "the most disturbing feature of the problem." But his view was at odds with the observations of most other scientists. Already in 1845, Jacques-Joseph Moreau noted that the effects of cannabis varied significantly from one individual to another—not from one "race" to the next. In the 1930s, most European pharmacologists understood that the effects of cannabis could be "very variable even on the same person, owing to individual dispositions and environment" (or "set and setting" in contemporary terms). Yet, Bouquet insisted on "race" as the primary concept for understanding this variability.[12]

Bouquet also believed that cannabis use led to insanity, violence, and crime—especially in non-Europeans. He quoted John Warnock's early Egyptian statistics to back up the claim that more than one-third of the men (and 0% of the women) hospitalized as "insane" in Tunis were "hashish addicts." He reported that long-term "hashish addicts" became dangerous and had "to be placed under restraint, as the result of some crime or at any rate of acts of violence." He wrote that old hashish users experienced "a general weakening of the intellect" and died "after only a few months of senile decay." Last, he minimized the medical utility of cannabis, writing: "Therapeutics would not lose much if it were removed from the list of medicaments."[13]

The Sub-Committee on Cannabis thanked Bouquet for his "outstanding, thorough, painstaking work," and American officials began referring to him as "the outstanding expert on cannabis in the world."[14]

* * *

In April 1937, American authorities held a Congressional hearing in Washington, DC, on a proposed national law called Taxation of Marihuana. They had found a way around the constitutional objections to a federal cannabis law, which consisted of imposing exorbitant taxes on the planting, harvesting, and transfer of the plant. The transcripts of these hearings reveal how the experts and politicians of the time understood and misunderstood cannabis.[15]

The first witness was Harry J. Anslinger, the Federal Bureau of Narcotics director. Usually known for his brutal tactics, sneering tone, and racist outbursts, Anslinger delivered a polished testimony to the congressional committee. He refrained from using racial code words and insisted instead on the susceptibility of young people to the corrupting influence of "marihuana." Drawing freely from Fossier's "Mariahuana Menace," Anslinger began with the story of the Assassins, who "derived their name from the drug called hashish which is now known in this country as marihuana." He deplored the drug's ready availability to schoolchildren and claimed that cannabis use led to insanity. "It is dangerous to the mind and body," he told the Congressional committee, "and particularly dangerous to the criminal type because it releases all of the inhibitions." He referred to Fossier's statistics concerning convicted criminals in the prison of New Orleans. He cited cases from other states as well, such as: "In Florida, a 21-year-old boy under the influence of this drug killed his parents and his brothers and sisters. The evidence showed that he had smoked marihuana. In Chicago, recently two boys murdered a policeman while under the influence of marihuana." Anslinger also quoted Bouquet on the medical obsolescence of cannabis and called him "the greatest authority on cannabis in the world today."[16]

The following testimony came from Dr. James Munch, a pharmacologist from Temple University in Philadelphia. Munch told the Congressional committee that "continued exposure to marihuana" tended to cause "the degeneration of one part of the brain," namely the cerebral cortex. He further testified that the medical use of *Cannabis* extracts was disappearing because the modern drugs for suppressing pain developed since 1880, such as morphine and novocaine, worked better and were easier to dose. Munch explained that these substances could be dissolved in water and injected with a syringe, but *Cannabis* extracts were resinous and oily and could not.[17]

The representatives of the United States Congress did not question the need to legislate against the "weed of insanity," which one citizen described as "possibly the greatest threat which confronts us today." However, they needed a clear understanding of the plant before writing an effective law. First, they wanted to know whether hemp, which seemed to grow like "an ordinary weed" in many parts of the country, belonged to the same species as the plant called "marihuana." Several experts testified that this was indeed the case. Botanist Lyster Dewey, who had spent his career working for the United States Department of Agriculture, declared that "the plant *Cannabis sativa*, so-called by Linnaeus in 1753, constitutes one species of hemp that has been known longer than any other fiber plant in the world."[18]

Considering *Cannabis* as a single botanical species seemed to simplify matters for the lawmakers. One congressman suggested that an effective way of "eliminating the evil" might be to suppress the plant's growth entirely and "kill the weed which kills people." But Congressman Daniel Reed remarked: "I can see a lot of trouble unless this is properly worked out because if you are going to start on a program of exterminating some weed, a weed that grows generally throughout the United States, you are undertaking a program that will be difficult and expensive."[19]

An additional complication for the lawmakers was that specific American industries used *Cannabis* fibers and seeds in their products. A National Institute of Oilseed Products representative testified that the hempseed industry imported an average of sixty-four million pounds of hempseeds yearly, which yielded a quick-drying oil used to make paints, soap, and linoleum. A representative of a birdseed company declared that a ban on hempseed would affect "all of the pigeon producers in the United States, of which I understand there are upwards of 40,000," adding that it would suffice to sterilize the seeds to raise the objection that people could use them for cultivating the plant. This last point carried particular weight because several experts associated the recent spread of wild *Cannabis* plants across the United States with feeding birds hempseed.[20]

One witness representing the American Medical Association, William C. Woodward, expressed concerns about the restrictions the law would impose on research, stating that "future investigation may show that there are substantial medical uses for *Cannabis*." Woodward also opposed the bill's terminology because "*Cannabis* is the correct term for describing the plant and its products. The term 'marihuana' is a mongrel word that has crept into this country over the Mexican border and has no general meaning except as it relates to the use of *Cannabis* preparations for smoking. It is not recognized in medicine, and I might say that it is hardly recognized even in the Treasury Department."[21]

The federal lawmakers did not take these objections into account. After the hearings, they created a law called the Marihuana Tax Act of 1937, which included the following definition: "The term 'marihuana' means all parts of the plant *Cannabis sativa* L., whether growing or not; the seeds thereof; the resin extracted from any part of such plant; and every compound, manufacture, salt derivative, mixture, or preparation of such plant, its seeds, or resin; but shall not include the mature stalks of such plant, fiber produced from such stalks, oil or cake made from the seeds of such plant . . . or the sterilized seed of such plant which is incapable of germination."[22]

The Marihuana Tax Act imposed a $100 per ounce tax on all transfers of "marihuana" (roughly equivalent to $2,000 per ounce today) and concerned anyone who "1) plants, cultivates, or in any way facilitates the natural growth of marihuana; or 2) harvests and transfers or makes use of marihuana." Failure to pay the tax could result in five years of imprisonment. The law aimed to discourage cannabis consumers and producers. Although it made exemptions for industrialists, physicians, scientists, and farmers, its cumbersome registration procedures dissuaded most legitimate uses of the plant. It also impacted landowners with wild *Cannabis* growing on their lands, who became "producers" subject to taxation if they did not kill the weeds.[23]

* * *

Several weeks after testifying at this Congressional hearing, Harry Anslinger traveled to the League of Nations headquarters in Geneva in June 1937 to represent the United States at the annual meeting of the Sub-Committee on Cannabis. Anslinger shared the new American definition of "marihuana" with the members of the subcommittee, who agreed that it solved most of the problems posed by the International Opium Convention's definition of "Indian hemp." However, they objected to the term "marihuana," which they considered too "local" for international legislation. The subcommittee decided to keep most of the American wording and replace the term "marihuana" with "cannabis"— and presto, they had a new definition to propose for the revision of the International Opium Convention.[24]

Progress was less forthcoming on other fronts, however. Many of the subcommittee's questions remained unanswered. Scientists had still not identified the plant's "active principle," and consequently, tests for detecting the presence of psychoactive *Cannabis* could not be infallible. The subcommittee asked Jules Bouquet to draw up a second questionnaire to be submitted to a new round of experts. However, the Second World War broke out in September 1939 and put an end to the subcommittee's activities by the following year.[25]

The Second World War lasted six years and resulted in the death of at least sixty million people. During the war, the League of Nations became inactive, and the preoccupation with *Cannabis* faded from international concern. In late 1945, after hostilities had ceased, representatives from fifty countries came together and founded the United Nations (UN) in the hope of preventing another global war like the one they had just experienced. In 1946, the League of Nations ceased to exist, having handed over its assets and responsibilities, including the control of international drug laws, to the UN.[26]

* * *

The newly founded UN Commission on Narcotic Drugs picked up where the League of Nations advisory committee had left off. Based in New York, the commission's first task was to simplify the existing international drug treaties and bring them together into a new single convention. In 1950, the secretary-general of the UN circulated several preliminary drafts of the future single convention containing radical *Cannabis*-control measures, such as prohibiting the plant's possession and cultivation except for scientific and medical purposes, uprooting wild plants, and requiring cultivators "to destroy, preferably by burning, the Indian hemp." One preliminary draft stated that establishing the medical obsolescence of *Cannabis* would facilitate the general prohibition of the plant.[27]

In 1952, the World Health Organization (WHO)—the UN agency responsible for international public health—published a report by its Expert Committee on Drugs Liable to Produce Addiction containing the following short paragraph about *Cannabis*: "The question of justification of the use of *Cannabis* preparations for medical purposes was discussed by the committee. It was of the opinion that *Cannabis* preparations are practically obsolete. So far as it can see, there is no justification for the medical use of *Cannabis* preparations."[28]

Though the WHO report did not cite a single scientific reference to back up this view, the UN Commission on Narcotic Drugs declared that it "attached special importance" to the opinion of the WHO Expert Committee. It drafted a resolution for the UN Economic and Social Council asking governments to explore the possibility of discontinuing *Cannabis* medicines. The UN Commission on Narcotic Drugs also decided to drop the term *Indian hemp* in favor of the word *cannabis* and asked the WHO to carry out an additional study on the physical and mental effects of cannabis use.[29]

Published in early 1955, the WHO study insisted that cannabis use caused insanity, violence, and crime. It included references to John Warnock's early work at the Cairo asylum. It quoted from an "important book" recently coauthored by Harry Anslinger containing accounts of atrocious murders committed in the United States by people reputedly intoxicated by cannabis. Such as: "a cotton-picker of 25 years of age drank, then smoked a 'reefer,' picked up a 17-month-old baby girl which had been left in the family car, violated and suffocated her; 'the real criminal, in this case, is marihuana,' said the murderer's own counsel." The WHO study concluded that "cannabis constitutes a dangerous drug from every point of view, whether physical, mental, social or criminological."[30]

With such clear statements from the WHO, the general prohibition of cannabis seemed at hand—as long as the UN Commission on Narcotic Drugs could reach an internal consensus on the plant's medical obsolescence. In April 1955, the commission's members considered adopting the position that "cannabis drugs have no further medical value." India, which had recently freed itself from British colonial rule, resisted this proposition and defended the use of *Cannabis* in Unani and Ayurvedic systems of medicine, but this was not sufficient to deter the prohibitionist forces at work.[31]

* * *

The Single Convention on Narcotic Drugs was signed in New York in March 1961. It remains the bedrock of contemporary international cannabis legislation, ratified by 186 countries. The Single Convention introduced a worldwide prohibitionist approach to "narcotic drugs," including penal obligations for signatory countries to criminalize the unlicensed production and trade of these substances, as well as the cultivation of plants such as the opium poppy, the coca bush, and *Cannabis*.[32]

This piece of international legislation superseded all previous drug treaties and placed *Cannabis* and its resin in the most restricted category called Schedule IV, reserved for drugs that are "particularly liable to abuse and to produce ill effects and that such liability is not offset by substantial therapeutic advantages not possessed by [some other drug]." Besides *Cannabis*, Schedule IV initially listed only three other substances, which were synthetic or semisynthetic opioids: heroin, desomorphine, and ketobemidone. The treaty encouraged signatory countries to prohibit Schedule IV drugs, including their production, manufacture, trade, possession, and use, "except for amounts which may be necessary for medical and scientific research only."[33]

In February 1961, during the negotiations leading up to the signature of the Single Convention, the Conference's Technical Committee discussed the listing of each substance one last time. Regarding Schedule IV, the representative of the United States, Dr. Nathan Eddy, declared that "only cannabis was extremely dangerous" compared to the other three drugs in the category. The representative of India, Mr. Tilak Raj, suggested, on the contrary, moving cannabis to a less restricted category and retaining only *Cannabis* resin in Schedule IV. The Technical Committee decided by eleven votes to one, with one abstention, "that both substances should be retained in Schedule IV." *Cannabis* was stuck on the short list of the world's most dangerous, harmful, and addictive drugs, those which "lack therapeutic value."[34]

India managed to extricate the plant's leaves from the Single Convention's most stringent measures. During the last discussions concerning the treaty's text, India's representative, B. N. Banerji, argued that *Cannabis* leaves were "far less harmful than alcohol and were used by the poorer people in India to make a mildly intoxicating drink [bhang] or as a substitute for analgesics and tranquilizers. There was no risk of addiction as they contained hardly any narcotic substance." The delegate from Pakistan agreed, and the two countries managed to sway discussions so that the Single Convention's final definition of "cannabis" omits the plant's leaves. It defines "cannabis" as: "the flowering or fruiting tops of the *Cannabis* plant (excluding the seeds and leaves when not accompanied by the tops) from which the resin had not been extracted, by whatever name they may be designated." This wording also excludes the plant's stalks and seeds to protect their industrial uses.[35]

With its definitions and categories, the Single Convention of 1961 imposed a worldwide legal interpretation of cannabis as a dangerous "narcotic drug" comparable only to heroin and two other opioid molecules produced by chemists. Yet scientists had still not identified the main psychoactive substance of the *Cannabis* plant at the time of the Convention's signature. The molecular structures of CBD and THC were only discovered in 1963 and 1964, respectively.[36]

The Single Convention placed cannabis in a more restricted category than cocaine and morphine. As the molecular structures of these two substances were well understood, it is possible that the unknown nature of the active principle of cannabis contributed to its listing among the world's most dangerous drugs.

The Single Convention went into effect in 1964. During the following decades, it obliged countries worldwide to enact laws resulting in the imprisonment of countless *Cannabis* cultivators and users. It also produced conditions discouraging the scientific study of the plant and its substances, including their therapeutic potential.[37]

# 8
# Spreading on the Wings of Music

In 1927, twenty-six-year-old American jazz musician Louis Armstrong smoked cannabis for the first time. He was backstage at a Chicago music club when a white musician named Mezz Mezzrow lit a reefer, took a few puffs, and passed it to him. At the time, cannabis was still legal in Illinois. "I had myself a ball," Armstrong later recalled. From then on, he openly praised the plant's benefits, viewing it as a "medicine" comparable to the wild herbs his mother fed him as a child in New Orleans. "It's an assistant, a friend, a nice cheap drunk if you want to call it that, very good for asthma, relaxes your nerves."[1]

In the months following this initial experience, Armstrong began smoking cannabis before his concerts and recording sessions. In 1928, he recorded an instrumental piece called "Muggles," a slang term for cannabis. The music evoked the way instruments sounded under the influence of the drug. This recording is considered one of the earliest in jazz history to feature musical improvisation.[2]

In 1931, Armstrong moved to California, where he played almost every night at a music club in Hollywood. However, California law prohibited cannabis. One evening, during the intermission of one of Armstrong's concerts, the police arrested him and a fellow musician after catching them smoking a joint in the parking lot behind the club. Facing a six-month prison sentence for possessing a half-smoked marijuana cigarette, Armstrong spent nine days in the downtown Los Angeles city jail while awaiting trial. Meanwhile, the club's owner and Armstrong's manager hired "a whole gang of lawyers," as Armstrong put it. On the day of the trial, the judge granted him a suspended sentence on the condition that he leave the state. Armstrong took a train back to Chicago, where his fans gave him a triumphant welcome. The arrest for smoking cannabis seemed only to increase his popularity.[3]

Armstrong did not enjoy this initial encounter with the law, but it did not dampen his enthusiasm for cannabis. He continued to smoke and share it with other musicians, often praising its benefits and offering advice. He once told pianist Harry Gibson, "I like blowin' on gage [smoking cannabis] a lot better than booze. But remember, man, it ain't the reefer or the liquor that's playing jazz. It's you, just feeling a lot better listening to it."[4]

Armstrong found that cannabis not only put him in the right mood for playing music but also helped him cope with the daily indignities of racial segregation, which limited where Black Americans could walk, talk, drink, rest, or eat in certain parts of the United States. Armstrong once told white music producer John Hammond, who disapproved of cannabis: "It relaxes you, makes you forget all the bad things that happen to a Negro."[5]

During the 1930s, several other Black American jazz musicians celebrated cannabis by composing and performing "reefer songs." Examples include Cab Calloway's "Reefer Man," Ella Fitzgerald and Chick Webb's "When I Get Low, I Get High," and Curtis Jones's "Reefer Hound Blues." After the passage of the Marihuana Tax Act of 1937, which effectively criminalized cannabis use across the country, the Federal Bureau of Narcotics began targeting jazz musicians who used cannabis. Here, too, the main result of treating musicians as outlaws was to enhance the rebellious allure of their music and of the plant they smoked.[6]

* * *

The growing international popularity of American jazz music contributed to the spread of cannabis. In Nigeria, a British colony and protectorate with no previously known usage of the plant, West African soldiers returning from service in India after the Second World War brought cannabis back with them, and young Nigerians took to smoking it in the music clubs of Lagos. They initially considered it a glamorous foreign drug associated with Black American music and jazz clubs.[7]

As a plant, *Cannabis* was not indigenous to West Africa, but during the early 1950s, people in Nigeria found that it grew profusely in the region's lush soils and tropical climate. *Cannabis* soon became a cheap and locally produced crop and intoxicant, which some Nigerians took to exporting and selling for many times its local price, especially in the United Kingdom. Enterprising individuals used cargo ships that called at the port of Lagos to smuggle Nigerian-grown cannabis out of the country, often hidden in luggage or parcels. Throughout this period, colonial authorities in Nigeria disregarded the drug's consumption and production, and British customs authorities paid little attention to individual cannabis smugglers.[8]

The first mention of illegal cannabis use in the UK dates back to the 1930s when a British newspaper reported that "a 'reefer' or dope habit" was spreading among jazz musicians in the clubs of London. The number of cannabis users surged after the passage of a law in 1948 allowing people from across the British Empire to migrate to the UK, some of whom brought with them the custom of smoking the drug. During the 1950s, most of the cannabis consumed in the UK was smuggled in on cargo boats from Burma, Nigeria, and Lebanon—and some of it ended up in the music clubs of London.[9]

In the late 1950s, a young Nigerian named Fela Kuti was studying at the Trinity College of Music in London when he first smoked cannabis in one of the city's jazz clubs. Inspired by the likes of Louis Armstrong and James Brown, Fela Kuti would go on to fuse West African music with American blues, jazz, and funk to create Afrobeat. He would also become Nigeria's most ardent proponent of cannabis.[10]

By 1960, imported cannabis had become easy to find in many parts of the UK. According to musician John Lennon: "People were smoking marijuana in Liverpool when we were still kids, though I wasn't too aware of it at that period. All these black guys were from Jamaica, or their parents were, and there was a lot of marijuana around. The Beatnik thing had just happened. Some guy was showing us pot in Liverpool in 1960, with twigs in it. And we smoked it and didn't know what it was."[11]

British authorities passed strict laws against cannabis use in 1928 in compliance with the Second Opium Convention but did not show a genuine interest in enforcing them before the early 1960s. As long as most cannabis use was limited to migrant communities, British police paid little attention to it. But in 1961, the Metropolitan Police of London noted for the first time that sizable numbers of native Britons had taken to smoking cannabis. According to one report: "Of the white users of the drug, they are mainly in their late teens or early twenties and are of the type frequenting jazz clubs and coffee bars of the West End. . . . The most dangerous trend is the interest shown by irresponsible young white people of both sexes in this drug."[12]

According to historian James Mills, cannabis attracted working-class and middle-class youths eager to challenge Britain's established order because it had a symbolic meaning. Unlike alcohol and tobacco, the traditional stimulants of the British, cannabis was associated more with the people who had been the subjects of the British Empire. British youths therefore regarded cannabis as "the drug of the 'colonized' rather than the 'colonizer.' At a time when the Empire was in rapid decline and the politics of imperialism were being rejected, to smoke cannabis and to refuse alcohol meant to identify with the oppressed of Britain's past and to dismiss the cultural and political relations of the old order. In this context, consuming the drug became a political act to be staged publicly, not simply a matter of taste that shaped private habits."[13]

In 1962, British authorities began instructing police officers to look out for cannabis users. Predictably, the number of convictions soared. In 1964, for the first time, British police arrested more "white" than "colored" people for cannabis infractions—these were the only two racial categories used in official statistics at the time. British authorities also passed stricter cannabis laws, as recommended by the UN's Single Convention on Narcotic Drugs. The UK's Dangerous Drugs Act of 1965 introduced new offenses relative to cannabis, including cultivating the plant and allowing the drug's use on one's premises. The revised law maintained heavy penalties for all cannabis infractions, including possessing small amounts of the drug. In one case, a person described as a "working-class black man" from Jamaica was arrested in London in 1965 in possession of three grams of cannabis and sentenced to thirty months in prison, which he served in full.[14]

British authorities also began targeting famous white musicians suspected of smoking the drug. In early 1967, the police raided the home of Keith Richards, the guitarist of the Rolling Stones, and arrested him after finding a small amount of cannabis in the possession of one of his guests. Richards faced up to ten years in prison for allowing his premises to be used "for the purpose of cannabis smoking." He stood trial and faced a judge and a prosecutor who expressed disdain for his "decadent" lifestyle. The jury found Richards guilty, and the judge sentenced him to one year in prison. However, Richards had hired one of Britain's most reputed and expensive lawyers, who appealed the sentence, arguing that the prosecution had failed to prove that anyone had, in fact, smoked cannabis on his client's premises. Richards spent twenty-four hours in prison and was released on bail. One month later, the judges of the Court of Appeal agreed that evidence was lacking and overturned the conviction and sentence. They ruled that Richards could not be convicted of "knowingly permitting" what the prosecution had failed to prove ever happened in the first place.[15]

The contrast between what a rich and famous white person and a working-class black person can expect from justice and the law is often stark; in the case of Keith Richards, having the means to hire a good lawyer made a big difference.[16]

Richards subsequently recognized that the judge who sentenced him to spend a year in prison for a cannabis infraction "managed to turn me into some folk hero overnight." As a plant prohibited by national and international laws, *Cannabis* tended to impart an aura of cool rebelliousness to its users. In the 1960s, consuming cannabis was a crime in most countries—but a victimless crime, meaning an illegal act involving only the perpetrator or occurring among consenting adults. People who used cannabis harmed no one except perhaps themselves, which made them seem inoffensive compared to most other criminals, at least in the eyes of some, mainly young, people.[17]

As for the British authorities, they continued targeting and arresting prominent musicians for cannabis possession, including Brian Jones, John Lennon, Yoko Ono, and George Harrison, usually releasing them within hours with a fine and a reprimand—and the guarantee that photographs of them emerging from police custody would be splashed across the nation's newspapers the following day. Many less fortunate Britons were sentenced to prison on similar charges, and by the end of the 1960s, the imprisonment of young people for cannabis infractions had become "one of Britain's most urgent social problems."[18]

* * *

The expansion of cannabis use in the UK after the Second World War and the subsequent crackdown against it have striking parallels with events in Nigeria during the same period. Nigeria became a sovereign nation in 1960 and signed the UN's Single Convention on Narcotic Drugs the following year. In 1962, Nigerian authorities began looking out for cannabis users and producers, and arrests soared. Nigerian psychiatrist Tolani Asuni reported in the *UN Bulletin of Narcotics* of 1964 that his country's cannabis problem had become "gigantic," and people from different walks of Nigerian life had taken to smoking the plant, writing: "The thug and criminal subculture believe that it increases their prowess and courage. Under the influence of cannabis, they become daredevils. Prostitutes and their clientele believe that it has a powerful aphrodisiac effect. Long-distance truck and lorry drivers and taxi drivers attribute to cannabis long staying power and sustained alertness, which they need to do very long hours. Dance-band musicians feel they can excel themselves under the influence of cannabis. Students and older schoolboys have smoked cannabis on festive occasions with the hope of achieving more mirth and enjoyment."[19]

According to Asuni, most Nigerian cannabis users were young city dwellers who belonged to "the fringes of society" and exhibited "disruptive and irresponsible behavior" as well as "disregard for society at large." Asuni noted that the policy of the Nigerian police was "to give as wide publicity as possible to cases associated with cannabis, in the hope that the heavy punishment usually meted out to culprits will deter others."[20]

In 1965, another Nigerian psychiatrist, Thomas Adeoye Lambo, published an article, also in the *UN Bulletin of Narcotics*, analyzing the medical and social problems of cannabis use in Nigeria. Lambo claimed that cannabis smoking caused "many forms of deviant behavior," even though some young Nigerian users "came from good homes." He argued that cannabis use attracted "aggressively minded adolescents temporarily alienated from the norms of the adult world," which led to "the gradual alteration of personality with complete lack of appreciation of the consequences of their acts."[21]

The ideas expressed by these two Nigerian psychiatrists were in line with the views held by most medical experts at the time, including those of the WHO and the UN. In 1966, Nigeria's Federal Military Government took up their warnings. It promulgated the Indian Hemp Decree, which ushered in the death penalty for cultivating the plant and a minimum sentence of four years in prison for smoking or possessing the drug.[22]

That same year, in 1966, Fela Kuti, now living in Lagos and trying to launch his musical career, had an epiphany after smoking cannabis. In his own words: "I used it a couple of times to relax. Then, one night, I went on stage stoned. Man, I used to stand there stiff as a stick. My feet were glued to the floor. That night, I just started jumping, dancing, flying. The music poured out. From now on, I said, we all turn on."[23]

In 1967, Nigeria spiraled into a civil war that lasted three years and caused more than two million deaths, particularly among Biafran civilians. These tragic events interrupted the ordinary course of life in the country, including the implementation of its new cannabis law. By the end of the civil war, some Nigerian soldiers had taken to smoking cannabis, which eventually led the Federal Military Government to reduce the death penalty for cannabis-related infractions to a maximum of ten years in prison.[24]

In the early 1970s, Fela Kuti became an internationally acclaimed musician and an ardent proponent of cannabis. Kuti did not hesitate to smoke cannabis while performing on stage as an act of civil disobedience, and Nigerian authorities did not fail to arrest him on several occasions—with each arrest seeming to enhance his status as a rebel. After one brutal crackdown on Kuti and his entourage in 1977, the musician emerged from prison with a broken leg in a cast and a mischievous smile. Several months later, *The New York Times* dubbed him "Nigeria's Dissident Superstar."[25]

* * *

As anthropologist Neil Carrier and criminologist Gernot Klantschnig observed: "While cannabis has many different cultures of consumption, there has been something of a globalization of its appeal over the twentieth century, especially through its link to various types of music. Long associated with jazz in the US, cannabis's popularity was also boosted by musicians such as Bob Marley and Fela Kuti. For Fela Kuti, cannabis had much symbolism as a symptom of defiance against authority, and this has long been a core of the herb's appeal for many consumers within various countercultures."[26]

Beyond its symbolic force, cannabis has a well-established reputation as a drug that intensifies the perception of sound. Mezz Mezzrow, the jazz musician who introduced Louis Armstrong to cannabis, explained in 1946 that the first time he played music after smoking cannabis, "I began to hear my saxophone as if it was inside my head, but I couldn't hear much of the band in back of me, although I knew they were there. All the other instruments sounded like they were way off in the distance; I got the same sensation you'd get if you stuffed your ears with cotton and talked out loud. Then I began to feel the vibrations of the reed much more pronounced against my lip, and my head buzzed like a loudspeaker. I found I was slurring much better and putting just the right feeling into my phrases. I was really coming on. All the notes came easing out of my horn like they'd already been made up, greased and stuffed into the bell, so all I had to do was to blow a little and send them on their way, one right after the other, never missing, never behind time, all without an ounce of effort."[27]

Present-day studies confirm that cannabis temporarily alters how people perceive music. Electroencephalogram (EEG) studies of brain activity show that cannabis stimulates certain areas of the brain involved in emotions, attention, and sensory perception. This increased stimulation leads to an intensification of the listening process, where sounds seem richer and deeper. It also alters the perception of time so that, on average, 15 seconds are perceived as 16.7 seconds, corresponding to the feeling that chronological time has slowed down—as if musicians under the influence of cannabis could observe music through a magnifying glass, perceiving each note and each silence in greater detail.[28]

Despite these effects, not all musicians find cannabis beneficial. Producer John Hammond, who frequently argued with Louis Armstrong over cannabis, complained that the drug "played hell with time," making music more difficult to control. Moreover, cannabis can hamper reading sheet music and playing a defined arrangement, and musicians need some experience with the drug before they can use it effectively.[29]

Still, contemporary research confirms several ideas initially expressed by Louis Armstrong. It does seem that cannabis can serve as an "assistant" for certain musicians, particularly those seeking to improvise and explore tonality and tempo. It does not change the music itself nor affect how our ears function. Instead, it influences how we perceive music. To paraphrase Louis Armstrong: "It ain't the reefer playing jazz. It's you, just feeling a lot better listening to it."

# 9

# A Drug for the Pursuit of Knowledge

In 1955, at the age of fifty-five, the Belgian writer and artist Henri Michaux began his first experiments with drugs. At the time, he lived in a fashionable neighborhood of Paris, where he was known for his courteous and composed manners. Over the next five years, he consumed mescaline, psilocybin, LSD, and cannabis, publishing several groundbreaking books recounting his experiences. Before embarking on these explorations, Michaux took care to study the scientific literature and consulted a doctor and a psychiatrist to learn about dosages and expected effects. When he spoke of "hashish," he sometimes referred to an edible form of cannabis and, at other times, to the plant's smokable resin.[1]

The first time Michaux tried cannabis, he was sitting in his apartment "waiting for the visions to appear" and flipping through some photographs to pass the time. Suddenly, he found himself noticing unusual details in the images—instead of focusing on the camel in the foreground, his gaze was drawn to the jagged peaks in the background. He realized that he could scrutinize the crevices in the rock with an "admirable optical dexterity." Photographs freeze a moment in time and space, but Michaux found that cannabis changed his gaze, writing: "Hashish de-photographs the places photographed so that *you can at last get in. The frozen is unfrozen.*" With cannabis, he discovered that he could explore the underlying relationships and movements within a static image. "I see in depth," he wrote, referring to both spatial and psychological depth and comparing the experience to "stereovision."[2]

To Michaux's surprise, cannabis seemed to act as an intelligent entity: "One can set hashish to work, ask it questions, give it problems to solve. It will have an answer, and perhaps most astonishing, it accepts the facts of the problem. But you have to be careful because its solution may make you and the facts of your problem look foolish."[3]

After trying cannabis several times, Michaux felt comfortable enough to test its effects beyond the confines of his apartment. The drug made him feel "pervaded by a feeling of kindness," which encouraged him to walk the streets of Paris, observing the world and himself looking at it. "When I go out after having taken hashish, I am another man with another gaze. Hashish designates, selects, observes, and penetrates with a blade that does not waver. Without it, I see the world like cattle do. . . . With hashish in me, I am a hunting falcon."[4]

He regarded cannabis as a mild hallucinogen "lacking in great scope, but manageable, convenient, repeatable, and without immediate danger." For him, it was an "admirable detector" and a valuable tool for understanding the world. He claimed that cannabis helped deepen his understanding of philosophical texts, allowing him to "hear the authors in person." Under its influence, he found that authors could no longer hide behind their words; they came "to the forefront," revealing their petty ambitions, self-satisfaction, and ignorance. "Hashish opens up the internal space of sentences, and hidden concerns emerge—it pierces them at once. It is curious that this hashish, when I test it on certain authors, never showed itself to be vain or eccentric. Set loose upon its prey, there was no turning back, no room for play. It was as focused as a falcon. Authors exposed this way never fully recovered their former mantle or retreat."[5]

Michaux acknowledged that cannabis could be deceptive and required effort to use effectively, noting: "One must learn to use it and patiently experience the mental upheavals it brings." While he considered the drug a more accessible means of acquiring knowledge than other hallucinogens, he emphasized the importance of personal effort in drawing lessons from it, writing: "Hashish—traitor, hunting dog, instructive guide. It sees faster than we do, pointing to what we have not yet understood. From the start, and each time, an effort is required. That is why it has not been used for this purpose."[6]

Michaux noted that history provided few examples of cannabis being used as a means of acquiring worldly knowledge. Historically, it has been used for many reasons: to relieve pain from various illnesses, to endure harsh physical conditions, to meditate on spiritual matters, to bolster courage before battle, to facilitate physical labor, to promote relaxation, to forget personal troubles, to enhance sensory pleasure, and to accompany music, singing, storytelling, and games. Perhaps the Arabic-speaking people living in the 1200s who nicknamed cannabis the "shrub of understanding" used it to deepen their perception of the world, and perhaps the Indian people living in the 1400s who described one of its effects as "inspiring mental power" did as well. However, history provides no further details in either case.[7]

Nor did the Europeans and North Americans who experimented with cannabis before Michaux dwell on it as a potential source of knowledge. The writer Charles Baudelaire, for instance, declared in 1860, in his work *Les paradis artificiels (Artificial Paradises)*, that cannabis users sought to "create paradise through pharmacy . . . like a lunatic who would replace solid furniture and real gardens with scenery painted on canvas and mounted on frames." Baudelaire relied on Silvestre de Sacy's interpretation of the myth of the Assassins, according to which the sect's leader drugged his disciples with hashish to make them believe in an artificial paradise. This illusion was then used to manipulate them and secure their suicidal loyalty. Convinced of the story's veracity, Baudelaire thought it proved that using cannabis led only to illusions and false knowledge.[8]

A century after Baudelaire, Michaux took a different view. He began one of his books on mescaline and cannabis with the statement: "Drugs bore us with their paradises. Let them give us some knowledge instead. The century we are in is not suitable for paradise." The contrast with *Artificial Paradises* was clear for French-speaking readers to the extent that Michaux did not even need to mention Baudelaire. In fact, he did not refer to him at all in his works.[9]

By breaking with Baudelaire in this indirect way, Michaux staked out a new intellectual position. It proposed that cannabis could lead to the acquisition of useful, reality-based knowledge rather than to an artificial paradise for dupes.

* * *

In 1958, Henri Michaux's writings on mescaline and cannabis caught the attention of two American writers living in Paris: Allen Ginsberg and William Burroughs. Both authors were part of the Beat generation, a literary movement that celebrated lived experiences, drug experimentation, sexual liberation, and jazz improvisation. The previous year, in 1957, US authorities had banned Ginsberg's book *Howl and Other Poems*, deeming it an endorsement of drug use and homosexuality. The collection contained lines such as "I smoke marijuana every chance I get" and descriptions of men who "screamed with joy" while having sex with one another. Ginsberg was prosecuted for "obscenity," but after a highly publicized trial, a judge ruled that his work had "redeeming social importance" and allowed its distribution in the United States. This decision made Ginsberg a national countercultural figure. However, his friend and mentor, William Burroughs, had not yet attained such notoriety.[10]

One day in July 1958, Allen Ginsberg sent a note to Henri Michaux, introducing himself as a young American poet with "much experience in the same hallucinogenic field." He proposed meeting to exchange ideas. Although Michaux was known as a recluse, he accepted the invitation and, a few days later, visited the run-down hotel where the Americans were staying. Dressed in a suit and tie, Michaux struck Ginsberg as "a sharp-eyed elderly man." According to Ginsberg, the encounter was significant: "He was the first substantial poet I had the opportunity to encounter who did know something about honorific use of drugs—all the younger American poets by that time, unbeknownst to their French contemporaries, had done extensive research into the fields of consciousness that could be catalyzed by peyote and hashish and mescaline. But we had no elder soul in America to check our senses with."[11]

Despite their cultural and generational differences, Michaux and Ginsberg spoke for hours, switching between broken French and English, and became friends. The following evening, Michaux returned to the Americans' hotel, bringing a chicken to cook for dinner, and met William Burroughs for the first time. A frail man in his forties, Burroughs had a deep interest in drugs and told Michaux about his quest for ayahuasca, a hallucinogenic vine he had sought in South America. The three writers spent the evening discussing poetry and hallucinogens.[12]

Their exchange proved mutually beneficial: Ginsberg helped Michaux find an American publisher for the English translations of his books. Burroughs, following Michaux's example, became a visual artist in addition to being a writer. Michaux, in turn, inspired his American colleagues to place their discussions of cannabis and psychedelics on higher ground by considering them as sources of knowledge.[13]

Burroughs's writings show a striking evolution on this count. Two years before meeting Michaux, he had published a letter on "dangerous drugs" in the *British Journal of Addiction,* in which he listed the effects of cannabis: "Disturbance of space-time perception, acute sensitivity to impressions, flight of ideas, laughing jags, silliness." He also emphasized its possible unpleasant effects, such as making bad situations worse and turning depression into despair. At no point did he suggest that cannabis could lead to any form of knowledge. However, three years after meeting Michaux, he took a different stance. In 1961, at the annual meeting of the American Psychological Association in New York, Burroughs presented a paper on "sedative and consciousness-expanding drugs," in which he stated: "It is unfortunate that cannabis (the Latin term for preparations made from the hemp plant, such as marihuana and hashish), which is certainly the safest of the hallucinogen drugs, should be subject to the heaviest legal sanctions. Unquestionably, this drug is very useful to the artist, activating trains of association that would otherwise be inaccessible, and I owe many of the scenes in *Naked Lunch* directly to the use of cannabis. . . . Cannabis serves as a guide to psychic areas, which can then be reentered without it. . . . It would seem to me that cannabis and the other hallucinogens provide a key to the creative process and that a systematic study of these drugs would open the way to non-chemical methods of expanding consciousness."[14]

Burroughs's groundbreaking novel *Naked Lunch*, published in 1959, chronicled the disjointed adventures of a man addicted to heroin. Its raw depiction of taboo subjects and use of the cut-up technique subsequently influenced numerous writers, musicians, and artists. In 1961, when he told the psychologists gathered in New York about how he used cannabis to his intellectual and creative benefit, he displayed considerable courage. His talk marked the first time an English-language author formally claimed to have gained valuable ideas by consuming cannabis, and he made this claim in the very city where, just months earlier, the United Nations had short-listed cannabis among the world's most dangerous and useless illegal drugs.[15]

* * *

In the early 1960s, Allen Ginsberg became an influential voice for young people in the United States and beyond. He traveled far and wide, visiting Peru, Morocco, India, and other countries. Whenever he passed through Paris, he made a point of contacting Henri Michaux. On one occasion, in June 1965, Ginsberg arrived in Paris at the end of a long journey through India and Eastern Europe. With no money, no place to stay, and only a plane ticket back to the United States a week later, he recalled: "Michaux like kindly father offered a few thousand francs which I was ashamed to accept. We made an appointment for lunch the next day. And went, of all places this time, to La Coupole in a taxi and ate shellfish and rare meat and talked about India." Ginsberg referred to Michaux as "the lovely teacher."[16]

Back in the United States, at the end of June 1965, Ginsberg embarked on numerous projects. He collaborated with musician Bob Dylan, spoke at peace rallies across the country, received a grant from the prestigious Guggenheim Foundation, and testified before the US Senate about LSD. He also began discussing cannabis as a means of gaining worldly knowledge.[17]

In November 1966, Ginsberg published a long article on cannabis in the prominent and usually staid *Atlantic Monthly*. The article was titled "The Great Marijuana Hoax: First Manifesto to End the Bringdown." Ginsberg kept the core of Michaux's argument while adopting a more direct and combative tone, declaring in the opening paragraph: "The actual experience of the smoked herb has been clouded by a fog of dirty language perpetrated by a crowd of fakers who have not had the experience and yet insist on degrading it."[18]

Ginsberg summarized his main argument as follows: "Although most scientific authors who present their reputable evidence for the harmlessness of marijuana make no claim for its surprising *usefulness*, I do make that claim: Marijuana is a useful catalyst for specific optical and aural perceptions. I apprehended the structure of certain pieces of jazz & classical music in a new manner under the influence of marijuana, and these apprehensions have remained valid in years of normal consciousness." He explained that smoking cannabis allowed him to view paintings by masters such as Rembrandt van Rijn, Paul Klee, and Paul Cézanne in a new light. He proposed that cannabis shifted attention in consciousness toward "a more direct, slower, absorbing, and occasionally microscopically minute engagement with sensing phenomena."[19]

These words, printed in a respected and widely circulated American magazine in late 1966, stood in stark contrast to the claims made by most experts during the preceding decades, which asserted that cannabis use led to crime, violence, and personal degradation. Ginsberg's claim that cannabis could enhance the acquisition of knowledge did not go unnoticed, nor did it stop many from trying it to see for themselves.

# 10

# World Sunlight Champion

During the late 1960s and early '70s, cannabis consumption surged in the United States and many other countries, propelling the plant to fame. Tens of millions of new users took to smoking its dried leaves, flowers, or resin. The jagged *Cannabis* leaf became an icon displayed on clothing, posters, pins, and flags. In a world where international authorities classified the plant as a dangerous and illegal drug, those who displayed this leaf image manifested its vegetal nature and showed their opposition to the official stance. The friction between aficionados and detractors turned *Cannabis* into "the world's most recognizable, notorious, and controversial plant."[1]

Botanists refer to *Cannabis*'s signature leaf as a "fan leaf" for its distinctive shape. Like the leaves of many other plants, it absorbs the energy of sunlight and uses it to split water and carbon dioxide molecules to produce glucose, the plant's primary energy source. This process, known as photosynthesis, also produces oxygen by-products that can damage other molecules. To guard against this, most plants produce antioxidant compounds. As a sun-loving species, *Cannabis* generates a variety of unique antioxidants. For example, cannflavins are compounds found in high concentrations in the fan leaves and have been shown to possess thirty times the anti-inflammatory activity of aspirin in human cells. In some parts of the world, *Cannabis* leaves have traditionally been used to dress wounds.[2]

The upper surface of a fan leaf is covered with microscopic hairs that contain calcium crystals, making it rough and unpleasant for herbivores to chew. The leaf's underside is dotted with tiny, sac-shaped glands that secrete bitter terpenes, which also deter grazing mammals. These glands produce small amounts of cannabinoids, although leaves contain about ten times less than the flowering tops of female plants.[3]

During the 1970s and '80s, many new users attempted to smoke *Cannabis* leaves, often with disappointing results. American writer Michael Pollan, for example, recounts an episode from 1982, when he germinated a few seeds from tropical Hawaii and planted the seedlings at the back of his garden in temperate Connecticut. Several months later, he found two tall, leafy plants that "looked like a pair of Christmas trees" but refused to flower as the autumn frosts set in. The seeds likely came from a variety adapted to Hawaii's long growing season, but they produced plants that were unable to reach maturity in Connecticut's shorter season. One day in early October, Pollan panicked upon learning that the man delivering his firewood was also the local sheriff. Within five minutes, he chopped down both plants, hacked off their branches, and "stuffed the fragrant mass of foliage into a pair of heavy-duty trash bags." A few days later, he smoked some of the dried leaves and found that the effect "had less in common with a high than with a sinus headache." He also noted that the leaves had lost their fragrance and "smelled like old socks."[4]

* * *

The aroma of *Cannabis* depends on the plant's variety, sex, and maturity, as well as the conditions under which it is grown, harvested, and stored. The potent scent of a mature female plant primarily comes from terpenes—aromatic compounds responsible for familiar fragrances such as lemon, lavender, and pine. As lightweight hydrocarbons, terpenes evaporate easily at room temperature and volatilize into the air, making them detectable to both humans and animals.

Many plant species produce terpenes for various purposes, such as repelling predatory insects, defending against microbes, and sending chemical signals to other plants. Terpenes are among the most energy-expensive substances that plants produce, requiring three times more energy to synthesize than glucose. This explains why plants often restrict terpene production to the floral tissues surrounding their vital reproductive organs. Many species, including *Cannabis*, coat their flowers with terpenes like limonene and pinene, which mimic the alarm pheromones of ants. Some ants produce substances chemically related to these terpenes to communicate danger to one another, and plants appear to use this chemical language to deter the insects from their flowers.[5]

Like many other plants, *Cannabis* secretes terpenes in tiny resin glands that cover its flowers. These glands form in dense clusters on female plants and sparser clusters on male plants. Under a magnifying glass, they resemble tiny mushrooms protruding from the plant's surface, each with a stalk and a bulbous head. *Cannabis* produces and stores most of its terpenes and cannabinoids in the small round structures at the tips of these glands.[6]

But unlike most plant species, *Cannabis* relies on the wind to spread its pollen. It doesn't need to attract animal pollinators, so it can dispense with colorful flowers, enticing aromas, and nectar. Instead, it can focus its terpene production on self-defense. By secreting aromatic, insect-repelling terpenes in its floral tissues and bitter terpenes in its lower leaves, *Cannabis* tailors its defenses to the most likely causes of damage—insects are more prone to harm the flowers while grazing mammals are more likely to eat its lower leaves.[7]

*Cannabis* stands out among nearly all plant species for producing an exceptionally wide variety of terpenes, including those that give off the aromas of lemon, mango, celery, carrots, jasmine, lavender, mint, hops, vanilla, bergamot, and pine. As a species, *Cannabis* generates more than 150 distinct terpenes. In comparison, a typical European pine tree produces only about fifteen, half of which smell like pine. Few plants rival *Cannabis* in terms of terpene diversity.[8]

*Cannabis* resin glands also produce cannabinoids, using the same chemical building blocks as terpenes. Despite this shared origin, cannabinoids are larger and heavier molecules than terpenes, making them less volatile and unlikely to contribute to the plant's aroma. Police dogs do not detect cannabis by sniffing THC or any other cannabinoid, but by identifying the oxidized form of one of the plant's spicy terpenes.[9]

*Cannabis* also produces a range of volatile sulfur compounds, which can influence the distinctive aromas of certain varieties—skunky, diesel-like, fruity, sweet, creamy, or savory. Research suggests that the plant secretes these powerful aromatic compounds in extremely low concentrations.[10]

The smell of "old socks" given off by some cannabis samples traces to a substance formed during the oxidation of the building blocks of terpenes and cannabinoids. Known as isovaleric acid, this by-product has a rancid and sweaty smell. Certain bacteria produce this same substance when they break down proteins in human sweat, while some fungi involved in the fermentation of blue cheese also generate it—meaning that isovaleric acid is also the major component of the smell of foot odor and overripe cheese.[11]

The presence of isovaleric acid in hastily harvested *Cannabis* plants, such as Michael Pollan's 1982 harvest, is understandable. Rough handling of *Cannabis* foliage can break individual leaves, exposing their contents to air and promoting oxidation. Since these leaves contain the building blocks of terpenes and cannabinoids, they will likely emit a cheesy aroma within a few days.

* * *

People in certain parts of South Asia have long understood that producing high-quality psychoactive cannabis takes know-how and skill. By the late nineteenth century, British colonial authorities in India became aware of this thanks to an Indian civil servant named Hem Chunder Kerr, whom they tasked with investigating ganja production in Bengal. Published in 1877, Kerr's report centered on local knowledge of the *Cannabis* plant's life cycle. He noted that the species had the "most remarkable peculiarity" of developing male and female organs on separate plants. He explained that male flowers produced pollen, which was carried by the wind, while female flowers received this airborne pollen and developed seeds. Kerr reported that Bengali farmers preferred to remove male plants before they released their pollen because unpollinated female plants developed more flowers and produced greater amounts of psychoactive resin. Kerr observed that in these unpollinated female flowers, "a slight exudation also takes place, which gives them a sticky, unctuous feel; but after impregnation, this disappears, and the ovaries become dry and comparatively inert: hence the anxiety to remove the male plants and enable the ovaries to secrete and retain the resin to perfection; and these ovaries, as also their surrounding appendages . . . constitute ganja."[12]

Kerr's report provided a detailed account of the expertise and intensive labor required to produce ganja. Bengali farmers cultivated *Cannabis* in open fields with maximum exposure to sunlight, enhancing the plant's resin production. They favored well-plowed, fertilized, sandy soil as the plant did best in rich, well-drained terrain. They hired "ganja doctors" to identify and remove male plants before they flowered and released pollen. Kerr noted that if female plants were pollinated and began producing seeds, the resulting ganja was "very inferior and scarcely saleable." Farmers waited until unpollinated female plants reached full floral maturity before harvesting them; otherwise, "the narcotic generally turns out to be of inferior quality." According to Kerr, Bengali farmers followed careful procedures when they cut and dried *Cannabis* flowers, such as removing "superfluous leaves," which ganja smokers considered "unfit for use."[13]

In 1893, sixteen years after Kerr's report, the British Parliament ordered the creation of the Indian Hemp Drugs Commission to study cannabis use in British India. This commission's report drew heavily on Kerr's observations of Bengali farmers, particularly regarding the definition of ganja. The report noted: "The advantage of isolating the females is known to some persons, the knowledge having been handed down from a time when ganja was openly cultivated [i.e., before colonial taxation], and it is difficult to believe that it should be confined to a very small number. The fakirs all over the country, who are mentioned as the principal secret cultivators, are certainly not ignorant on this point and would disseminate the knowledge."[14]

Fakirs probably disseminated ganja cultivation techniques across

British India during this period, but the reports commissioned by British authorities disseminated them internationally. In 1915, when the United States Department of Agriculture published a booklet advising American farmers on the cultivation of drug plants, its recommendations on *Cannabis* cultivation closely mirrored those in Kerr's report, placing a particular emphasis on removing male plants before they released pollen.[15]

Half a century later, in the 1960s and '70s, as the plant's cultivation and use had become illegal worldwide, growers in Mexico and the United States rediscovered the technique of growing seedless psychoactive *Cannabis*. After appreciating the stimulating effects of the drug made from such plants, several Americans appear to have learned this technique in Mexico. When they began importing seedless Mexican marijuana to the United States, they dubbed it *sinsemilla*—two Spanish words, *sin* ("without") and *semilla* ("seed"), stuck together to form a new word in American English. Sinsemilla refers to a production technique, like fine wine-making, as well as to the cannabis it produces, rather than to a specific variety of the plant.[16]

The "sinsemilla revolution" in California during the 1970s was actually the rediscovery of an ancient, labor-intensive technique first practiced by farmers in South Asia. The words *ganja* and *sinsemilla* are synonymous and refer to the same growing technique and the same product—seedless female *Cannabis* inflorescence.[17]

* * *

In 1977, an American published a book called *The Primo Plant: Growing Sinsemilla Marijuana*. Writing under the nickname "Mountain Girl," Carolyn Garcia (née Adams)—one of the original members of Ken Kesey's Merry Pranksters and a partner and future wife of Jerry Garcia, the singer and lead guitarist of the Grateful Dead—was a well-known figure in the counterculture movement. Mountain Girl drew on her experience to describe how to grow seedless *Cannabis*. The daughter of a botanist, she had a way with plants and a gift for sharing her knowledge about them. "Knowledge is the key to success," she wrote, recommending that readers study their local climate and consider using a greenhouse "anywhere that is cold in November." She also advised choosing seeds from *Cannabis* varieties that could mature in local conditions. "You get what you plant," she wrote, noting that tropical varieties usually failed to mature at temperate latitudes. Mountain Girl encouraged growers to select south-facing planting sites because "top-to-toe all-day sunshine is essential for perfect plants; no plant should shade another." She also explained why unpollinated female plants produce more resin: "Seeds are the final, ultimate purpose of marijuana life, so the strong female plant will make as many as it can (if you let it). Pot seeds are high in proteins, fats, and stored sugar to start that baby plant. The plant will spend its entire store of nutrients and energy making pounds of seeds, not resin." She observed that unpollinated female plants "will continue to produce new flowers every day at the tops of the flower cluster until the tops are huge. Plants that have been heavily pollinated seem to stop growing new flowers right away and throw all their energy into seed making."[18]

Carolyn "Mountain Girl" Garcia noted that growers who succeeded in cultivating mature seedless *Cannabis* plants risked not having seeds for the next season. She advised them to preserve pollen from a carefully selected male plant—one with a "healthy appearance that is also an early bloomer"—and to manually pollinate a small number of flowers on a selected female plant. "Seeds from homegrown of past years will have improved over the original imported strain only if the grower has deliberately selected his very best plants each year to use for seed," she wrote. "Fifth-generation domestic grass may have degenerated considerably if no selective process was used. In some countries where families have been cultivating marijuana for generations, breeding carefully for their ideal plant, the seeds are jealously guarded." With these few sentences, Mountain Girl outlined the basic principles of *Cannabis* breeding.[19]

During the 1970s, several knowledgeable growers in California began crossbreeding early maturing varieties from northern Mexico, Afghanistan, and South Africa with highly psychoactive but slower-maturing plants from equatorial regions, such as India, Thailand, and Colombia. Their goal was to create potent hybrids with faster flowering times. By the early 1980s, seeds from these new hybrid varieties became widely available through international mail-order sales from seed banks in the Netherlands. This allowed people living in temperate countries to grow potent psychoactive *Cannabis*, especially if they applied the sinsemilla technique.[20]

* * *

As a wind-pollinated species with male and female flowers on separate plants, *Cannabis* mixes its genetic material over long distances. Few plant species use this reproductive strategy, which relies on extensive pollen dispersal. Male *Cannabis* plants release vast amounts of pollen grains that can travel more than one hundred kilometers (sixty-two miles) on favorable winds. A single female plant can capture airborne pollen—often from numerous male plants—and produce thousands of genetically diverse seeds. All living members of the *Cannabis* species can cross-pollinate this way, enhancing the species' genetic diversity and adaptability. Botanists who breed *Cannabis* in research centers recommend an isolation distance of about five kilometers (three miles) from other plants—whether wild or cultivated—to minimize the risk of wind pollination by unknown male plants. One of the first difficulties in growing seedless *Cannabis* is finding a planting site sufficiently secluded from airborne *Cannabis* pollen.[21]

Humans have long selected and bred *Cannabis* plants, some for their strong fibers or nutritious seeds, others for their psychoactive or medicinal resins. While *Cannabis*, as a species, is amenable to cultivation, plants that revert to the wild soon "turn weedy" and lose the traits enhanced through human selection. As Mountain Girl noted, maintaining desired traits in given varieties takes careful, ongoing selection and breeding.[22]

During the 1970s, growers in California and other parts of the United States—all of whom operated outside the law—worked out how to select, breed, and cultivate potent, seedless *Cannabis*. By 1978, demand for sinsemilla was such that it commanded, ounce for ounce, the price of gold. *The New York Times* printed the word *sinsemilla* for the first time in 1979, describing it as a cultivation technique and a highly potent form of cannabis. The newspaper spoke of a new "gold rush" in California, explaining: "The key is the astronomical price being paid for the local marijuana, which many connoisseurs consider among the most potent in the world." The article quoted a district attorney from Northern California, who described a recent and massive increase in violent crimes related to sinsemilla cultivation and claimed that growers feared assaults by armed criminals more than arrests by police officers. To combat this growing phenomenon, the newspaper reported, police officers had taken to flying over remote areas in small planes during the late summer, "using $5,500 gyrostabilized binoculars to scan the mountainsides for bright green plants."[23]

Finding a planting site isolated from *Cannabis* pollen was one thing, but growing plants so they remained undetected by criminals and police officers in low-flying planes was another. An American horticulturist named Kayo summed up the difficulties facing growers in such a context: "As farmers, they need sunshine, and as outlaws, they need seclusion; but where they find sunshine, there is no seclusion, and where they find seclusion, there is no sunshine." Kayo called this contradiction "the cultivators' dilemma."[24]

In his 1982 book *The Sinsemilla Technique*, Kayo elaborated on the importance of sunshine for *Cannabis*. He explained that plants cultivated under intense sunlight developed dense flower clusters, which improved productivity and offered a solution to the cultivators' dilemma. By growing a small number of plants on a secluded, south-facing plot, he argued, cultivators could use the sinsemilla technique and produce cannabis that was potent enough—and therefore lucrative enough—to compensate for the small size of the harvest. With sinsemilla, less could be more.[25]

* * *

During the 1970s and '80s, while outlaw cultivators like Carolyn "Mountain Girl" Garcia and Kayo gained hands-on knowledge about the plant, most scientists kept their distance. International laws defined *Cannabis* or certain parts of the plant as an illegal drug, making it difficult for scientists to study it.[26]

More recently, the move to re-legalize *Cannabis* in certain countries has renewed scientific interest in the plant and drug. Today, researchers publish thousands of studies on the topic every year. As it turns out, this growing body of knowledge validates many practices and observations of nineteenth-century ganja farmers and twentieth-century sinsemilla growers. For instance, studies show that higher light levels increase the size and density of unpollinated female *Cannabis* flowers, resulting in higher yields of harvested plant material.[27]

Research also confirms that *Cannabis* has an exceptional capacity to use sunlight. Most plant species reach a saturation point beyond which they cannot absorb additional light energy for photosynthesis. However, three leafy species reportedly do not exhibit a clear light saturation point: sunflower, maize, and *Cannabis*. Under ideal conditions, these fast-growing, sun-loving annual herbs can continue absorbing light even at the highest natural intensities. Known as "peak sun," this upper limit of solar radiation corresponds to midday sunlight on a clear day at the equator. While the yields of sunflower and maize level off under these conditions, laboratory research shows that *Cannabis* yields continue to increase beyond peak sun. This astonishing capacity distinguishes *Cannabis* as the plant that makes the most of intense sunlight. It also suggests that cultivators seeking to increase yields do well to expose their plants to as much light as possible.[28]

*Cannabis*'s remarkable ability to harness intense light relies in part on the photoprotective properties of its cannabinoids. These molecules can absorb and partially neutralize the most harmful wavelengths of ultraviolet (UV) radiation that reach Earth's surface. Since cannabinoids are primarily secreted around the floral tissues, one of their functions may be to protect the plant's vulnerable reproductive organs from UV-induced damage.[29]

Most of the plant's roughly 125 cannabinoids arise through a cascade of degradations, in which each UV-absorbing molecule transforms into another, smaller molecule with distinct properties. Collectively, these compounds span the most hazardous regions of the UV spectrum and help form a protective shield against photodamage. The plant's resin glands secrete five primary acidic cannabinoids, including THCA (delta-9-tetrahydrocannabinolic acid) and CBDA (cannabidiolic acid). These compounds accumulate at the tips of the glands, just above the plant's surface, where they absorb UV photons and undergo photochemical transformations. For example, UV exposure causes THCA to lose a molecule of carbon dioxide ($CO_2$) and degrade into THC. Like its precursor, THC also absorbs UV radiation, though less efficiently, and it too degrades, ultimately forming CBN, the cannabinoid capable of absorbing more UV photons at harmful wavelengths than any other.[30]

A similar sequence occurs with CBDA, which can absorb significant UV energy before converting into CBD. Although CBD is less UV-absorbent, it can also break down into a degradation product, HU-331, which again demonstrates stronger UV absorption than its precursor.[31]

In addition to providing photoprotection, cannabinoids act as antioxidants and likely help maintain the chemical integrity of other compounds in the resin, particularly the terpenes. Cannabinoids neutralize highly reactive, oxygen-rich molecules—by-products of photosynthesis that can strip electrons from surrounding substances. By donating electrons, cannabinoids diffuse this oxidative threat, undergoing degradation in the process, while helping to keep the terpenes in a volatile and functional state.[32]

Cannabinoids also possess a variety of other protective properties. A recent study found that several primary acidic cannabinoids deter certain insects from feeding on the plant's leaves. However, their effectiveness as repellents seems limited, as many herbivorous insects continue to consume *Cannabis*.[33]

Cannabinoids may also help protect the plant from thermal stress and drought. At high temperatures, resin glands can rupture and release their contents, coating the plant's surface with a cannabinoid-rich glaze. This layer helps retain moisture and protects the plant from heat, similar to the waxy cuticle found in cacti.

Last but not least, cannabinoids contribute to an insect-trapping mechanism. When resin glands burst upon contact, the volatile terpenes evaporate, and the remaining cannabinoids solidify into a sticky adhesive that immobilizes the insect, similar to flypaper.[34]

Cannabinoids serve a variety of protective roles, shielding the plant from UV radiation, oxidative stress, herbivores, heat, and drought. While they may not be the most potent natural compounds in any single role, their broad and overlapping functions make them notably multifunctional.[35]

In brief, cannabinoids are versatile compounds—Swiss Army knife–molecules created by a Swiss Army knife–plant for its own defense.

* * *

To date, scientific research on cannabinoids has primarily focused on their potential as medicines and intoxicants for humans. Human cells share similarities with plant cells in various ways. For instance, both are vulnerable to oxidative damage, and both utilize similar antioxidant defenses. In humans, oxidative damage contributes to numerous serious illnesses. Thankfully, plant-based antioxidants are also effective in human cells—which is why consuming fruits and vegetables is an effective way to boost cellular defenses with plant-derived antioxidants.[36]

As *Cannabis* produces a variety of unique antioxidants, research has increasingly focused on their therapeutic potential, especially that of CBD. In addition to absorbing UV radiation, CBD helps protect lipids and proteins from oxidation. These antioxidant properties have anti-inflammatory effects on human cells, which are composed mainly of lipids and proteins. Oxidative stress and inflammation are linked to certain cancers, neurodegenerative diseases, immune disorders, intestinal inflammation, cardiovascular diseases, and arthritis. Preliminary research suggests that CBD may have anticancer properties, support immune regulation, offer neuroprotection, and improve conditions such as epilepsy, autism, psoriasis, and Alzheimer's and Parkinson's diseases. One of CBD's degradation products has also shown strong anticancer activity in animal cells, shrinking tumors without the adverse effects associated with conventional chemotherapy. Neither CBD nor its breakdown products induce euphoria, which presents an additional advantage from a medical standpoint, as patients can benefit from these compounds without feeling intoxicated.[37]

Only four or five of the plant's cannabinoids, aside from THC, show psychoactive properties. THC's precursor, THCA, is nonpsychoactive. Since the living plant contains significantly more THCA than THC, consuming raw cannabis usually results in little to no intoxicating effect. To activate THCA by turning it into THC, users must heat cannabis to a sufficient temperature, either by toasting, cooking, vaporizing, or smoking the plant material.[38]

THC shares the same chemical composition as CBD but has a slightly lower capacity to absorb UV radiation. The key difference between the two is in their atomic structure, which gives them distinct effects. Like CBD, THC exhibits antioxidant, anti-inflammatory, neuroprotective, and anticancer properties. However, unlike THC, CBD does not produce euphoric or intoxicating effects and may even reduce them in some cases.[39]

* * *

All in all, *Cannabis* produces a range of unique substances, several of which may have protective or health-enhancing properties in humans. The evolutionary origin of these unusual compounds remains speculative. Plant scientist David Potter offers an intriguing hypothesis: *Cannabis* may have originated in regions with intense sunlight and nitrogen-poor soils, similar to present-day Afghanistan. Unlike many other plants that produce nitrogen-based alkaloids for defense—such as tobacco, which synthesizes the natural insecticide nicotine—*Cannabis* may have lacked access to sufficient nitrogen. As a result, it evolved to rely on its most abundant resource, sunlight, to produce a variety of nitrogen-free compounds—terpenes, flavonoids, and cannabinoids—that help shield it from herbivores, insects, molds, heat, and UV radiation. In this view, the plant became chemically creative out of necessity.[40]

The events outlined in Potter's hypothesis may have taken place millions of years ago, long before the emergence of humans. While determining the exact conditions that led *Cannabis* to develop its distinctive compounds remains beyond the reach of current scientific knowledge, it is likely that human selection over the past ten thousand years has also shaped the plant's chemistry, particularly by promoting the diversity of its terpenes.[41]

One thing is certain: The *Cannabis* plant, as it exists today, has a complex and intriguing chemistry rooted in its remarkable capacity to harness intense sunlight.

# 11

# Indoor Outdoor Hyperaccumulator

During the 1960s and '70s, American *Cannabis* breeders began selecting plants for their potency and developing new varieties with increasingly higher THC content. One simple procedure involved harvesting all male plants, sampling them for potency, using the pollen from the strongest ones to fertilize the females, and then comparing the seeded females for potency and using seeds from the most psychoactive plants for the next generation.[1]

Scholars argue that the US government's prohibitionist crackdown on cannabis, initiated in 1981 as part of the ongoing War on Drugs, contributed to driving up cannabis potency, just as the prohibition of alcohol had favored the production of whiskey over beer. They refer to the Iron Law of Prohibition, which posits that the potency of forbidden substances tends to rise as law enforcement intensifies. According to this theory, the greatest cost of selling an illegal product is evading detection, and more concentrated forms of the product take up less space, weigh less during transport, and command higher prices. Whether the focus of *Cannabis* breeders on increasing potency was entirely due to the prohibitionist context is unclear. Still, the crackdown on cannabis drove up prices, leading customers to expect a bigger bang for their buck, and such circumstances likely motivated breeders and growers to produce ever-stronger products.[2]

During the early 1980s, the War on Drugs gave rise to the nationwide Domestic Cannabis Eradication/Suppression Program (DCE/SP), a joint effort involving local police, National Guard troops, and FBI agents. Between 1982 and 1998, according to official estimates, DCE/SP operations eradicated more than three billion *Cannabis* plants and led to the arrest of more than one hundred fifty thousand individuals. Cultivating *Cannabis* outdoors in the United States became fraught with danger, so cultivators began to grow their plants indoors under artificial lights. Indoor growing offered several advantages. It was hard to detect, even from a helicopter. It kept out unwanted pollen. It allowed growers to produce mature flowers without having to worry about autumn rains and frosts. It could also be practiced anywhere, even in regions where *Cannabis* did not grow well outdoors; all one needed was the know-how, the right seeds and technology, and access to electricity.[3]

Indoor growing was also a boon for *Cannabis* breeders. By exposing young plants to intense light for up to twenty-four hours a day to accelerate growth and then switching to just twelve hours a day, producers could create an "artificial autumn equinox" that induced the plants to flower. Instead of maturing in four to five months and producing seeds that could only be planted at the start of the following outdoor growing season, indoor *Cannabis* plants could be enticed to produce mature seeds in less than four months. This enabled breeders to work with three generations a year instead of just one, significantly accelerating their ability to imprint selective preferences on the plant.[4]

In the 1980s, several American *Cannabis* breeders moved to the Netherlands, where cannabis laws were more lenient and law enforcement was less strict. There, they joined forces with Dutch breeders, benefitting from a legal framework that allowed *Cannabis* cultivation for seed production and commerce. Together, American and Dutch breeders developed varieties that did well indoors. Tall plants are less economically productive under lamps because only the top parts receive sufficient light. Starting with seeds from Afghanistan, the breeders selected squat plants that matured quickly and produced potent, resinous flowers. In 1985, a Dutch seed company began advertising in the American cannabis magazine *High Times* for "Indoor indicas, potent and heavy-yielding," available by mail order. Within a year, the company had shipped hybrid seeds to fifteen thousand customers in the United States alone.[5]

The Netherlands provided not only a more tolerant context for *Cannabis* breeding but also a long tradition of greenhouse horticulture that included experience with hydroponics and atmospheric $CO_2$ enrichment. Growing *Cannabis* outdoors in the Netherlands was challenging because of the country's northern latitude and consequent lack of intense light, and its dense population made it hard to find grow spots undetectable to thieves. For practical reasons, growing *Cannabis* indoors or in greenhouses made sense in the Netherlands. However, Dutch growers initially did not use lights powerful enough to produce quality indoor *Cannabis* until their American counterparts introduced them to high-intensity discharge lamps, which had only recently become commercially available. In brief, a fusion of American and Dutch know-how during the early 1980s yielded a new cultivation method called Sea of Green. This technique involves growing small female plants, often clones, packed closely together under intense lighting in hydroponic systems with constant access to water, nutrients, and $CO_2$-enriched air, all designed to speed up growth and boost yields. With only four weeks of vegetative growth before flowering is induced, these plants remain small, producing a single large central inflorescence on a tender stem that requires support. These short, stick-like plants are ready for harvest in just three months. Their speeded-up life cycle allows four to five harvests per year or even six with multiple rotating grow rooms.[6]

Proponents of the Sea of Green favor clones—rooted branch cuttings taken from a mature mother plant—for several reasons. Clones are genetically identical to the mother plant and guaranteed to be female, eliminating the need to remove male plants. They also mature at the same rate, ensuring a uniform, reliable, and predictable product.[7]

Horticulturalists have long practiced cloning to keep an appealing plant variety unchanged, meaning unaffected by the variability caused by sexual reproduction. Granny Smith apples, seedless yellow bananas, navel oranges, and seedless grapes all come from cloned plants and are copies of copies dating back several decades or even centuries. But cloning comes at a cost. A field of clones is particularly vulnerable to illnesses and pests since all the plants share the same weaknesses. Outdoors, cloned crops suffer from debilitating diseases, and commercial growers use synthetic pesticides to combat them. Indoors, strict hygiene is required to prevent bacterial, fungal, or pest infestations.[8]

Initially, many indoor *Cannabis* growers wanted to avoid detection and make the most of small, discreet spaces like garages, basements, and attics. Over a year, a grower practicing the Sea of Green method could produce five pounds (or two kilograms) of dried sinsemilla in a space the size of a small closet. One study published in 2006 by the *UN Bulletin on Narcotics* estimated indoor *Cannabis* crops to be "between 15 and 30 times as productive per square meter of cultivation space as outdoor crops."[9]

Indoors, everything can be intensified. The light levels can reach five hundred times those recommended for reading, matching those found in hospital operating rooms. High-intensity discharge lamps left on for long hours can deliver significantly more daily light than the sun provides anywhere on Earth. The temperature can be kept consistently high, and plants can have unlimited access to water, nutrients, and $CO_2$. In indoor grows, air often lacks $CO_2$, slowing plant growth. To counter this, cultivators can use ventilation systems to expel stale air and moisture transpired by the plants. They can also enrich the incoming air with $CO_2$ concentrations up to four times higher than usual, allowing *Cannabis* plants to grow about twice as fast. Speeding up the vegetal life cycle in this way requires sealing off the grow room and using additional technology.[10]

By pushing for maximum productivity, many indoor *Cannabis* growers set themselves up for trouble. Pests such as aphids, mites, and molds could have a field day in a warm, humid room filled with juicy plants. Outdoors, these pests had natural predators; indoors, they went unchecked and could quickly turn a grow room into a thicket of sickly plants. Early guidebooks on indoor *Cannabis* growing acknowledged this problem and usually recommended spraying synthetic pesticides. Among the suggested solutions were "bug bombs" or total-release aerosol insecticides. A manual published in 1983 explained: "Bug bombs are very strong insecticides that essentially exterminate everything in the room. They were developed to kill fleas and roaches. Bug bombs are used *between crops* to rid the grow room of all bugs before introducing the next crop. Many manufacturers produce bug bombs under many brand names containing a menagerie of toxic chemicals. Place the bug bomb in the *empty* room. Turn it on. Then, leave the room. The chemicals are very toxic! Follow directions to the letter!"[11]

By the 1990s, growing *Cannabis* indoors had become the new norm in many countries. Countless individuals set up clandestine grow rooms, drawn by the possibility of producing potent sinsemilla, hidden from view and out of reach of authorities. This new practice went largely unstudied, as nearly all these growers operated outside the law. Police reports concerning the chemical composition of sinsemilla-type cannabis seized in the 1990s and 2000s focused primarily on its THC content, ignoring the possible presence of toxins and pesticide residues—perhaps because authorities considered cannabis unfit for consumption in any case.[12]

* * *

Botanists describe *Cannabis* as a plant that has relatively few problems with pests or diseases when it grows naturally. As a wind-pollinated species with male and female flowers on separate plants, it has the resilience conferred by the sexual reproduction of distant individuals. There tends to be considerable variation from one plant to the next, both genetically and in terms of chemical composition. This pervasive diversity limits the ability of pathogens to attack, as only a small number of plants are likely to be simultaneously affected by a specific disease. Under favorable conditions, outdoor *Cannabis* plants grow vigorously and produce a wide range of terpenes and other substances that repel most bacteria, fungi, insects, and spider mites. Botanists refer to *Cannabis* as "a hardy plant" capable of thriving in many different places—essentially, a true weed.[13]

Outdoor *Cannabis* plants are not pest-free but rather pest-resistant or pest-tolerant. When problems arise, they are rarely catastrophic. Wild animals such as deer, monkeys, rabbits, raccoons, rats, and other rodents may eat young *Cannabis* plants. Beetles, butterflies, snails, and stink bugs can also cause damage. The most common disease that afflicts the plant is a fungus causing gray mold, particularly in humid, temperate climates. Some *Cannabis* varieties are less resistant to this fungus than others. In particular, hybrid varieties bred from plants that initially evolved in arid conditions like Afghanistan have almost no resistance to gray mold, especially during the development of flowers. When grown in humid environments, either indoors or outdoors, the tightly packed inflorescences of these plants hold moisture and rot quickly. John McPartland and colleagues point out in their book *Hemp Diseases and Pests*: "This unfavorable trait even appears in hybrids with a small percentage of *afghanica* heritage. What marijuana breeders gained from the introduction of *afghanica* (potency, short stature, early maturity), they paid for with extreme susceptibility to bud rot."[14]

Despite these vulnerabilities, most domesticated *Cannabis* varieties require minimal attention when grown outdoors, even over large areas of mono-cropped plants. In the words of botanist Ernest Small, who dedicated much of his career to studying the plant: "*Cannabis sativa* is known to be significantly resistant to most harmful organisms and rarely requires protective treatment. Indeed, the most valid claims for the environmental friendliness of hemp relate to its very limited need for agricultural biocides."[15]

A female *Cannabis* plant that grows outdoors in favorable conditions can mature into a bush with a hard, woody stem as thick as a wrist. In contrast, a mature indoor clone resembles the rooted branch of an outdoor plant and typically has a slender stem as thin as a finger. The clone's sped-up life cycle makes it a chimera-like plant, featuring a large, mature inflorescence on a short, immature stem. When grown under intense lighting, these clones have a significantly higher proportion of floral material, which increases the strain on their stunted stems while maximizing yields. In most cases, indoor *Cannabis* clones require support to remain upright—unlike their bushy, outdoor relatives, which become strong through exposure to winds and the elements.[16]

* * *

In the 1990s, scientists discovered that *Cannabis* plants excel at absorbing toxic substances and storing them in their tissues. This unusual capacity became apparent when Ukrainian scientists planted *Cannabis* in fields near the site of the Chernobyl nuclear accident and found that the plants extracted notable amounts of radioactive heavy metals from the soil. Subsequent research confirmed that *Cannabis* roots absorb many kinds of toxic substances with no apparent ill effect on the plant. Besides heavy metals and radioactive elements, *Cannabis* can pull up and store poisonous hydrocarbons and synthetic pesticides. For this reason, scientists classify it among "hyperaccumulator plants"—species capable of extracting and trapping environmental toxins.[17]

Few plant species tolerate heavily contaminated soils. Besides *Cannabis*, hyperaccumulator plants include willow, sunflower, maize, and bamboo, each possessing specific absorption capabilities. *Cannabis* does well with zinc, lead, nickel, and cadmium. As a plant, it grows fast, sets down deep roots, tolerates high concentrations of toxins, has a low susceptibility to diseases and pests, and quickly generates large quantities of aboveground biomass. These qualities make it suitable for the bioremediation of contaminated sites.[18]

*Cannabis*'s capacity to act as a hyperaccumulator stems from its evolution as a pioneer species. It is adapted to rapidly colonizing disturbed soils, such as landslides and eroded riverbanks, and reestablishing vegetal cover. These soils can contain heavy metals brought up from deeper layers of subsoil. *Cannabis* has probably been pulling toxic metals out of the ground for millions of years.[19]

All this is good news for cleaning up contaminated sites—but less so for cannabis consumers. Like a sponge for poisons, the plant absorbs the substances its roots encounter, including those harmful to most other living organisms. *Cannabis* roots can penetrate the soil up to six feet (two meters) deep in some areas, extract toxins, and expedite them to the plant's aerial parts, including the leaves, seeds, and flowers.[20]

This spongelike quality means that the contents of a *Cannabis* plant depend on the conditions in which it grows, including the soil, water, fertilizers, and inputs such as pesticides. *Cannabis* not only has a complex and poorly understood chemistry of its own, but individual plants tend to absorb the substances in their immediate environment. As a result, *Cannabis* products like sinsemilla flowers, compressed resin (hashish), and concentrated oils may contain a wide range of substances in addition to those specific to the plant.

* * *

In 1996, the citizens of California voted in favor of a law permitting medical patients to use cannabis. Local activists and patients had lobbied for years to decriminalize the plant's therapeutic use. People with AIDS found that smoking cannabis reduced their suffering and stimulated their appetite in ways that other medicines did not; similarly, cancer patients undergoing chemotherapy reported that smoking cannabis reduced nausea and increased their appetite, while multiple sclerosis patients discovered that it lessened spasms and made their condition more manageable. The new law, known as the Compassionate Use Act, made California the first jurisdiction in the world to re-legalize *Cannabis* and some of its products. It also ushered in a new era of scientific interest in the plant's therapeutic potential.[21]

In the subsequent years, a dozen American states followed California's lead and re-legalized the medical use of cannabis. Several countries, including the Netherlands and Canada, did the same. In most of these places, patients preferred smoking cannabis rather than ingesting it in an edible form or as a pharmaceutical extract. Smoking cannabis allows users to experience its effects almost immediately and adjust their consumption accordingly. In contrast, an ingested edible or extract takes much longer to take effect, so users cannot adjust the dosage once they have swallowed the product and must wait at least thirty minutes before they get an inkling of its strength. This preference for smoking raised a question: Could a dried herb be considered a medicine? Experts agreed that medicines must be "standardized, efficacious, and safe preparations." They also pointed out that herbal cannabis is "a highly variable product with respect to composition," not only from one plant to another but even from one inflorescence to the next on the same plant. Could such contradictions be reconciled?[22]

In 2003, the Dutch Ministry of Health, Welfare and Sport outlined a solution to the problem. According to its "Guidelines for Cultivating Cannabis for Medicinal Purposes," botanically trained personnel could grow cloned plants in indoor facilities "under such standardized conditions that the content of the constituents is constant." By respecting proper hygiene, using uncontaminated soil, and avoiding pesticides "as far as possible," Dutch authorities stated that it was possible to grow "standardized plants" and supply patients with medical cannabis of "highly standardized quality."[23]

The idea of a "standardized plant" appeared to contradict the inherent biological variation of living beings. In most cases, individual plants of the same species have genetic differences and grow in diverse conditions, so they cannot have identical chemical compositions. Still, the Dutch method of growing genetically identical plants in a strictly controlled indoor environment seemed to solve the problem. It certainly proved sufficient to establish herbal cannabis as a "modern medicine."[24]

The first signs of trouble concerning the quality of medical cannabis arose in the United States. In 2009, the Los Angeles City Attorney's Office tested three samples of medical marijuana purchased in local dispensaries and found two samples containing extremely high levels of a neurotoxic insecticide called bifenthrin. One sample had sixteen hundred times the legally acceptable digestible amount, and the other eighty-five times the legal limit. At the time, California had no regulations concerning the production of medical cannabis. Five years later, researchers in Oregon found pesticide residues on close to half the medical cannabis sold in the state's dispensaries, often at "concentrations exceeding the allowable levels on any agricultural product." They noted that bifenthrin was the most frequently detected insecticide at excessive levels. Local growers used it to "bomb" indoor grow rooms between cultivation cycles, and most believed that doing so would not contaminate their next crop. The researchers examined the grow rooms and found that "bifenthrin applications lead to long-term contamination of workspaces, tools, lights, and ventilation systems that result in chronic cross-contamination of later crops."[25]

Bifenthrin blocks nerve cell signal transmission in insects, causing immobility and death. Prolonged or excessive exposure in humans can lead to respiratory paralysis, tremors, and seizures. Like many other synthetic insecticides, bifenthrin has a chemical structure designed to last. The longer these substances persist, the more efficacious they can be. Many are so resistant to degradation that they survive combustion. When present in cannabis and smoked along with it, about 70% of bifenthrin and other synthetic insecticides end up intact in the mainstream smoke. By comparison, only about 50% of the cannabinoids in the herbal material survive combustion and make it into the smoke.[26]

The dangers of smoking pesticide-tainted cannabis are speculative because existing studies of pesticide toxicity in humans primarily focus on ingestion rather than inhalation. An ingested pesticide passes through the digestive system and undergoes considerable metabolic degradation before entering the bloodstream. In most cases, only a small portion survives digestion. However, when the same substance is inhaled, it goes directly from the lungs into the bloodstream. This direct route makes inhalation a potent means of exposure to pesticide residues. Despite the lack of toxicity data concerning the dangers of smoking or vaporizing such substances, experts agree that consuming medical cannabis contaminated with neurotoxic insecticides poses a threat to patients, particularly those suffering from neurological issues. As for all other consumers, they are probably better off using untainted plants and products.[27]

* * *

The illegal status of *Cannabis* pushed the plant indoors, but its relegalization as a medicine has done little to bring it back outside. At present, dozens of countries across five continents—as well as a majority of US states—have legalized medical cannabis, generally requiring the cultivation of cloned plants in enclosed spaces under strictly controlled conditions. Health authorities concur that medical-grade cannabis must be standardized and dependable, and they tend to believe that only indoor cultivation can yield such a product and ensure its constancy from one crop to the next.[28]

Security is an additional concern motivating authorities to prefer indoor production. In the United States, the federal government's National Center for Cannabis Research and Education in Mississippi protects its outdoor plantations behind tall steel fences overlooked by security turrets and employs armed guards and electronic surveillance systems to secure its premises. In the Netherlands, the original guidelines for producing medical cannabis recommended hiring security guards to protect cultivation facilities and restricting access to authorized personnel. Many countries have adopted similar rules. For example, when Greece legalized medical cannabis in 2018, the law required cultivation "under enclosed conditions" in fenced-off production areas protected by security guards. It also stipulated that production sites had to be located at least one kilometer (0.6 miles) from schools, archeological sites, and other sensitive locations.[29]

Why take such drastic measures when cultivating a weed-like plant? Because its illegal status has made it expensive, attracting the interest of thieves. As plant scientist David Potter explains: "This highly regulated plant has a large retail cash value and a unique cachet that makes it vulnerable to those with a curious, acquisitive, or malicious intent."[30]

Growing *Cannabis* indoors eases security concerns and reduces the risks of unwanted pollination from distant plants and crossover pesticide contamination from adjacent fields. It also allows cultivators to control light, temperature, $CO_2$ concentration, humidity, soil or growing medium, nutrient supply, and ventilation. Each new crop can be cultivated in precisely the same conditions as the previous ones, ensuring a homogenous and uniform product. Experts agree that the risk of contamination by bacteria, fungi, insects, and spider mites is highest in sealed grow rooms, somewhat less so in greenhouses, and less still in living soil in the open air. Currently, authorities in different countries who oversee the production of medical cannabis understand the importance of growing contaminant-free plants and have formulated regulations to ensure product quality.[31]

Cultivating uncontaminated indoor medical-grade *Cannabis* typically requires costly infrastructure, extensive knowledge, meticulous work, and copious amounts of electricity. One recent approach, widely regarded as exemplary, was developed in the Netherlands and Great Britain, and involves growing cloned plants in a computerized greenhouse equipped with lamps, blinds, and ventilation. Over one year, about half the light energy within the greenhouse comes from electric lamps. The plants are small and have accelerated life cycles, usually reaching maturity within three months. Here, the goal is to avoid developing conditions that favor infections and infestations rather than maximize productivity—crop uniformity and quality are more important than yield. Plant density is reduced compared to other indoor methods, thus improving air circulation. Constant ventilation helps prevent the establishment of plant diseases. Animal pests are kept in check by introducing beneficial insects that predate on them, such as ladybugs and parasitic wasps. Temperatures are kept relatively low to limit pest proliferation. Only organic and biodegradable pesticides are used as a last resort. Lastly, the water, soil, and nutrients are certified free of contaminants, and rigorous cleaning protocols are applied. By respecting all these measures, experts have succeeded in cultivating *Cannabis* in enclosed conditions that consistently yields "compliant, uniform, and safe cannabis medicine."[32]

* * *

At present, *Cannabis* remains illegal in most countries, and illicit growers continue to produce crops of unverified quality, mainly in clandestine grow rooms. The rare toxicological studies conducted on illegal sinsemilla-type cannabis reveal alarming levels of synthetic pesticides: 44% of tested samples in Austria, 64% in Belgium, and 92% in Canada contained them. Environmental journalist Nate Seltenrich observes that "growing plants indoors to escape detection often increases the risk that insect infestations and harmful microbes will spread quickly. For illicit growers with little knowledge of other methods and no regulatory oversight, it is easier and cheaper just to spray."[33]

A small number of countries and about half the states in the United

States have legalized *Cannabis* for "recreational" use. In most cases, the law allows adults to grow a few THC-rich plants for personal consumption. Such plants require no standardization, which makes their cultivation in the open air possible. However, regulations vary from one jurisdiction to the next. For example, in Luxembourg, the law states that "the plants must not be visible from public spaces," whereas in South Africa, they can be grown in a "private place," including "the garden of one's residence." In Canada, individuals can grow up to four plants in their home or garden, while in Uruguay, the limit is set at six, as long as the plants remain out of public sight.[34]

Some of these laws mean that psychoactive *Cannabis* may once again grow in living soil under sunlight. However, this does not guarantee contaminant-free plants. A recent example in Switzerland illustrates this challenge: Swiss authorities launched a pilot project allowing citizens to purchase organic, THC-rich cannabis; by legal prescription, the plants were to be grown in living soil and contain no bacteria, heavy metals, or pesticides. Cultivation was entrusted to a private company operating under high security: double fences, surveillance cameras, alarms, and restricted access. The soil in which the *Cannabis* grew had remained fallow for years, but the previous owner had used it for ornamental flower production. A preliminary analysis detected no pesticides in the soil. However, when the first dried *Cannabis* inflorescences were tested, they showed traces of a synthetic pesticide at levels just above the limit of detection. Authorities deemed the harvest unfit for consumption and postponed the distribution of recreational cannabis planned under the pilot project. A thorough investigation revealed that the plants had extracted long-buried residues and stored them in their aerial tissues at concentrations ten to thirty times higher than in the soil, making them detectable. A spokesperson for the company concluded that the only way to ensure that a plot of living soil lends itself to the cultivation of pure, organic *Cannabis* is to grow *Cannabis* plants on it beforehand and check the resulting crop for residues.[35]

In brief, the hyperaccumulator plant is so proficient at drawing synthetic pesticides out of the ground that growing organic *Cannabis* can be challenging, even with the best intentions.

# 12

# Beyond THC

When *Cannabis* plants escape cultivation and propagate through wind pollination, they regain the qualities of wild-type plants within a few generations. Free-living *Cannabis* tends to produce the main cannabinoids, THC and CBD, in similar amounts, with mature female flowers containing between 1% and 2% of each substance.[1]

However, humans can alter this balance by selecting and breeding plants that favor a specific cannabinoid. Breeding for higher THC levels necessarily drives down CBD content and vice versa. This trade-off occurs partly due to the competition between the enzymes responsible for producing these cannabinoids, which rely on the same chemical precursor. THC and CBD are cut from the same cloth, so producing more of one means making less of the other. Some intensively bred varieties can produce flowers containing over 30% THC and almost no CBD. However, without human intervention, these high THC levels quickly drop off over subsequent generations.[2]

Contemporary *Cannabis* breeders have also created varieties with unnaturally *low* levels of THC. During the 1990s, several European countries decided that *Cannabis* with flowering parts containing less than 0.3% THC was, in fact, "industrial hemp" and could be legally cultivated for fiber and seed production. This arbitrary threshold was established for political, not scientific, reasons, as there is no correlation between THC levels and fiber or seed quality. Industrial hemp plants that escape cultivation and return to the wild typically see their THC levels increase over subsequent generations.[3]

The fixation with driving THC levels up or down stems from the scientific and pharmaceutical focus on THC after its discovery as the "main active compound" of *Cannabis* in 1964. By then, scientists had already perfected the extraction of active ingredients from medicinal plants, the analysis of their chemical structures, and their modification to create synthetic versions that enhanced the desired effects. In the nineteenth and twentieth centuries, several major painkillers, such as aspirin and synthetic opioids, were developed using this approach.[4]

Scientists working in the second half of the 1960s found they could synthesize THC from scratch by combining certain chemicals in the laboratory. Producing THC as a "pure substance" in this way allowed them to do away with the plant and circumvent legal obstacles in certain jurisdictions. In the United States, in particular, the Controlled Substances Act of 1970 placed "all parts of the plant *Cannabis sativa* L., whether growing or not," and every one of its compounds in the most restricted category of drugs "with no accepted medical use"—but did not concern cannabinoid-like substances not found in the plant, unless regulatory authorities decided otherwise. This loophole gave American scientists and pharmaceutical companies leeway to produce and study synthetic THC and develop variants of the substance.[5]

In the 1970s, American pharmacological authorities assigned a generic name to synthetic THC, *dronabinol*—a contraction of the chemical name tetrahy*dro*can*nabinol*. Dronabinol was initially developed as an experimental treatment for medical conditions such as glaucoma and chemotherapy-induced nausea. It is a viscous resin and usually comes combined with sesame oil in a gelatin capsule ready for ingestion.[6]

Though dronabinol was supposed to have the same effects as cannabis, clinical trials soon revealed differences between the two. Ivan Silverberg, a physician and professor of clinical oncology, prescribed dronabinol as an anti-vomitive to chemotherapy patients in the late 1970s. Since each patient to whom he prescribed THC had smoked cannabis before, he did not expect any of them to suffer any ill effects.

To his surprise, every single one of them refused to take it after the first or second dose because the psychoactive effects were too extreme: "Several said the THC drug was intensely anxiety-provoking. Others said they were heavily sedated by synthetic THC. A couple reported the equivalent of 'bad trips' on THC. . . . I had several patients who accused me of trying to poison them with an inferior product. One woman actually threw her bottle of THC back in my face."[7]

Studies found dronabinol less effective therapeutically, less predictable, slower acting, and marred by debilitating side effects when compared to smoked herbal cannabis. One report issued in 1981 by the New York Department of Health listed dronabinol's most frequent side effects as sedation, disorientation, dizziness, hallucinations, anxiety, paranoia, rapid heartbeat, headache, blurred vision, slurred speech, and depression. Another study conducted at Johns Hopkins University with cannabis-using volunteers who received either dronabinol or a placebo found that "the subjective effects of the active drug were very different from standard descriptions of somebody who had eaten (or smoked) an equivalent amount of cannabis. On standard scales to rate a marijuana 'high,' our subjects reported variously that the dronabinol made them sleepy, interfered with their ability to think clearly, and made them dizzy or slightly anxious, but definitely not 'high,' 'buzzed,' 'mellow,' or 'feeling good.'"[8]

During the 1970s and early '80s, several studies demonstrated that cannabis contained pharmacologically active substances besides THC. Clinical trials with healthy volunteers suggested that CBD, in particular, could reduce anxiety, depersonalization, and confused thoughts sometimes induced by THC. However, these studies did little to shift the scientific focus away from the "active principle" of cannabis. On the contrary, university and pharmaceutical laboratories began using manufactured chemicals to create synthetic substances with a molecular structure analogous to THC. One such substance, nabilone, showed useful anti-nausea and anti-vomitive properties and helped manage glaucoma. However, animal studies revealed that this molecule was seven times more psychoactive than THC by weight. Developed by US pharmaceutical company Eli Lilly and Company, nabilone became the first cannabinoid to receive approval for medical use. It was marketed in 1982 in Canada, followed by Mexico, the United Kingdom, and Germany—despite frequent reports of hard-to-manage side effects, such as drowsiness, rapid heartbeat, hallucinations, and intense feelings of unhappiness.[9]

During this period, scientists working for pharmaceutical companies also produced "nonclassical synthetic cannabinoids," meaning compounds that lack a core, ring-shaped component of the THC molecule but that mimic its effects. These substances, also known as "cannabimimetic agents," were designed to harness the medicinal properties of THC while reducing its psychoactive effects. However, the more these molecules showed therapeutic potential, the more they proved to be highly mind-altering. Pfizer Inc., for example, developed a series of these compounds with exceptional painkilling effects but accompanied by unmanageable levels of psychoactive intensity. One of these substances, CP 55,940 (CP being the initials of Charles Pfizer, cofounder of the company), was found to be more potent than morphine for pain relief but twenty-five times more psychoactive than THC. Due to these excessive side effects, Pfizer halted its research on "nonclassical cannabinoids" in the mid-1980s.[10]

Even so, these experiments led to significant scientific advancements. They revealed which parts of the THC molecule were essential to its psychoactive effects and highlighted the inseparable link between its pain-relieving and psychoactive properties. These studies also suggested that THC acted on a specific, yet unidentified, biological receptor. In addition, CP 55,940 would soon become a crucial tool in identifying this receptor—a discovery that would revolutionize the understanding of how THC works in the human body.[11]

* * *

By the 1970s, scientific research had established that nerve cells, or neurons, possess receptors on their outer surface to which certain brain chemicals bind, like keys fitting into locks, triggering cascades of internal reactions. Most psychoactive substances exert their effects because they have molecular profiles like those of the brain's natural chemicals, allowing them to fit into the receptors like skeleton keys. By studying how nicotine, LSD, and morphine work in the brain and body, scientists discovered the receptors that these molecules activate; then, they identified the brain chemicals that usually fit into these receptors. For example, by tracing the pain-relieving action of morphine, researchers identified a specific docking site on the surface of neurons: the opioid receptor. Further investigations in the mid-1970s revealed that the brain itself produced molecules capable of binding to this receptor. They called them *endorphins*, a contraction of *endogenous* (meaning *made from within*) and *morphine*.[12]

Ten years after the discovery of opioid receptors and endorphins, the mechanism of action of THC continued to stump scientists, who still had no idea how the active principle of *Cannabis* exerted its effects. Cannabinoids are oily and stick to the fatty membranes of neurons and other cells, making their binding sites difficult to detect. Even when researchers labeled THC with a radioactive marker—a standard method for pinpointing binding sites—they were unable to identify an anchor point. The breakthrough came in 1988 when Pfizer provided researchers with its potent nonclassical cannabinoid CP 55,940. Scientists at Saint Louis University Medical School in Missouri, in collaboration with Pfizer, attached a radioactive marker to this molecule and tracked it to its binding site in the brain. They then discovered a previously unknown receptor, which they named the *cannabinoid receptor*, or *CB receptor*.[13]

This discovery set off a wave of research. Scientists found that these CB receptors were mainly concentrated in brain regions involved in coordination, memory, cognition, and emotions—all functions influenced by cannabis. They also discovered that all vertebrates and some invertebrates possess these receptors, suggesting that the first vertebrates already had them. Given that vertebrates appeared around 500 million years ago, while *Cannabis*, as a plant species, is estimated to be no more than 28 million years old, scientists concluded that CB receptors existed long before the plant.[14]

These discoveries led an international team of researchers based at the Hebrew University of Jerusalem to hypothesize that the brain itself produces a cannabinoid-like substance capable of binding to these receptors. In 1992, they launched a research project to find it. First, they devised a novel, ultra-potent synthetic cannabinoid, HU-243 (HU being the initials of the university). This compound, structured like THC, had an affinity for CB receptors 1,000 times higher and a psychoactive potency 125 times greater. To identify a potential natural molecule, the researchers placed neuronal membranes in a solution containing radioactively labeled HU-243 and then added various natural brain chemicals to see if any displaced the radioactive probe. This approach led to the discovery of a new molecule, which they named *anandamide*, from the Sanskrit *ananda*, meaning bliss.[15]

In the years that followed, scientists identified a second cannabinoid receptor, which they named CB2, thus renaming the first one CB1. They also developed synthetic molecules capable of blocking these receptors to explore their functions, and they discovered other endogenous cannabinoids beyond anandamide, which bound to and activated the same receptors. By the late 1990s, scientists had established a distinction between endocannabinoids, naturally produced by the human body and animals; phytocannabinoids, produced by plants; and synthetic cannabinoids, manufactured by chemists.[16]

The intensive research conducted throughout this period revealed that CB1 receptors are primarily present in the central nervous system, digestive tract, and cardiovascular system. They are responsible for the psychoactive effects of cannabis, THC, and synthetic cannabinoids. CB2 receptors, widely distributed throughout the body—including in the brain—do not induce psychoactive effects. However, they play a crucial role in immune response, bone and blood cell formation, inflammation regulation, and energy balance.[17]

Research conducted in the 1990s also showed that anandamide binds briefly to the CB1 receptor, acting as a partial agonist, much like THC. In contrast, some synthetic cannabinoids act as full agonists, binding tightly and activating the receptor with intensity. This interaction can have devastating effects. In 1999, a researcher who ingested an unspecified dose of HU-210—a synthetic cannabinoid one hundred times more potent than THC—reported feeling "prostrate, mute, tachycardic, panicked, and hallucinating for 48 hours."[18]

* * *

In the final decade of the twentieth century, scientists discovered that endocannabinoids are produced by neurons "on demand." Once released, they travel to CB receptors on neighboring neurons to activate them. As soon as anandamide transmits its signal, a specific enzyme breaks it down into components that neurons can reuse to rebuild their membranes. In other words, unlike neurotransmitters such as dopamine or serotonin, these substances are not stored but are created and recycled as needed.[19]

When anandamide binds to a CB1 receptor, it triggers a cascade of reactions inside the neuron that generally inhibits the release of neurotransmitters, thereby limiting the transmission of nerve signals. In most cases, anandamide calms the activity of nerve cells that may be overreacting due to illness or injury. It also binds to receptors on immune, limbic, digestive, cardiovascular, and muscular cells and on organs such as the heart and bladder. It helps regulate inflammation, soothe muscle spasms, and relax blood vessels. It can also stimulate appetite, reduce nausea, alleviate pain, and exert an anti-inflammatory effect. Its presence in breast milk initiates the feeding process of newborns. In 1998, a team of Italian researchers summarized the primary functions of endocannabinoids in five verbs: "relax, eat, sleep, forget, and protect."[20]

By the early 2000s, scientists referred to the endocannabinoid system as a central biological system in all vertebrates and some invertebrates. It consists of endocannabinoid messenger molecules, enzymes responsible for their production and degradation, and CB receptors. This system has many functions, including regulating all other bodily systems. Its scope explains why cannabinoids can have many different effects on users and offer therapeutic potential for a wide range of medical conditions.[21]

In 2001, a research team supported by the French pharmaceutical company Sanofi demonstrated that the CB1 receptor controlled a well-known effect of cannabis: appetite stimulation. Sanofi was developing a CB1 blocker called rimonabant, intended to function as an appetite suppressant. Five years later, in 2006, despite concerning side effects such as depression, anxiety, and nausea, European regulators approved its sale as an "anti-obesity medication." In the United States, however, health authorities denied approval, deeming the risks greater than the benefits. Rimonabant enjoyed initial success in Europe, where medical doctors prescribed it as a pill to be taken daily for a year. At first, most rimonabant users lost weight, slimmed down, and improved their cardiovascular health profiles in measurable ways. However, clinical trials had failed to anticipate a major issue: In the long run, users tended to get depressed, feel nauseous, and think about suicide. Since cannabis stimulates appetite, alleviates nausea, and produces a euphoric sensation, some critics nicknamed rimonabant the "anti-pot pill." By late 2008, after several suicides among users, European regulators withdrew the drug from the market and prohibited it.[22]

Rimonabant's short tenure as a "medication" ended in disaster. Nevertheless, its failure revealed a key point: Blocking cannabinoid receptors can be just as dangerous as overstimulating them.

* * *

From the 1960s to the '90s, scientists and pharmaceutical companies focused almost exclusively on THC rather than on the other cannabinoids present in *Cannabis*. However, their quest to synthesize a THC variant did not yield a new drug with the groundbreaking therapeutic properties some had hoped for. Instead, it resulted in a significant discovery: the existence of the endocannabinoid system. This breakthrough eventually reignited scientific interest in the plant's other compounds.[23]

In the 1990s, researchers tested CBD to see if it activated CB1 and CB2 receptors. Since it did not, they initially classified it as "inactive." However, in the early 2000s, further studies revealed that CBD indirectly influences cannabinoid receptors by inhibiting the enzyme that breaks down anandamide. This extends the usual effects of anandamide, which may, in turn, reduce THC's psychoactive effects since anandamide and THC compete for the same receptor. CBD functions as a "weak antagonist" of both cannabinoid receptors. More precisely, it seems to bind to an inactive part of the CB1 receptor, which may prevent THC from attaching to it, reducing its psychoactive effects. Additionally, CBD interacts with other receptors, including specific opioid receptors, which may explain its pain-relieving properties.[24]

Research during this period also discovered that cannabinoids like THC and CBD exhibit "biphasic activity." Their effects vary depending on dosage: Low doses of THC tend to reduce anxiety, while high doses can increase it. Likewise, low doses of CBD reduce anxiety, but at high doses, it can cause "distinct feelings of depersonalization, derealization, and altered internal and external perceptions." However, a high dose of CBD taken alongside THC seems to reduce THC's psychoactive effects. Neuroscientist Godfrey Pearlson comments that these two cannabinoids "have complicated and separate effects of their own that blend and combine unpredictably when they are administered together."[25]

THC and CBD also have notable effects on the human heart. By stimulating the CB1 receptor, THC can trigger the constriction of the veins, which increases blood pressure and accelerates heart rate—indeed, a growing body of research links cannabis use to an increased risk of heart attacks, strokes, and other cardiovascular diseases. Yet, paradoxically, ultralow doses of THC appear to be cardioprotective. CBD, in contrast, acts as an anti-inflammatory and antioxidant, which relaxes blood vessels, helps stabilize heart rate, and offers cardiovascular protection.[26]

The two main cannabinoids seem to work in tandem, but only sometimes.

* * *

In the 2000s, as scientists and pharmaceutical companies moved beyond their exclusive focus on THC and THC-like substances, several synthetic cannabinoids "escaped the lab" and became drugs of abuse. Produced by clandestine manufacturers and sold online in powder form, these chemicals were then dissolved in solvents and sprayed onto dried plant material to create "herbal products" mimicking the appearance and effects of cannabis. These products were marketed as "incense" or "smokable blends" under trade names such as Spice and K2. Since the synthetic cannabinoids they contained were chemically distinct from THC, they did not show up on tests designed to detect cannabis—an added selling point for some. When authorities first identified these substances in commercial products in 2008, they primarily found CP 47,497-C8, a nonclassical cannabinoid developed in the 1970s by Pfizer, and JWH-018, a cannabimimetic agent structurally unrelated to THC that scientist J. W. Huffman designed as a research tool in the 1990s. Both compounds were several times more potent than THC and easy to produce, thanks to detailed descriptions available in scientific journals.[27]

At the time, these synthetic cannabinoids were not illegal, and banning them proved more difficult than expected in certain countries. Since most of the compounds were chemically distinct from THC, they could not be considered synthetic THC analogs and banned as such under existing drug laws. Instead, each new substance had to be treated individually and according to its specific chemical makeup. This slowed down the prohibition process. It also gave clandestine manufacturers more time to produce other, yet-to-be-banned substitutes, chosen among the hundreds of synthetic cannabinoids previously described in scientific publications. These relatively obscure substances were then added to Spice or K2 products until they, too, were detected and banned.[28]

In 2012, researchers from Japan's National Institute of Health Sciences discovered two potent "cannabimimetic substances" in herbal products purchased online. One was unknown to science. The other was a compound designed by Pfizer as a potential medication targeting the CB1 receptor. Pfizer had abandoned research on this substance shortly after patenting it in 2009. The patent referred to it as: "N-(1-amino-3-methyl-1-oxobutan-2-yl)-1-(4-fluorobenzyl)-1H-indazole-3carboxamide." With Pfizer's permission and following scientific conventions, the Japanese scientists named it *AB-FUBINACA,* and authorities in different countries promptly added it to the growing list of prohibited synthetic cannabinoids. In 2014, a few months after this substance's prohibition in the United States, analyses of K2 products marketed in US convenience stores revealed the presence of a slightly modified version of the original molecule. This new substance, named *AMB-FUBINACA*, was unknown to science, and, as such, its production, distribution, and use had yet to be prohibited. Experts noted that patents, such as the one established by Pfizer for AB-FUBINACA, are published in the public domain and contain all the information a competent chemist needs to produce and modify the substance.[29]

AMB-FUBINACA was easy to make and eighty-five times more potent than THC. It was available online as a powder, costing about $3 per gram. According to some calculations, one gram of this substance diluted in a solvent and sprayed on cheap, smokable herbal material could yield a "cannabis substitute" with a retail value of more than $500. The business model was lucrative, but a slight miscalculation or the uneven spraying of the drug could have devastating effects. In 2016, AMB-FUBINACA caused a well-documented "zombie outbreak" in a New York City neighborhood. Emergency responders found dozens of people standing immobile, lying prostrate on the ground, or wandering the streets in a "trance-like state, groaning and moaning, their eyes lifeless and their movements slow and seemingly mechanical." The victims had all smoked K2 laced with AMB-FUBINACA. Fortunately, after enduring debilitating effects for about nine hours, they all recovered.[30] Other victims of this ultra-potent synthetic substance were not so lucky. Between 2017 and 2019, sixty-four people in New Zealand died after smoking herbal materials containing AMB-FUBINACA at three times the dosage of the New York outbreak.[31]

This synthetic substance acts as a full agonist of CB1 receptors, but the exact mechanisms by which it causes death are not fully understood. Recent studies suggest that when a molecule of AMB-FUBINACA binds to a CB1 receptor, it locks into place completely, triggering an intense and rapid reaction. At low doses, its effects resemble those of cannabis, producing euphoria and relaxation but also anxiety, drowsiness, dizziness, rapid heartbeat, dry mouth, nausea, and red eyes. However, at higher doses, its effects go far beyond those of cannabis and can include depression of the central nervous system, cardiac or respiratory arrest, brain or kidney dysfunction, and death. AMB-FUBINACA appears to exacerbate preexisting cardiovascular diseases, and the presence of other synthetic drugs can further amplify its effects. During the New Zealand outbreak, nearly half of the victims had preexisting heart conditions.[32]

By the late 2010s, AMB-FUBINACA became the most widely used synthetic cannabinoid in the world, according to drug seizure reports. Unfortunately, its low cost and its potency mean that it is especially attractive to vulnerable populations, including displaced Indigenous people, people experiencing homelessness, and prisoners. In the New Zealand case, the vast majority of victims were Indigenous Maori people.[33]

* * *

After the initial "zombie outbreak" caused by AMB-FUBINACA, *The New York Times* ran the headline 33 SUSPECTED OF OVERDOSING ON SYNTHETIC MARIJUANA IN BROOKLYN. However, the term "synthetic marijuana" is misleading. AMB-FUBINACA is manufactured from industrial chemicals, and its molecular structure has nothing in common with plant-based THC. When burned, it releases cyanide, a highly toxic compound with severe neurological and cardiovascular effects. Unlike high-THC cannabis, which poses no risk of fatal overdose even at extremely high doses, AMB-FUBINACA can be lethal. The only similarity between the two substances is their capacity to activate cannabinoid receptors in the body and brain—and even then, in significantly different ways.[34]

Synthetic cannabinoids raise several questions, starting with terminology. Scientists refer to them under various names: *THC derivatives*, *nonclassical cannabinoids*, *cannabimimetic agents*, or *synthetic cannabinoid receptor agonists*—all of which may seem opaque to nonspecialists. Some experts argue that substances like AMB-FUBINACA are not even "true" synthetic cannabinoids, as they are "neither cannabinoid in structure nor analogs of natural phytocannabinoids." Instead, these experts prefer the term *cannabimimetic agent*, meaning "any substance that is a CB1 receptor agonist as demonstrated by binding studies." From a scientific point of view, the term "synthetic cannabinoid" is becoming increasingly imprecise, now encompassing an ever-growing range of diverse chemicals. Nevertheless, these substances pose a serious public health threat, and referring to them as synthetic cannabinoids may help public understanding.[35]

International health authorities report a growing number of poisonings and deaths linked to synthetic cannabinoids. Some deplore the lack of an effective antidote, while others suggest reintroducing rimonabant—the CB1 antagonist once used as an appetite suppressant before being banned due to its long-term depressive and suicidal effects. Research indicates that rimonabant can reverse the activity of the most potent synthetic cannabinoids without significant risks if administered in emergencies. As a onetime antidote against overdoses, rimonabant could save lives rather than imperil them. As the well-known adage goes, the difference between remedy and poison is almost always dosage.[36]

During the final decades of the twentieth century, scientific and pharmaceutical interest in THC led to the creation of potent synthetic derivatives—souped-up versions of THC—some of which have become potentially lethal poisons used in "cannabis substitutes." However, labeling them as such unfairly implicates *Cannabis*. The plant has been blamed for many problems over the years, mainly due to its primary psychoactive component, THC. It is true that *Cannabis* is the only known natural source of this molecule. But synthetic cannabinoids are human creations, concocted in laboratories by chemists. They do not come from the plant.

# 13

# Potencies Rising

The potency of alcoholic drinks is generally well understood. The higher the alcohol content, the stronger the drink and the greater its capacity to intoxicate. Spirits like vodka, whiskey, or rum often contain ten times more alcohol than beer does, so people usually consume them in much smaller quantities. Most consumers know that drinking several shots of vodka is more intoxicating than having a beer.[1]

After scientists identified THC as the "active principle" of cannabis in the 1960s, they used it by analogy with alcohol to measure potency. In some countries, authorities began testing THC levels in confiscated cannabis samples, and by the 1980s, with the rise of potent sinsemilla, records showed a marked increase in THC content. In the United States, the average THC concentration in seized cannabis more than tripled between 1980 and 2005, rising from about 2% to 7%. During this period, emergency services observed a significant increase in visits related to "unexpected effects" following cannabis consumption.[2]

Research conducted in the early 2000s confirmed that using high-THC cannabis could trigger panic attacks, paranoia, and psychotic symptoms in otherwise healthy users, precipitate psychosis in vulnerable individuals, and worsen symptoms in diagnosed schizophrenic patients.[3]

Rising international concern led the United Nations *World Drug Report 2006* to focus on the health risks associated with high-THC cannabis. It stated that "the plant itself has been transformed into something far more potent than in the past" and noted that this "new cannabis" was rapidly gaining market share. It also mentioned research suggesting that CBD might counteract THC's effects and called for including CBD content in cannabis potency assessments.[4]

The following year, the British government conducted a nationwide study on cannabis potency, taking both THC and CBD into account. An analysis of several thousand cannabis samples "confiscated from street-level users" revealed that imported cannabis resin, or hashish, contained an average of 6% THC and 3% CBD, while domestically produced sinsemilla had an average THC content of 16%, with virtually no CBD. Known locally as *skunk* due to its pungent odor, sinsemilla had recently replaced hashish as the most popular form of cannabis smoked in the UK.[5]

At the time, most hashish imported into the UK came from Morocco, Pakistan, and Afghanistan, where producers neither selected their plants for potency nor removed male plants before pollination. In these mixed fields, the resulting resins contained relatively balanced proportions of THC and CBD. To produce hashish, growers harvested mature female plants, dried them, and then rubbed or beat the flowers against a sieve to extract the resin glands. The resin was then compressed into a solid, sticky paste: hashish. The first round of sieving produced the purest resin, while subsequent extractions contained more plant material. Like olive oil, sieved cannabis resin can be graded based on whether it comes from the first or later extractions. Some hashish from the first sieving could contain up to 45% THC, but the average THC content of resins exported to Europe in the early 2000s ranged between 6% and 8%.[6]

Until then, British cannabis smokers were accustomed to mixing this relatively mild hashish with tobacco, which they rolled into joints. Within just a few years, however, the majority had shifted to skunk, which they also mixed with tobacco. Authorities did not explain this shift, but they considered its impact on public health.[7]

In 2009, a team led by psychiatrist Marta Di Forti published a study showing that nearly half of the patients hospitalized in south London for a "first episode of psychosis" were regular smokers of high-THC skunk. The study also found that healthy cannabis users living in the same area were more likely to smoke hashish. In the following years, Di Forti and her colleagues conducted larger-scale studies, confirming a strong correlation between the use of high-THC, low-CBD skunk and the development of schizophrenia. They demonstrated that skunk users had a higher risk of developing a psychotic disorder compared to nonusers, with the risk increasing with frequency and quantity of consumption. A 2015 study estimated that daily skunk users had five times more risk of developing a psychotic disorder compared to nonusers, and that risk was even greater for those who started smoking skunk before age fifteen. Surprisingly, no similar trend was observed among hashish smokers, regardless of their level of use. In fact, daily hashish users appeared to have a lower risk of developing psychotic disorders than those who had never used cannabis. Drawing on studies suggesting that CBD "ameliorates the psychotogenic effect" of THC, Di Forti and her team hypothesized that the relatively high CBD content in imported hashish "might explain our results."[8]

Critics responded that the vast majority of cannabis users do not develop schizophrenia or other forms of psychosis. Fewer than one in a hundred people in the general population develops schizophrenia during their lifetime, and schizophrenia rates do not vary significantly between countries where cannabis use is widespread and those where it is rare. In most cases, the psychosis-like states that cannabis can induce subside within a week of abstinence and do not develop into chronic psychiatric disorders. The correlation between cannabis use and psychosis does not necessarily prove causation. It is possible that people genetically predisposed to psychosis or schizophrenia are also more likely to use cannabis and to use it more frequently. Some research suggests that schizophrenia is linked to alterations in the endocannabinoid system, which could explain why certain at-risk individuals react differently to cannabis. In other words, cannabis may accelerate the onset of schizophrenia in predisposed individuals, but it does not appear to cause the disorder on its own.[9]

Both Di Forti et al. and her critics had valid points. Neuroscientist

Godfrey Pearlson puts it this way: "One is that people who develop a short-lived psychotic illness related to cannabis use that lasts a month or more should quit using cannabis, because their risk for developing schizophrenia is ten times higher than that of the general population. The other is that if somebody has a family history of schizophrenia, then they are taking a calculated risk if they use cannabis, and probably shouldn't be tempting fate."[10]

* * *

The studies on the mental health risks associated with using high-THC, low-CBD cannabis raised a question: Was the increased risk due to the high THC content or the absence of CBD? To answer this, several research teams conducted experiments, administering these two cannabinoids in pure form to healthy volunteers under strictly controlled conditions.[11]

Intravenous injection of cannabinoids is neither a practical nor a recommended method of consuming cannabis. Producing injectable cannabinoids is expensive and complex, using syringes is repulsive to many, and establishing and administering precise doses is challenging. However, for scientific research, this method minimizes variables and improves the reliability of the experimental results.[12]

In real life, most people inhale cannabis by smoking or vaporizing it. When smoked, about half of the cannabinoids are destroyed by combustion, while the other half ends up in the smoke. The proportion of THC or CBD reaching the bloodstream varies significantly, usually between 10% and 35%. This percentage, known as bioavailability, depends on numerous factors: lung capacity, depth of inhalation, duration of retention, and loss through sidestream smoke. Once inhaled, THC rapidly enters the bloodstream via the lungs, reaching the brain within seconds. Experienced users can adjust their consumption based on the effects they feel. However, this method remains imprecise and subject to considerable variations.[13]

Others prefer to consume cannabis in the form of edibles or infusions. With oral consumption, the effects typically appear between thirty minutes and two hours after ingestion. About half of the cannabinoids are destroyed by stomach acids before they reach the intestine, where the other half enter the bloodstream and are metabolized by the liver. The liver converts some of the THC into 11-hydroxy-THC, a metabolite that is at least as psychoactive as the original molecule, if not more potent. The bioavailability of ingested cannabinoids varies greatly, ranging from 2% to 20%, depending on cooking methods, the contents of the consumer's digestive tract, and individual genetic differences that affect cannabinoid absorption, digestion, and metabolism. For researchers, oral administration is the least reliable and most unpredictable method for measuring THC and CBD uptake.[14]

Unlike inhalation and ingestion, intravenous injection delivers a precise dose of a cannabinoid directly into the bloodstream, ensuring total bioavailability. The effects of injected THC appear within seconds, peak within twenty to thirty minutes, and then gradually fade depending on the dose. Volunteers who receive a low intravenous dose of THC report the typical sensations associated with cannabis: well-being, dizziness, relaxation, hunger, racing thoughts, enhanced appreciation of food, and heightened sensitivity to sounds. However, a high dose of injected THC consistently induces transient psychosis-like symptoms, including hallucinations, paranoia, panic attacks, and a severe distortion of time perception. In various experiments, volunteers believed that the laboratory clock had been intentionally set back to confuse them, that the experimenter was a fake doctor conducting a fake interview, or that they were still being administered THC through improbable means—such as through the bedsheets. However, when these same volunteers received a high oral dose of CBD three hours before the THC injection, they exhibited far fewer psychosis-like symptoms.[15]

In 2017, several scientists called for further investigation into whether increasing CBD levels in cannabis products could improve consumer safety. Subsequent studies yielded mixed results. One experiment showed that a high dose of vaporized CBD reduced the psychoactive effects of THC, whereas a low dose enhanced them—particularly in occasional users. Another study found that a high dose of ingested CBD taken simultaneously with THC actually intensified THC's psychoactive effects. CBD and THC are biphasic substances, meaning their effects differ depending on the dose. When CBD is administered at levels generally found in cannabis or hashish, there is no conclusive evidence that it mitigates THC's psychoactive effects.[16]

The researchers who initially proposed studying CBD's potential to reduce THC-related risks ultimately concluded that its impact is "considerably less important than the dose of the primary intoxicating cannabinoid, THC." According to their revised view, the benign effect of low-potency cannabis is primarily due to its low THC content rather than any protective role played by CBD.[17]

* * *

Determining a low-risk and effective dose of THC for the general population is challenging, as individual reactions to the substance vary considerably. Even when taking body weight into account, there is no universal dose at which most consumers experience intoxication. A dose suitable for one person may be too strong or even uncomfortable for another. An occasional user may experience severe reactions, such as paranoia, anxiety, or vomiting, at doses that regular users consider "standard." This variability in effects sets cannabis apart from other intoxicants, particularly alcohol.[18]

In many countries, public health authorities have defined "standard drinks" to help consumers manage their alcohol intake. A mug of beer, a glass of wine, and a shot of whiskey all contain roughly the same amount of pure alcohol, with volume differences compensating for alcohol strength. The amount of pure alcohol in a "standard drink" varies slightly from country to country, but the goal remains the same: to establish serving sizes that are low-risk, effective, and easy for consumers to track. Most people are reluctant to calculate their pure alcohol intake in grams but can easily keep track of the number of drinks they've consumed.[19]

A similar system for cannabis seems unfeasible. While potency can be measured in terms of THC percentage and doses expressed in milligrams, there is no universal quantity that is both effective and low-risk for all. Consequently, there can be no THC measure equivalent to a "standard drink."[20]

The variability of responses to cannabis is further amplified by differences in gender. Until recently, scientific research on the effects of cannabis tended to focus exclusively on men. As a result, most studies that include women are preliminary (none to date consider nonbinary people). Still, existing research suggests that women are more sensitive than men to THC and require smaller amounts to experience similar effects. This may be due to differences in the number, location, and sensitivity of cannabinoid receptors between the sexes. Research indicates that the female sex hormone estrogen influences the body's production of endocannabinoids and the availability of its cannabinoid receptors. Among women with a menstrual cycle, a low estrogen concentration at the start of the cycle can make their receptors more sensitive to THC, whereas higher estrogen levels later in the cycle may reduce its effects. As a result, the same dose of cannabis may produce different effects throughout the menstrual cycle.[21]

After menopause, women no longer have menstrual cycles, and their estrogen levels decrease. Some studies suggest that this hormonal change increases the number and availability of cannabinoid receptors, making postmenopausal women more sensitive to THC than younger women.[22]

Increased sensitivity to a substance implies that a lower dose achieves a similar effect. In practice, a "low dose" of THC varies from person to person depending on multiple factors, including metabolism, age, sex, and frequency of use. Recreational users usually consume between 1 and 20 milligrams of THC per session, illustrating just how much a "standard dose" can vary. For beginners, the advised procedure is to "start low, go slow, and stay low."[23]

* * *

During the 2010s, the potency of confiscated cannabis continued to increase worldwide. In the European Union, the THC content of imported hashish nearly tripled in ten years, reaching close to 30% on average. Authorities attributed this sharp increase to changes in production methods in traditional hashish-producing countries, including the introduction of high-THC plant varieties, intensive cultivation practices, and more efficient resin-extraction techniques.[24]

To explain this increase in cannabis potency since the 1980s, commentators often cite the Iron Law of Prohibition, which suggests that strict law enforcement leads to an increase in the concentration of illegal substances. The argument is that, under the pressure of illegality, traffickers aim to reduce the size and detectability of their products while maximizing their market value, leading them to prefer higher potency. However, developments in newly legalized recreational cannabis markets challenge this view. In jurisdictions that have adopted a liberal model without THC limits, the potency, diversity, and market share of high-THC products have increased rather than decreased—as consumers flocked to high-potency products and turned away from milder ones. In 2020, a leading California-based cannabis producer told *Forbes* that he had been forced to "throw in the trash" an exceptional sinsemilla "with an amazing terpene profile, the best smoke I've ever had," simply because its THC level was below 20%.[25]

Consumers appeared drawn to increasingly potent cannabis products, and producers and retailers responded to this demand. There was certainly money to be made in the high-potency cannabis market, but this alone does not explain what initially motivated consumers. Most people do not associate potency with quality when purchasing beer, wine, or spirits, so why do cannabis consumers seem to favor products with higher and higher THC levels?[26]

This question remains understudied, yet several of its elements are understood. First, cannabis has a "biological ceiling," which some experts estimate to be around 35% THC. It seems that no other plant produces such a high concentration of a single compound; however, dried *Cannabis* flowers still have limitations, as they consist of leaves, stems, pistils, and other plant material, not just THC-containing resin.[27]

Second, high-potency products are not necessarily more dangerous when consumed in moderation. However, measuring and controlling small amounts is more challenging than managing larger quantities, which increases the risk of serving and consuming more than intended. Studies on alcohol have shown that consumers tend to overpour strong spirits, while they pour more precise amounts when serving beer or wine. Whether users of cannabis concentrates have similar tendencies has yet to be researched.[28]

Third, people who drink spirits can use their senses to detect high potency thanks to the distinct smell and burning taste of alcohol. In comparison, THC is tasteless and odorless and provides few visual cues, making it hard for consumers to gauge the potency of cannabis products. When cannabis is smoked or vaporized, the effects appear within seconds, but it generally takes three to ten minutes to feel their full impact. This means that even regular smokers of medium-potency cannabis can be caught off guard by the intense effects of ultra-potent cannabis, leading to anxiety, paranoia, nausea, and severe vomiting, similar to what a beginner might experience at a lower dose. In brief, when individuals consume significantly more THC than they are accustomed to, they risk adverse effects, particularly nausea and vomiting.[29]

In some cases of repeated exposure to high doses, vomiting can become severe. Medical doctors refer to this condition as cannabis hyperemesis syndrome (CHS), defined as a disorder characterized by "intractable nausea, cyclic vomiting, abdominal pain, and hot bathing behavior associated with ongoing THC exposure." While hot baths and showers may provide temporary relief, the only effective cure appears to be discontinuing cannabis use. Clinicians describe CHS as "unique in presentation" due to a specific biphasic effect of THC, which acts as a vomitive at high doses but serves as a medically recognized anti-vomitive at lower ones.[30]

* * *

In the 2010s, a cannabis concentrate called butane hash oil (BHO) became popular because it was potent, cheap to make, and easy to consume. Butane gas, used as a solvent, can strip nearly all cannabinoids and terpenes from cannabis trimmings without removing much other plant material. This produces an oily mixture that can be filtered and heated, and once the butane evaporates, the result is BHO. Commonly called *dabs*, *wax*, or *shatter*, this concentrate typically contains 60% to 90% THC but can be refined further. In its pure form, THC is a sticky oil that requires high temperatures to volatilize. The dabbing method involves placing a small amount of concentrate—often the size of a pinhead—onto a red-hot metal element, then inhaling the fumes produced. This method allows users to absorb the THC content of several joints in one breath. As cannabis expert Ed Rosenthal put it: "Dabs get you very stoned, very quickly."[31]

As dabbing gained popularity among consumers, some commercial *Cannabis* growers began converting their entire crops into oily concentrates. In 2013, journalist Bobby Black explained in *High Times* that these extracts were "more potent and profitable, and easier to conceal and consume." According to Black, growers increasingly viewed their plants as "nothing but raw material to be transformed into a variety of waxes, oils, and shatter." He also described the main risk of dabbing as "passing out and falling over after doing a dab." Black consequently advised readers to consider "dabbing in a seated position whenever possible or if you are in line for a dab at an event, asking the person behind you to be your spotter. . . . Be ready to perform a 'dab grab' in case the person in front of you faints!"[32]

Inhaling cannabis concentrates to the point of fainting is neither advisable nor everybody's cup of tea. However, studies conducted on chronic dabbers highlight a crucial difference between THC and alcohol. Some dabbers consume very high and repeated doses of THC throughout the day, resulting in daily intakes more than a hundred times higher than a basic intoxicating dose for an infrequent cannabis user. With alcohol, drinking one hundred standard drinks in a single day would be fatal, as the lethal dose is ten to twenty times the effective dose. Even heavy drinkers risk death if they consume two dozen shots of spirits in quick succession. By contrast, THC has no known lethal dose, even at the high levels reached by chronic dab users. They may vomit or pass out, but no cases of death have been directly attributed to a THC overdose.[33]

Preliminary studies indicate few differences between the short-term health concerns of chronic dabbers and chronic users of potent cannabis. Although dabbers have, on average, twice the THC levels in their blood, research suggests that they do not necessarily experience stronger psychoactive effects. Prolonged exposure to high doses appears to induce significant tolerance, meaning users reach a threshold beyond which additional consumption has minimal effect. The precise mechanism behind this tolerance remains unclear, but experts suggest that cannabinoid receptors become saturated with THC at these elevated levels.[34]

In the late twentieth century, cannabis was seen as much less addictive than other popular intoxicants, as most users could stop consuming it without notable consequences. However, today's high-THC cannabis products are a different story. Users develop dependence, and when they quit, they experience withdrawal symptoms, including cravings, irritability, sleep disturbances, and weight loss. Health professionals refer to this combination of dependence and withdrawal as cannabis use disorder (CUD) and report a growing number of affected patients.[35]

While high doses of THC can cause dependence, low doses do not appear to have anti-addictive properties. In this case, the biphasic action of cannabinoids is reflected in THC's sibling molecule, CBD, which is known for its anti-addictive properties. Some studies indicate that CBD can help reduce tobacco consumption, decrease heroin cravings, and alleviate withdrawal symptoms related to CUD.[36]

Regular use of high-THC cannabis and its products poses specific risks for adolescents and young adults. Endocannabinoids are crucial in forming connections between neurons, particularly in reward circuits, which THC can alter in a way that increases a young person's likelihood of using drugs later in life. Additionally, regular use of high doses of THC seems to affect cognition in young people, although studies on this issue remain inconclusive. What is certain is that the younger a person starts consuming cannabis, and the more they consume, the higher the risk of experiencing adverse effects. By limiting themselves to occasional use of low- or medium-potency cannabis until age twenty-one, young people allow their brains to develop fully with minimal interference from THC. Of course, the most effective way to avoid any risk of harm is to abstain from using cannabis altogether.[37]

Once again, THC's biphasic action is evident. Preliminary research suggests that low doses may enhance cognition, particularly in older adults, whereas high doses impair cognition across all age groups, especially among young people.[38]

* * *

Currently, some producers distill and purify concentrates to obtain pure THC, an oily substance that is easy to consume in vape pens. Since THC is odorless, users can take discreet puffs throughout the day and end up with a considerable daily intake of this substance.[39]

This focus on a single substance stands in stark contrast to the ways of the *Cannabis* plant itself. *Cannabis* does not produce THC in isolation but alongside hundreds of other bioactive compounds, including cannabinoids, terpenes, and flavonoids. The interaction of these substances, known as the "entourage effect," influences the overall impact of consumed cannabis. Terpenes, in particular, have well-documented psychoactive effects. For example, limonene enhances alertness and reduces anxiety, myrcene promotes relaxation, and linalool acts as a sedative. Several terpenes seem to modulate the intensity of THC's effects. However, scientific research on this complex phenomenon is still in its early stages. Until a reliable method exists to quantify the entourage effect, cannabis potency will continue to be measured exclusively by its THC content.[40]

When cannabis resin is concentrated into an oil containing around 80% THC, its chemical composition remains faithful to the original plant, including both its qualities and its flaws. On the positive side, concentrated terpenes contribute intense aromas and flavors, which dabbing connoisseurs appreciate. On the negative side, concentrates may also contain pesticides, heavy metals, and other toxins present in the original plant material. Producing high-quality concentrates means starting with contaminant-free and healthy *Cannabis* plants.[41]

At 90% THC or higher, concentrates lose most of their terpenes and become less aromatic. They can be further distilled and purified into pure THC—a translucent oil that no longer reflects the plant's mixed chemical composition because it has become a single substance. This phenomenon is similar to the distillation of alcohol from wine.[42]

Wine made from fermented grape juice generally contains 10% to 15% alcohol and small amounts of terpenes and flavonoids. Distilled, it becomes brandy, which contains trace amounts of terpenes and can reach 90% alcohol, though most brandy for human consumption contains between 40% and 50%. When distilled close to purity, it becomes a "neutral spirit," a colorless liquid that looks like water with no palpable connection to grapes, wine, or brandy. At this stage, alcohol becomes a dangerous poison: Consumed in its pure form, it burns internal tissues and can cause organ failure. The risk comes not from the nature of the substance but from its extreme concentration—the danger arises from the substance's purity.[43]

Although the effects of alcohol and THC differ significantly, both substances play similar roles. They are the central agents responsible for the psychoactive effects of their respective intoxicants—cannabis without THC does not make users high, just as alcohol-free beer or wine does not make consumers drunk. Both molecules are the primary drivers of dependence for their respective substances. And both can be distilled and purified into increasingly potent and hazardous forms.[44]

While pure THC does not pose the extreme dangers associated with pure alcohol, its regular use suggests risks with minimal apparent benefits. The health impacts of dabbing or vaping pure THC remain understudied. Anecdotal evidence and initial research indicate that beyond the usual risks associated with potent cannabis products, long-term use of pure THC may increase the risk of depression and suicidal thoughts, especially among young users.[45]

Psychoactive and biphasic, THC is a tricky substance. Those who use it gain by knowing it.

# 14

# New Ideas

Studying the positive effects of nonmedical cannabis use has long been challenging for scientists. Its impact on creativity, in particular, remains debated. To understand why, it helps to revisit research conducted in the 1970s.

A study funded by the US government in 1970 set out to establish the full range of effects experienced by regular cannabis users. The research involved anonymous participants completing a questionnaire that listed several hundred possible effects. The goal was to determine how cannabis intoxication influenced the senses, body, mind, perception of space and time, interpersonal relationships, and sexual desire, among others. Participants were asked to indicate how frequently they experienced these effects over the past six months and specify the level of intoxication needed to feel them. They were also invited to add any relevant comments at the end of the questionnaire.

The study was led by psychologist Charles Tart, an associate professor at the University of California, Davis, and a recognized specialist in altered states of consciousness. In the letter accompanying the questionnaire, Tart emphasized that scientists knew very little about the experience of smoking cannabis and encouraged participants to respond as precisely as possible to help advance research.[1]

One hundred and fifty experienced users participated in the study. Based on the biographical information, Tart described the group as "a predominantly young, highly educated group of Californians, primarily students, with a fair number of older individuals and professionals among them." After analyzing the results, he published *On Being Stoned: A Psychological Study of Marijuana Intoxication* in 1971, a book that became a reference in the scientific study of cannabis's psychological effects.[2]

Tart's anonymous respondents confirmed many effects already associated with cannabis. At low to moderate levels of intoxication, they reported enjoying music and food more, feeling increased empathy, having deeper conversations, experiencing heightened sexual desire, and laughing more easily. At higher levels, they became easily sidetracked, forgot the beginning of their conversations, perceived time as moving more slowly, and had difficulty estimating distances.[3]

By asking participants to rate their level of intoxication for each effect, Tart's study highlighted the biphasic nature of cannabis. At low doses, users felt more sociable and talkative, less anxious, and sometimes more capable of recalling past events. However, at higher doses, they became less sociable and communicative and more anxious, and they experienced short-term memory difficulties. The biphasic nature of effects was particularly noticeable when it came to thought processes and creativity. At low to moderate levels, many respondents found their minds worked more efficiently, allowing them to solve problems more easily by switching to intuitive thinking. However, at higher levels of intoxication, their minds worked less efficiently, and clear thinking and problem-solving became arduous.[4]

The majority of participants reported having original ideas under the influence of cannabis. Tart cited the testimony of a forty-year-old physicist who explained how occasional cannabis use had stimulated his professional thinking: "I smoke marijuana once or twice a week for recreation, but a couple of times, I've started thinking about my work and had real breakthroughs as a result. Once, when I had been in the process of setting up a new laboratory for several months, I got stoned one evening and started thinking about things at the lab and suddenly had all these ideas popping into my mind of little things I had to do if the laboratory was to function on schedule, little details about equipment that were unspectacular but essential. I listed about twenty ideas in an hour, and every one of them checked out the next day. They were all sorts of things that had been pushed to the back of my mind by more obvious problems in setting up the laboratory. Another time, I got thinking about a problem area in my work, and all sorts of theoretical ideas came popping into my head. They fit together into a coherent theory which looked damned good the next morning—I have since published the theory and organized a lot of research around it, to my great advantage."[5]

Tart made no claims about the validity of such statements. Instead, he viewed the creative effects of cannabis intoxication on highly trained individuals as "an intriguing research question." He stated plainly that cannabis could cause decreased thinking efficiency but also stressed that the difference between cognitive impairment and enhancement mainly depended on intoxication levels—impairment occurring at high levels and enhancement at low to moderate ones. Tart concluded his book by calling for further research to evaluate both the benefits and the physiological and psychological costs of cannabis use so that people could make informed decisions about whether the benefits outweighed the costs.[6]

Four months after Tart's book was published, in June 1971, US president Richard Nixon declared drug abuse to be "public enemy number one," launching the War on Drugs, an international campaign that would last for decades. From then on, most scientific research on cannabis and cognition focused on negative effects, relegating the study of potential benefits to the background.[7]

One notable exception was a 1975 report by a team of Canadian researchers led by psychiatrist Thaddeus Weckowicz from the University of Alberta. At that time, several studies had already demonstrated that cannabis impaired three main cognitive functions: short-term memory, learning, and time perception. However, Weckowicz and his colleagues noted that these studies used doses of cannabis or THC far higher than those usually consumed by occasional users. Furthermore, many occasional users stated that cannabis enhanced their thinking, perception, and creativity rather than impairing them. Intrigued by these discrepancies, the researchers decided to examine the impact of low-dose cannabis on cognitive abilities.[8]

They recruited eighty-four occasional cannabis smokers and divided them into four groups based on their treatment: high-dose, low-dose, placebo, and control. The first group smoked cannabis containing 6 mg of THC, the second group smoked cannabis with 3 mg of THC, the placebo group smoked cannabis with no THC, and the control group consumed nothing. All participants then took tests evaluating two forms of thinking central to creativity: divergent thinking, which involves generating new solutions to a problem, and convergent thinking, which narrows down ideas to arrive at a single optimal solution.[9]

The results showed that participants in the low-dose group had the highest scores on divergent thinking tests and two out of three convergent thinking tests. In contrast, those in the high-dose group had the lowest scores. The researchers concluded that "low doses of cannabis may improve performance on some cognitive tests," whereas consuming 6 mg of THC or more "may interfere with such higher cognitive functions as reasoning and concept formation." They emphasized the need for future cannabis research to consider a wide range of doses.[10]

Over the following decades, these findings were largely ignored, and most studies focused on cognitive impairments associated with cannabis, often at high doses. In 1992, one study showed that occasional users who smoked a cigarette containing 19 mg of THC exhibited impairments in nearly all cognitive tasks: They responded more slowly to questions and reading tasks, struggled to understand and remember what they read, had difficulty detecting spelling mistakes, and performed poorly in arithmetic. Other studies confirmed these negative effects, noting that non-intoxicated daily cannabis users had short-term memory and learning deficits, while occasional users who received a high oral dose of THC showed significant impairments in word recall and learning tasks.[11]

However, some research produced seemingly contradictory results. In 2001, a study found that heavy users who smoked multiple times a day showed no cognitive impairment after consuming cannabis containing 30 mg of THC, nearly ten times the effective dose for an occasional user. Another study revealed that while sober, heavy users had slower information processing speeds compared to nonusers, but after consuming their usual dose of cannabis, these deficits disappeared—suggesting that using the drug normalized their performance while abstinence impaired it.[12]

By the mid-2000s, researchers realized that the disparate results of these studies were not solely due to differences in dosage or administration methods but also to participants' consumption history. Increasing evidence indicated that regular users developed tolerance to many of cannabis's impairing effects so that an occasional user was more likely to be impaired by a given dose than a daily user. In turn, this suggested that overlooking this human variable when assessing cognitive performance among cannabis users amounted to comparing apples and oranges.[13]

* * *

People who consume cannabis by choice rather than for medical reasons cite a wide range of motivations. They primarily seek relaxation, a sense of well-being, stimulation of appetite and sexual desire, enhanced thinking, better sleep, or a boost in creativity. They are much more interested in its perceived benefits than its negative effects. However, scientific research has historically focused on the risks of cannabis rather than its potential benefits. At the end of the twentieth century and into the early 2000s, scientific research barely considered its positive effects on cognition and only did so occasionally and in passing. One study noted that people who had just smoked cannabis made "more uncommon associations" when responding to prompt words. Another reported that regular users who had abstained for a day outperformed nonusers in fuzzy logic, which is the ability to make decisions based on uncertain or incomplete information. A third study found that THC improved participants' verbal fluency but impaired their performance in all other measures. These rare references to cognitive enhancement were otherwise lost in a sea of impairments.[14]

It was only in the late 2000s that the question of cannabis and creativity reemerged as a subject of scientific discussion. The rise of medical cannabis in several countries led doctors to prescribe it for various conditions, including anxiety, headaches, insomnia, premenstrual syndrome, and writer's block. These evolving medical practices generated a need for research into its efficacy, including its potential to help people overcome creative slumps.[15]

In 2012, Irish psychologist Gráinne Schafer and her team at University College London conducted an experiment to measure cannabis's effect on creativity and divergent thinking, building on the work of Weckowicz and his colleagues in the 1970s. For this study, they recruited 160 occasional cannabis users, all of whom had abstained for 24 hours. On the first day, participants completed a series of tests and answered a questionnaire about their creative achievements in different fields. The researchers then classified them into three groups: high, intermediate, or low creativity. The following day, participants returned to the research center and, in a calm and comfortable room, smoked their own cannabis. They were then given tests assessing their divergent and convergent thinking. The results showed that cannabis significantly improved divergent thinking in participants from the low-creativity group. In a verbal fluency task—designed to measure real-time creativity—they had to generate as many words as possible in sixty seconds starting with a given letter (either *B* or *M*). When sober, they performed significantly less well than those in the high-creativity group. However, their performance improved to match the scores of the high creativity group after smoking cannabis. Schafer and her colleagues concluded: "This increase in verbal fluency for those low in trait creativity on the intoxicated day supports the notion that cannabis can enhance aspects of creativity."[16]

The researchers acknowledged that their study did not measure originality, a key criterion in most definitions of creativity. They also recognized that their findings provided limited insight into how cannabis affects individuals believed to have high creativity since the verbal fluency scores for the high-creativity group remained constant across both days.[17]

As early as 1971, Charles Tart suggested that research on cannabis and creativity should focus on individuals whose professions require creative thinking. Tart also emphasized the influence of other factors, such as the user's mindset and the environment in which they smoke, which can modulate the drug's effects. In his view, a "cold and sterile" laboratory setting risks stressing participants and skewing observations. He recommended that future studies focus on experienced users, allowing them to manage their own dosage and timing of consumption in similar conditions to their habitual and ordinary use.[18]

Fifty years later, in 2021, a multidisciplinary research team from Washington State University implemented many of Tart's recommendations while incorporating methodological insights from more recent studies. Led by Benjamin Warnick, an associate professor of entrepreneurship, the team combined expertise from business studies and psychology. Their goal was to study cannabis's impact on entrepreneurs, a group whose profession depends on creative thinking to develop new products or services. The researchers noted that several successful entrepreneurs, including Steve Jobs, had already attributed a positive influence on their creativity to cannabis. Jobs himself had stated that smoking cannabis helped him feel "relaxed and creative."

To test this hypothesis, the researchers conducted an anonymous survey of 254 established entrepreneurs via a data collection platform. About half of the participants used cannabis several times a week, while the other half abstained entirely. All participants received a description of an emerging technology—virtual reality—and had three minutes to brainstorm as many business ideas as possible related to new products or services based on this technology. They then selected the best idea and developed it in detail. Participants also completed a questionnaire regarding their entrepreneurial passion and experience while providing biographical information that allowed researchers to control for variables such as age, gender, education, personality, and familiarity with virtual reality.[19]

Two independent experts, who had no information about the participants, then evaluated the originality and feasibility of the selected ideas. The results showed that the cannabis-using entrepreneurs generated significantly more original ideas than the nonusers, though these ideas were somewhat less feasible. The study also showed a link between entrepreneurial passion and originality, while entrepreneurial experience was more strongly associated with the feasibility of ideas. These findings confirmed that motivation and knowledge influence creative thinking, particularly among cannabis users.[20]

Warnick and his colleagues noted that cannabis users were generally more impulsive, less inhibited, and more adept at spotting connections between seemingly unrelated concepts. They were also less likely to criticize or filter their ideas prematurely. The researchers suggested that these traits, combined with the typical cognitive impairments of cannabis use—such as short-term memory and attention deficits—may have contributed to the lower feasibility of their ideas.[21]

Warnick's team carefully weighed the cognitive benefits and drawbacks of cannabis. They acknowledged its adverse effects, such as the risk of addiction and tolerance, a possible link to psychosis, and an increased risk of traffic accidents. They also recommended alternative strategies to boost creativity, such as mindfulness meditation, sufficient sleep, or regular breaks. Nonetheless, their findings indicated that cannabis use can enhance creativity in professions that require it, particularly among individuals who are passionate about innovation and possess a strong foundation of knowledge to draw from when forming new connections.[22]

* * *

Beyond psychological studies and cognitive tests, recent advances in neuroscience support the idea that cannabis can promote creative thinking. Research shows that cannabinoid receptors play a key role in brain areas linked to creativity. When activated, they cause neurons to reduce their release of neurotransmitters. These CB receptors are particularly abundant on neurons that produce glutamate and GABA (gamma-aminobutyric acid), two substances that regulate many other neurons. Glutamate acts as an accelerator of neuronal activity, while GABA serves as a brake. By reducing the availability of these two substances, cannabinoids modulate brain chemistry. More specifically, THC reduces the release of GABA, which usually inhibits dopamine-producing neurons. This reduced inhibition leads to an increase in dopamine release, a neurotransmitter essential to many brain functions, including creativity.[23]

Magnetic resonance imaging (MRI) scans conducted on individuals engaged in creative tasks—such as poetic writing, sketching images, or musical improvisation—show similar patterns of brain activity. Three major brain networks collaborate during these tasks: the default mode, salience, and executive control networks. Each of these networks consists of several brain regions that include dopamine-producing neurons as well as other neurons equipped with CB receptors.[24]

The default mode network handles spontaneous thinking, imagination, mind-wandering, and memory retrieval. It also generates "candidate ideas." The salience network filters these ideas and identifies those most relevant. The executive control network then refines these ideas and directs them toward practical and achievable solutions. This network also manages decision-making, short-term memory, problem-solving, and goal-directed behavior.[25]

Dopamine plays a key role in all three networks. It facilitates memory formation, stimulates divergent thinking, and encourages the exploration of new ideas. Studies on Parkinson's disease have shown that patients treated with dopaminergic medications often experience increased artistic and creative drives, as well as improved verbal and visual creativity.[26]

Preliminary research suggests that THC indirectly stimulates dopamine production, encouraging the emergence of self-generated thought and broadening the range of ideas available for creative activity. Under the influence of THC, people may notice details they would otherwise overlook, and their "trains of thought" may include more unusual material, novel ideas, and unexpected associations.[27]

Several well-known effects of cannabis are linked to its impact on dopamine levels, particularly the distortion of time perception. The brain's internal clock relies on brief bursts of dopamine to track time. By increasing available dopamine, cannabinoids speed up this clock, making external time feel slower. The sensation of time slowing down is a classic effect of cannabis.[28]

Dopamine-producing neurons also regulate the pleasurable sensations derived from food and sex. Elevated dopamine levels can lead to feelings of euphoria, creativity, and heightened concentration, but too much dopamine can cause distraction, confusion, and states close to psychosis. All these effects align with those that can be experienced under the influence of cannabis, suggesting that THC's capacity to increase dopamine levels underlies several effects generally attributed to cannabis.[29]

These explanations simplify ongoing neuroscience research. THC's influence on brain chemistry is not limited to dopamine—it also affects serotonin, acetylcholine, and endorphins. Further research is needed to reach a better understanding of how THC boosts dopamine and how dopamine enhances creative thinking. Nonetheless, neuroscientists studying creativity now consider cannabis use a "promising research area" for the experimental examination of specific brain regions.[30]

More than fifty years after Charles Tart called for in-depth studies on cannabis and creativity, cannabis has become a tool for exploring how the brain works during creative activity. Sometimes, science advances, slowly but surely.

# 15

# Intoxication and Aromatic Pleasure

Understanding how cannabis intoxicates those who consume it requires considering the effects of THC on the body and brain. For public safety reasons, much recent research on the psychoactive effects of cannabis has focused on its impact on driving ability.

In countries or states where cannabis has been re-legalized, authorities are faced with the complex issue of setting tolerance thresholds for driving under its influence. Discouraging heavily intoxicated users from getting behind the wheel makes sense, particularly when they experience disorientation, distraction, or drowsiness. But what about those who are only slightly under the influence, such as individuals who have taken a prescribed dose of medical cannabis? To what extent are they still fit to drive? Where is the line between a legally acceptable level of impairment and pronounced, dangerous intoxication?

When it comes to alcohol, the guidelines are clear. Drunk driving is widely recognized as dangerous, and most countries have set legal blood alcohol limits for drivers. These thresholds are well understood by the general public and typically correspond to a specific number of standard drinks that a person can reasonably consume before driving. Law enforcement officers use breathalyzers, which reliably estimate blood alcohol content by analyzing the concentration of alcohol in exhaled air.[1]

This system of control works because of how ingested alcohol interacts with the human body. As a water-soluble substance, alcohol quickly and evenly distributes itself throughout the body, which is largely composed of water. Blood and breath alcohol concentrations accurately reflect those in the brain and correlate with the level of psychoactive effects and impairment experienced. Additionally, the body breaks down alcohol at a constant rate, allowing for predictable elimination; within twenty-four hours, even after excessive consumption, nearly all the alcohol is cleared from the body in most cases.[2]

Almost none of this applies to cannabinoids like THC and CBD. Unlike alcohol, these substances are fat-soluble, meaning they dissolve in fats rather than water. When a person smokes or vaporizes cannabis, cannabinoids enter the bloodstream through the lungs and bind to fat-transporting proteins. They are then rapidly distributed to lipid-rich organs, particularly the brain, liver, and fatty tissues. Following inhalation, THC levels in the blood peak within about fifteen minutes before quickly decreasing as the substance spreads throughout the body and the liver begins metabolizing it. As THC leaves the bloodstream and enters the brain, psychoactive effects and impairment increase. Approximately thirty minutes after inhalation, blood THC concentration reaches a low plateau that can last for several hours, during which consumers experience the most pronounced impairing effects.[3]

When cannabinoids are ingested rather than inhaled, they are absorbed more slowly and result in considerably lower blood THC levels, but the effects can be more intense and prolonged. After ingestion, it takes about two to three hours for THC blood levels to peak. Then, its concentration gradually declines over about eight hours, paralleling the subjective effects and observed impairment, particularly when driving.[4]

The time THC remains detectable in the bloodstream depends on how often and how much a person uses cannabis. In occasional consumers, THC usually clears from the blood within twelve hours, and its metabolites are eliminated through urine and feces within two to five days. However, in regular users, THC tends to accumulate in fat-rich tissues, where it is gradually released over days and even weeks. Although these residual traces are too low to produce psychoactive effects, they can be detected in the bloodstream for an extended period before being eliminated. Studies have shown that some chronic users may test positive for THC up to thirty days after their last consumption, yet they experience no psychoactive effects during this period. As a result, the blood THC level of a regular user can be similar to that of an occasional user who has recently smoked and is genuinely under the influence. In other words, the mere presence of THC in the body does not necessarily indicate an altered state of consciousness or impaired faculties.[5]

Another challenge lies in detecting THC in exhaled air. Unlike alcohol, which is expelled in large quantities through water vapor in each breath, THC is exhaled in minute amounts as tiny particles coated in lung fluid. In a single breath, a person who has consumed alcohol exhales roughly twelve million times more alcohol molecules than a cannabis user exhales THC molecules. Even with a sophisticated breathalyzer capable of measuring THC precisely, the amount present in a regular user's breath on days they have not consumed can be similar to the amount detected an hour after consumption. This ambiguity makes it difficult to differentiate between recent and past consumption and complicates the enforcement of strict limits for driving under the influence of cannabis.[6]

Despite the complexity of THC metabolism, which presents numerous challenges for authorities seeking to establish legal limits for cannabis-related driving impairments, there is no doubt that the drug can affect driving performance. Under its influence, drivers tend to drift within their lane, react more slowly, and struggle to manage multiple tasks simultaneously. Often, they are aware of their altered state and generally adopt a more cautious driving style by reducing speed, increasing following distances, and avoiding overtaking. This behavior contrasts with alcohol-impaired drivers, who often drive faster, overtake more frequently, and leave less space between vehicles. However, these cautious adjustments do not fully offset the driving deficits caused by the mind-altering effects of cannabis.[7]

Experts agree that drivers under the influence of cannabis have a "slightly but significantly increased risk of crashes." Some estimates suggest this risk is 1.4 times higher than that of a sober driver. By comparison, drivers under the influence of alcohol are five to thirty times more likely to crash, depending on their blood alcohol concentration. While cannabis poses considerably less risk to driving safety than alcohol, this is not saying much, given the extreme dangers of drunk driving.[8]

Authorities in some jurisdictions take a zero-tolerance approach, even when cannabis is legal for medical or recreational use. In Uruguay, the first country to legalize recreational cannabis in 2013, any detectable presence of THC in a driver's blood is illegal and can result in license suspension. In practice, this means that regular users, including patients on daily medical cannabis, risk breaking the law if they drive, even if they are not intoxicated or experiencing impairment.[9]

Most countries and states that have re-legalized cannabis have set legal blood THC limits for drivers, ranging from 1 to 5 nanograms per milliliter: 1 in Luxembourg, 2 in Canada, 3.5 in Germany, and 5 in several US states. However, none of these thresholds reliably correspond to actual driving impairment levels.[10]

With a limit set at 5 nanograms, most occasional cannabis users escape detection since their blood THC level usually drops below this threshold within thirty minutes of smoking or vaporizing cannabis and then remains low for two or three hours before going down to zero. However, these same users may still experience intense intoxication and measurable driving impairment at these low levels. Experts refer to such cases as "false negatives," where drivers are impaired but not detected by a blood test.[11]

Conversely, with a limit as low as 1 nanogram, some frequent users who have abstained for days or even weeks continue to test positive due to the persistence of THC in their system. This varies based on metabolism, prior consumption, and body composition. Women are particularly disadvantaged, as their bodies appear to eliminate THC more slowly than men's. On average, heavy cannabis users take about three weeks to clear all traces of THC. At a threshold of 1 nanogram, these individuals run the risk of being classified as "false positives," meaning that they test positive without being under the influence of cannabis.[12]

A driving simulator study involving occasional users who had vaporized a high dose of cannabis revealed that at both the 1-nanogram and 5-nanogram thresholds, more than half of the participants were misidentified—either as "false positives" or "false negatives."[13]

In another study, individuals who had abstained from cannabis for three months consumed a potent edible containing either 10, 20, or 50 mg of THC. All participants reported "significant subjective drug effects" at all doses. Yet, none exceeded 5 nanograms of THC in their blood, and some—including two who had consumed the highest dose—did not even reach 1 nanogram. When ingested, THC must pass through the liver, which releases it slowly, in small amounts, and over a long period—preventing rapid accumulation in the bloodstream. Nevertheless, ingested THC can produce powerful and long-lasting mind-altering effects, significantly impairing driving ability without being associated with a high blood concentration.[14]

Ultimately, all current legal blood THC limits misidentify a significant number of drivers in relation to their actual levels of intoxication and impairment. In contrast, blood alcohol limits are effective, as nearly all drivers show impairment once these limits are reached.[15]

* * *

Scientists warn that blood THC limits "cannot reliably discriminate between impaired and unimpaired drivers" and "cannot be scientifically supported" because "drug concentration in blood does not correlate to driving impairment." Nevertheless, political authorities in different countries continue to define cannabis-related driving offenses in terms of blood THC levels.[16]

Germany provides a clear illustration of this issue. In the early 2020s, as lawmakers worked on the country's new cannabis law, they asked the Federal Ministry of Transport and Digital Infrastructure to establish a blood THC limit comparable, in terms of road safety, to Germany's legal blood alcohol limit of 0.05%. In response, the ministry's Limit Value Commission issued a 2022 report explaining why it could not establish such a value. THC can persist in the blood for days and even weeks, its metabolism varies considerably between individuals, and it behaves so differently from alcohol in the body, the commission wrote, that a direct comparison of their concentrations in body fluids is "fundamentally not possible."[17]

This up-to-date position clarified the challenges but did not satisfy the lawmakers. In response, they inserted a clause into the cannabis bill requiring the Ministry of Transport and Digital Infrastructure to appoint a working group tasked with "proposing, by March 31, 2024, a blood THC concentration threshold beyond which the safe driving of a motor vehicle is no longer guaranteed, based on current scientific knowledge."[18]

As a result, the ministry convened a new panel of experts, including several researchers who had previously made comparisons between the effects of THC and alcohol on driving. Their report recommended a limit of 3.5 nanograms per milliliter of blood. While they did not dispute the previous commission's conclusion that it was impossible to determine a clear threshold at which safe driving "is no longer guaranteed," they instead proposed a limit at which an impact on safe driving "would not be unlikely." According to their findings, the first signs of driving impairment appear at blood alcohol levels of 0.02% and blood THC levels ranging between 2 and 5 nanograms per milliliter. They chose the median value within this range—3.5 nanograms—as an indicative threshold "from which it would not be unlikely for occasional users to experience effects relevant to road safety."

The panel's report acknowledged that regular users "who had not consumed cannabis and who showed no loss of driving performance" could still have baseline blood THC levels above the proposed legal limit and "would consistently be considered under the influence of cannabis when driving." Their report also suggested that occasional users who "maintain an adequate waiting period between consumption and driving" would not be affected. Arguing that the primary goal of a blood THC limit was to "guide cannabis users toward appropriate behavior," the panel's report suggested that this median threshold could encourage regular users to reduce their intake.[19]

A few days before the legislative deadline, while Germany's new cannabis law was not yet in effect, the Ministry of Transport and Digital Infrastructure announced that an "independent and interdisciplinary" working group had established a "cannabis limit" for drivers. In an official statement, the ministry asserted that this blood THC limit was based on "a conservative approach, presenting a risk comparable to a blood alcohol concentration of 0.02%."[20]

Two months later, the German Parliament approved the cannabis driving limit. Presented as less than half the legal alcohol limit, the THC limit suggested that authorities were taking a firm stance on the dangers of driving under the influence of cannabis. The limit made sense politically, if not scientifically.[21]

Several members of the ministry's Limit Value Commission promptly voiced their objections in online interviews. Toxicologist Volker Auwärter reiterated that "there is no scientific data to justify" a direct comparison between THC and alcohol concentrations in blood. Auwärter described Germany's THC limit as the result of a "theoretical comparison" between the "abstract risks of accidents when driving under the influence of alcohol or cannabis" while noting that the studies required to determine those risks for cannabis did not yet exist in sufficient number.

Professor of medicine Matthias Graw, president of the German Society for Traffic Medicine, emphasized that THC, unlike alcohol, "quickly disappears from the blood as it accumulates at the site of action. This is why we initially see an increasing effect as blood concentrations fall. Blood concentration and effect are contradictory and go in opposite directions, which is why a blood THC limit cannot be reliably defended."

Graw emphasized that blood THC levels vary depending on how regularly a person consumes cannabis. The new limit would fail to detect many occasional users, he said, even though they posed a higher risk of "causing accidents because they are less accustomed to cannabis." At the same time, it would penalize regular users, who tend to "experience fewer symptoms at these concentrations" and might even be sober and unimpaired at the time of their arrest.

When asked about the recommended waiting time before driving after cannabis use, Graw gave a nuanced answer. For most occasional users, he explained, blood THC levels drop below the legal limit of 3.5 nanograms within three to five hours. However, their driving abilities may still be impaired, and he therefore recommended waiting at least twelve hours before getting behind the wheel. For regular users, whose blood THC levels may remain above the legal limit even after several days without consuming cannabis, he suggested a "period of abstinence lasting several weeks" before driving.[22]

Graw's recommendations made clear the trade-off in the new regulations: Just as cannabis use was becoming legal in Germany, driving was becoming practically illegal for many regular users. "They give you this, but you pay for that," as the song goes.[23]

Since blood THC limits do not reliably correlate with cannabis intoxication or its effect on driving ability, experts are currently exploring alternative assessment methods. One possible solution involves roadside sobriety tests, which enable police officers to evaluate a driver's concentration, coordination, and balance.

Research indicates that even dabbers—regular consumers of high-potency concentrates who tend to develop a tolerance to THC's impairing effects—exhibit balance issues after use. When asked to stand with their eyes closed, dabbers experience the same difficulties as occasional or regular cannabis smokers. These impairments generally disappear within an hour of consumption, suggesting they could serve as a reliable indicator of recent cannabis use, even among heavy users who are otherwise tolerant to other effects.[24]

A recent US government–funded study found that roadside sobriety tests accurately identified 81% of drivers under the influence of cannabis. However, these tests also incorrectly classified nearly half of the participants who had smoked a placebo cigarette as unfit to drive. This issue was resolved by cross-referencing results with toxicological analyses that were used to eliminate false positives and confirm that these drivers were, in fact, sober. After all, a person with no detectable THC in their system cannot be deemed intoxicated by cannabis—just as someone with no alcohol in their blood cannot be considered drunk.[25]

* * *

Among all alcoholic beverages, wine is undoubtedly the one most often compared to cannabis. Both are plant-based intoxicants that contain a wide range of aromatic substances, such as terpenes, flavonoids, and volatile sulfur compounds. However, despite these similarities, scientific research has historically focused on wine rather than cannabis. In the twentieth century, legal restrictions hindered studies on cannabis, while research on wine flourished into the scientific discipline of enology. Hundreds of books have been published on wine quality, often written by experts who refined their understanding through blind tasting, a process in which they assess wines without knowing their identities. No comparable body of knowledge exists for cannabis, which suggests that methods and insights from wine science may offer untapped guidance for contemporary cannabis studies.[26]

Research on the subjective effects of wine focuses on the pleasure it arouses, which scientists attribute more to its aromatic compounds than to its alcohol content. While alcohol is an inherent component of fermented grape juice, excessive concentrations can mask a wine's aromatic complexity and reduce its pleasure-giving qualities. This is why winemakers sometimes halt fermentation to limit alcohol levels, and the most esteemed wines—as judged by professional tasters—rarely exceed 15% alcohol by volume.[27]

Scientists consider the pleasure associated with wine to be generated by the brain, which constructs smells and tastes from sensory input. Drinking wine is a complex multisensory experience that engages multiple brain areas. Among the five human senses, smell plays the most significant role in shaping flavor, as olfactory signals travel directly from the inner nose to the centers of memory and emotion. This explains why certain scents instantly evoke memories. Specifically, when wine aromas stimulate olfactory receptors, they send signals straight to the cerebral cortex, triggering emotions and pleasure and influencing taste expectations. In contrast, inputs from other senses, such as taste, are filtered through the thalamus, a brain structure that regulates sensory information before transmitting it to the cortex. This difference in brain wiring explains why aromas have such a powerful, evocative effect.[28]

Although this brain architecture is the same in all healthy individuals, the appreciation of wine remains subjective: Each person perceives aromas based on their physiology, emotions, and personal memories. In other words, the perceived quality of a wine is deeply personal—just as beauty lies in the eye of the beholder. The pleasure and enjoyment of wine are not merely reactions to quality—they are the experience of quality itself.[29]

Wine experts generally deplore intoxication and drunkenness resulting from excessive alcohol intake and emphasize that wine enjoyment requires moderation. When people drink too much wine, their ability to appreciate its quality diminishes, underscoring that wine quality exists in individual perception and appreciation.[30]

The alcohol in wine interacts with several neurotransmitters, notably endorphins. By stimulating their release in the nucleus accumbens, a brain region associated with pleasure and reward, alcohol can enhance feelings of pleasure.[31]

In sum, the pleasure associated with wine arises from a subtle interaction between its aromatic compounds and alcohol content. While alcohol contributes to the neurological stimulation of pleasure, it is primarily the aromas that shape the sensory richness and tasting experience. The combined effects of wine's aromatic compounds and alcohol likely produce the beverage's capacity to elicit multifaceted enjoyment and pleasure.

Although cannabis research is still in its early stages, initial studies, including a blinded trial of cannabis samples, have drawn conclusions similar to those in wine science, particularly concerning the central role of aroma in subjective enjoyment. In a groundbreaking 2021 study led by American phytochemist Jeremy Plumb, researchers recruited several hundred participants, most of whom were daily cannabis users. Each participant received a kit containing *Cannabis* flower samples in sealed glass jars, labeled only with codes. After abstaining from cannabis for forty-eight hours, they tested a sample and rated its quality and its effects on enjoyment and mood.

The results were striking: Aroma, not THC content, was the primary determinant of appeal and pleasant effects. Although most participants reported appreciating cannabis for its capacity to induce "a prominent shift in consciousness," they rated the most pleasantly scented samples as the most enjoyable, even when the THC content was low. The researchers suggested that a pleasant aroma "may have primed individuals for a pleasant consumption experience," a phenomenon also observed in wine appreciation.[32]

One key difference between cannabis and wine is that cannabinoids, including THC, have no odor, while alcohol has a distinct scent that can overpower a wine's aromatic complexity. In wine, limiting alcohol content helps preserve aroma. In cannabis, there's no comparable reason to limit THC content. Yet Plumb and colleagues found that "high-THC cannabis may cause people to feel high, but high-THC cannabis is not always enjoyable."

This conclusion suggests that intoxication can, at times, diminish both pleasure and enjoyment. Nevertheless, many contemporary cannabis users associate high THC content with better quality. Plumb and colleagues described this assumption as a "carryover from cannabis prohibition" and argued instead that aroma may serve as a more reliable indicator of quality.[33]

These preliminary findings align with informal observations from cannabis competitions, where blind evaluations consistently show that people prefer fragrant *Cannabis* flowers over merely potent ones. This does not imply that THC plays no role in the pleasure cannabis provides. Rather, it suggests that the combined effects of the plant's aromatic compounds and THC likely influence the user's overall experience.

If wine science is any indication, it may take decades of research before experts reach consensus on how THC and aromatic pleasure contribute to cannabis quality. In the meantime, if you need to gauge some *Cannabis* flowers, the safest bet may be to trust your nose.

# 16

# The Future

*This is the best reason to learn history: not in order to predict the future, but to free yourself of the past and imagine alternative destinies.*[1]

—Yuval Noah Harari

Cannabis prohibition during the twentieth century hindered scientific research and created significant gaps in knowledge. Two questions, in particular, remain unanswered, both of which are vital for future generations. The first concerns male fertility, and the second relates to pregnancy, breastfeeding, and early brain development.

Research indicates that cannabis use may reduce sperm count, alter sperm shape and motility, and even affect the genetic material that sperm carry. Scientists have identified cannabinoid receptors on sperm cells that are usually regulated by the body's own endocannabinoid, anandamide. When sperm enter the female reproductive tract, they encounter anandamide, which helps coordinate essential biochemical changes in their membranes. These changes influence how sperm swim and enable them to acquire the ability to fertilize the egg. By binding to the same receptors as anandamide, THC may disrupt these processes. However, due to the lack of large-scale, long-term studies, the risks remain unclear.[2]

The second area of concern is cannabis use by mothers during pregnancy and breastfeeding, which may impact early brain development in fetuses and newborns. In this context, anandamide plays a crucial role in facilitating the formation of connections between neurons. THC consumed by the mother can cross the placenta or be transferred through breast milk, potentially disrupting anandamide's normal function. Research indicates that THC accumulates in breast milk by binding to its fat-rich components. Although the amount of cannabinoids an infant receives in a single feeding is relatively small, repeated exposure over time raises concerns. However, the risks remain uncertain, as reliable population-level studies are just beginning to emerge.[3]

Current medical guidance errs on the side of caution: Men trying to conceive are advised to abstain from cannabis for at least three months, while women are encouraged to avoid using it throughout pregnancy and breastfeeding. These are precautionary recommendations, and many cannabis users have healthy children. Future research will hopefully provide a clearer understanding.[4]

The risks that alcohol, tobacco, and opioids pose to male fertility, pregnancy, breastfeeding, and early brain development are well-documented and supported by decades of research. In contrast, the lack of reliable data regarding cannabis is striking.

Over 200 million people worldwide use cannabis, yet relatively little is known about how cannabinoids interact with the human brain and body. In the coming decades, science is likely to seek more knowledge about cannabis—if only to make up for lost time.[5]

* * *

One way to consider the future of cannabis is to extrapolate from today's most significant trends. Three major developments stand out: the rise of industrial manufacturing, the growth of craft cultivation, and the emergence of nonprofit production within a regulated framework. This chapter examines each in turn.

The first—and perhaps most visible—of these is the rapid expansion of the legal cannabis industry. Already generating tens of billions of dollars in annual sales, this industry is growing at an extraordinary pace. In countries with a free-market approach to legalization, such as Canada and the United States, large companies now produce a wide and growing range of products, some of which are redefining how cannabis is consumed.[6]

For example, nano-THC drinks are a new type of beverage sold in standard soft drink cans that contain nano-emulsified THC. In these products, THC oil is mixed with water and emulsifiers, then broken down into microscopic, water-soluble droplets using high-pressure blenders. This process makes THC more absorbable by the body because, as a spokesperson for one company explains, the THC nanoparticles "can bypass the digestive system and be absorbed directly into the bloodstream through the mucous membrane in the mouth and throat. This means that users can feel the effects of a nano-THC drink within minutes, rather than waiting up to an hour or more for traditional edibles to take effect."[7] Nano-THC not only acts more quickly but is also significantly more potent. Studies indicate that it can deliver up to 85% absorbable THC, compared to only 6% to 20% from conventional edibles.[8]

The shift toward more potent legal cannabis products is especially evident in markets with few or no limits on THC content. For example, one US-based manufacturer offers, among other items, large pre-rolled joints infused with cannabis concentrate, each containing 1,000 mg of THC—roughly fifteen to twenty times what a typical daily user might consume in one session. According to the product description, these joints are "designed for highly experienced users." As they contain 42% THC, even a single puff may exceed what inexperienced users can tolerate. The company underscores its "ability to meet customer demand with better quality, hard-hitting products."[9]

High-dose customers are particularly valuable to cannabis companies because they represent a disproportionate share of sales. In free-market environments without THC caps, around 20% of users account for more than half of all retail sales, primarily by purchasing high-potency products. While these customers may drive the market, little is known about the long-term health effects of regular, high-dose THC use. Research indicates that an increasing number of cannabis users experience issues such as tolerance, withdrawal, cravings, loss of control over use, and continued consumption despite adverse consequences.

Since nearly all serious health risks associated with cannabis rise with higher THC intake, it is reasonable to think that frequent use of high-potency products may carry additional risks. However, for now, those risks remain unmeasured.[10]

The THC in many modern products—drinks, vape juices, concentrates, extracts, edibles, gums, and skin patches—is typically extracted from plants grown indoors in "state-of-the-art cannabis cultivation and manufacturing facilities." In these factories, *Cannabis* flowers are treated as raw material for producing "highly formulated products," including some with therapeutic applications.[11]

As Ellen Holland, author of *Weed: A Connoisseur's Guide to Cannabis*, observes: "The race to capitalize on cannabis as a pharmaceutical medication lies in breaking down the plant elements and reformulating them to perform in their optimal capacity."[12]

While some focus on increasingly refined industrial extractions, others envision a more radical future: genetically modified *Cannabis* plants designed to produce concentrated cannabinoids, or algae engineered with *Cannabis* genes that generate these compounds in vats. In their book *Weed of Wonder*, Jules Marshall and Floris Leeuwenberg go so far as to suggest that, by 2050, cannabinoids may serve as "the oil that lubricates the human robot and provides a means to manipulate attention and consciousness."[13]

* * *

A second trend that may shape the future of cannabis is emerging in California. It involves establishing regulatory frameworks that support craft cultivation in specific regions.[14]

In 2024, California became the first jurisdiction in the world to adopt a law allowing cannabis from certain parts of the state to carry a protected designation of origin. This "appellation of origin" system mirrors the model that has long been used in the wine industry. Just as only sparkling wine produced in France's Champagne region can legally be called "champagne," California's law seeks to confer similar protections and prestige on cannabis grown in areas like Humboldt and Mendocino Counties—regions already celebrated for producing sinsemilla with distinctive aromas and effects. The goal is for appellation labeling to enable these craft products to command a market premium, much like fine wines whose higher prices reflect perceived quality and uniqueness.[15]

Supporters of cannabis appellations draw on the world of wine to explain why regional identity matters: the concept of *terroir*. This French term refers to the combination of local climate, soil, sun exposure, and farming practices that form the character of agricultural products in a specific place. Terroir gives wine its "taste of place"—the unique flavor and aroma that set Bordeaux apart from Burgundy, or Napa from Rioja. Scientific research confirms that grapevines absorb minerals from the soil and that terroir has a significant influence on the terpene content of grapes.[16]

*Cannabis* is also regarded as a terroir-expressive plant—sometimes to a problematic degree, as it can hyperaccumulate heavy metals from contaminated soil, unlike grapevines. Still, some connoisseurs claim that sun-grown *Cannabis* cultivated in healthy, living soil develops more complex and pleasing aromas than the same varieties grown indoors. Preliminary research appears to support this view: When grown outdoors under natural sunlight, *Cannabis* tends to produce a broader range of terpenes, often at notably higher levels. Its cannabinoid profile remains largely unchanged, although indoor-grown plants often contain higher concentrations of oxidized or degraded cannabinoids.[17]

Terroir-based cannabis is the opposite of a standardized product; it embraces natural variation and relies on environmental conditions rather than complete human oversight. However, it can still benefit from certified quality control, like any other professionally produced commodity.[18]

How this model will develop in the coming decades remains uncertain. Some believe it represents the future of *Cannabis* farming, or at least a viable alternative to large-scale indoor production. Whether an appellation system could take root in other historic cannabis-producing regions—such as Kashmir in India, Mazar in Afghanistan, Michoacán in Mexico, or Santa Marta in Colombia—mainly depends on how legalization unfolds worldwide. For small farmers in these areas, an appellation system could serve as a lifeline, just as it has for small-scale wine producers.[19]

One day, global consumers may be able to purchase Kashmiri charas, Thai sticks from Northern Thailand, kief from Morocco's Rif Mountains, or sinsemilla from Humboldt County, each with certified origin and quality, a distinct character, and a story of place.

* * *

A third trend that may influence the future of cannabis involves noncommercial cultivation by individuals and private clubs for personal use within a nationally regulated framework. This approach was initially introduced in Uruguay, the world's first country to legalize cannabis for "recreational" or "nonmedical" use. Since 2017, adult residents of this South American country have had three options for accessing legal cannabis. First, individuals can grow up to six *Cannabis* plants per household per year, with a total yield not exceeding 480 grams of dried flower, provided that the plants remain out of public view. Second, individuals can join a Cannabis Social Club, which can have up to forty-five members and cultivate up to ninety-nine plants per year. Each member is entitled to a share of the harvest, also limited to 480 grams per year. Third, individuals can purchase cannabis from government-licensed pharmacies, subject to the same yearly limit, and containing no more than 9% THC—a restriction not imposed on homegrown or club-grown plants.[20]

To safeguard public health, the Uruguayan system prohibits advertising and promotion, requires consumers to choose only one supply method, and enforces strict operational rules for clubs, including proximity restrictions to schools. Furthermore, cannabis sold in pharmacies must be quality-controlled and packaged in plain, resealable pouches that display only essential information, such as potency and consumption regulations.[21]

In 2024, when Germany legalized recreational cannabis, it adopted a framework inspired by the Uruguayan model, with its own modifications. In Germany, residents aged eighteen and older may cultivate up to three *Cannabis* plants simultaneously at their residence for personal use only. Private possession is limited to 50 grams of dried cannabis, and sharing between individuals remains prohibited. This law also permits adults to establish nonprofit *Cannabis* cultivation associations, where each member can obtain up to 50 grams of dried cannabis per month. For members aged eighteen to twenty, the monthly limit is reduced to 30 grams and capped at 10% THC. Adults aged twenty-one and older face no THC limit on cannabis obtained through clubs or grown at home.[22]

The German cannabis law features numerous provisions, several of which impose severe penalties. For example, the production and possession of cannabis edibles remain illegal due to concerns about potential risks to children, and violations can lead to prison sentences of up to three years. Public consumption of cannabis is prohibited in and within sight of schools, playgrounds, and public sports facilities, as well as in pedestrian zones between 7 A.M. and 8 P.M. It is also forbidden in the presence of minors and at certain public events, such as Oktoberfest, the annual beer festival held in Bavaria, where regional authorities have imposed restrictions and heavy fines.[23]

Critics point out inconsistencies in the German cannabis law. For instance, it is unclear how an individual can cultivate three plants without eventually possessing more than 50 grams of dried cannabis, considering that an average indoor plant yields more than that amount, and a mature outdoor plant can produce ten to twenty times as much. While the law prioritizes cultivation, it appears to overlook—or misunderstand—its basic principles. What is a grower supposed to do when all three plants reach maturity? Harvest only what the law allows and destroy the rest? Sharing the surplus is prohibited, and leaving the plants unharvested is also not a viable option, as they will soon dry out and exceed the legal limit for possession. This specific conundrum is more than a quibble, given that individuals found with more than 60 grams at home face up to three years in prison.[24]

Some experts describe this law as "the dawn of a new era for cannabis policy in Europe." At the same time, they acknowledge that the law's ultimate impact will depend on how the reforms are implemented across Germany's federal states and the level of political resistance they face. In short, the decisions made from this point onward will shape the future of cannabis in Germany, and perhaps beyond.[25]

* * *

In addition to the three emerging trends discussed above, a fourth, long-standing dynamic is likely to persist: the illicit production of cannabis and its derivatives. This includes everything from small-scale cultivation for personal use or informal profit to large-scale operations controlled by organized criminal networks.

Despite the recent shift toward legalizing recreational use in several countries, the majority of cannabis consumed worldwide still comes from unregulated sources. Accurately measuring this is challenging due to the covert nature of these operations. The UN's *World Drug Report 2022* estimates that illicit cannabis is produced in over 190 countries and territories. Even in places where recreational cannabis is legal, black markets continue to thrive. For instance, their share of total consumption is estimated to be around 40% in Canada, 50% in Uruguay, and 60% in Germany and California. These illegal markets remain attractive despite legalization, mainly due to lower prices and easier access.[26]

A global consensus on cannabis legalization seems unlikely anytime soon, and even if such an agreement were reached, there is little reason to believe that it would eliminate unregulated production. *Cannabis* is relatively easy to grow, and the high cost of many legal products continues to favor illegal alternatives.

The four trends or dynamics outlined here are likely to continue coexisting in some form, though no one can know precisely what the global landscape will look like twenty-five years from now.[27]

Uruguay may continue to require that *Cannabis* plants grown for personal use remain out of public sight, or it may not. Canada might impose potency limits on cannabis products, or it might not. Countries like Egypt, Pakistan, Nigeria, France, the UK, Finland, and El Salvador may uphold bans on recreational use, or each could individually change course. Consumers may become less willing to buy products made from factory-farmed *Cannabis* plants, or they may remain indifferent. Certified cannabis from the world's finest terroirs might one day be sold in specialty shops modeled on today's cigar stores, or perhaps not. The future of the global cannabis landscape depends on countless interacting variables. Predicting it is like trying to forecast the weather in Tokyo on a specific day twenty-five years from now.[28]

Historian of ideas Yuval Noah Harari suggests focusing less on predicting the future and more on imagining alternative destinies. Viewed from this perspective, *Cannabis* presents an intriguing possibility. If industrialized societies choose to shift toward a more sustainable, solar-powered economy, the plant's remarkable ability to flourish in intense sunlight becomes increasingly significant. Realizing this potential will require a deeper understanding of its cannabinoids—many of which have UV-absorbing and antioxidant properties that remain understudied, particularly regarding how they function within the living plant.[29]

A recent discovery highlights how much there is still to learn. Researchers have identified a cannabinoid produced by the plant in trace amounts that is considerably more potent than THC. This compound, called THCP, likely contributes to the psychoactive effects of cannabis, yet its properties remain poorly understood. Nonetheless, breeding plants for higher THCP levels is already an active area of research. The future role of this substance remains uncertain, but it now stands as the most psychoactive naturally occurring cannabinoid.[30]

* * *

The future can seem uncertain, unwritten, and unpredictable, but cannabis itself can be hard to pin down. The rise of nano-THC "cannabis beverages" illustrates this ambiguity. These drinks may appear futuristic, but are they truly cannabis?[31]

The *Cannabis* plant produces cannabinoids from compounds derived from fatty acids. As a result, cannabinoids have an oily or resinous texture. They readily dissolve in fat and are described by experts as lipophilic, meaning "fat-loving." Even when cannabis is smoked, the cannabinoids that survive combustion and enter the body retain their affinity for fat, which influences how they are distributed and stored. In contrast, when THC is nano-emulsified to become water-soluble, it behaves more like alcohol. This transformation in how the substance is absorbed and experienced is entirely engineered. Calling nano-THC drinks "cannabis beverages" obscures how far removed they are from the plant.[32]

Consider this analogy: If a beverage company distilled wine into pure alcohol and then mixed it with juice and fruit flavorings, the result would no longer be wine but an alcopop. Now, imagine that the alcohol in this liquid was chemically altered to become fat-soluble instead of water-soluble, changing how it's absorbed in the body and giving it effects more like a cannabis edible—slower-acting, longer-lasting, and more intense than standard alcoholic drinks. If such a beverage were marketed as a "wine product," connoisseurs would likely object, and regulators might prohibit the labeling, because the distance from the original product would be too great. The same is true for a cannabinoid extracted from *Cannabis* and designed to act like alcohol; it has strayed too far from its origin to retain its name.

Cannabis has been described as "many things to many people"—but water-soluble is not one of them. The plant and the drug have limits that chemistry helps define.[33]

* * *

As one of the world's most tightly regulated plants, *Cannabis* can seem entirely subject to human control. Yet time and again, it has escaped cultivation and reverted to a weedy, self-reliant existence. Rewilded *Cannabis* proves as resilient as wild horses—animals whose ancestors were once domesticated but now thrive independently.[34]

Biologists refer to plants and animals that prosper in human-dominated landscapes as "weedy species." Some foresee a future dominated by such species, given the well-documented global trends in deforestation, species extinction, and expanding human land use. These trends favor fast-growing, disturbance-tolerant plants—such as grasses, vines, brambles, and thistles—over slower-growing species, like trees. The same is true for animals: Species that are prolific, versatile, opportunistic, and mobile—such as rats, sparrows, cockroaches, ants, termites, carp, and iguanas—increasingly have an advantage.[35]

Drawing on such patterns, science writer David Quammen famously predicted that we may soon find ourselves living on the *Planet of Weeds*, an impoverished world of our own making, populated by a short list of weedy species. Writing in 1998, Quammen forecast: "Earth will be a different sort of place—soon, in just five or six generations."[36]

This scenario is not an outcome I hope for or wish to promote, but it remains a plausible one, especially if current trends continue unchecked. The *Planet of Weeds* deserves attention as a worst-case scenario—one that can encourage reflection on human responsibility and inspire the search for solutions.

Quammen assigned humans a dominant role in this imagined world. "*Homo sapiens* itself is the consummate weed," he wrote, quoting a paleontologist who described humans as "one of the most bomb-proof species on the planet." Although Quammen did not mention *Cannabis*, it seems improbable that people in such a world would choose to live without it. Like other weedy plants, *Cannabis* helps detoxify soil and restore vegetation on damaged land. Additionally, it offers something rare among weeds: the capacity to ease sorrow, relieve pain, and spark new ideas.[37]

Besides, *Cannabis* does not depend on humans to endure. Its track record as a hardy species stretches back at least nineteen million years, long before the evolution of human beings.

All the weedy species mentioned above appeared at least ten million years ago. Research indicates that termites have existed in their current form for over 150 million years, cockroaches for approximately 125 million, and one species of grass for more than 60 million. In contrast, *Homo sapiens* has been around for only about 300,000 years. Which species deserves the title of "weediest of them all" depends on the criteria. But in the league of professional weeds, a rookie species like ours may need to prove its staying power before laying claim to the crown.[38]

Plant biologist Robin Wall Kimmerer proposes that humans view weedy plants as partners in ecological restoration. These species often emerge first on degraded land, helping to jump-start nutrient cycling, form soil, and renew biodiversity. Kimmerer suggests that people can learn valuable lessons in resilience and renewal by observing how weeds grow, where they flourish, and how they respond to change. As she writes: "Plants tell their stories not by what they say, but by what they do."[39]

From this perspective, we can start paying attention to and learning from weedy species without delay. One plant, commonly called *weed*, may even help us imagine alternative destinies.

# Conclusion

# Knowledge for People, Freedom for Plants

Plants can sense light, gravity, touch, temperature, moisture, and sound, and respond to these stimuli in complex ways. They also communicate through shape and color, and by releasing a wide range of chemicals that serve as signals to other plants, animals, fungi, and microbes.[1]

Many scientists and philosophers refuse to describe plants as conscious, since standard neurological definitions of consciousness require a brain or nervous system, which plants do not have. But some argue that plants interact with their environment in ways that challenge conventional human notions of thought and awareness. For instance, philosopher Michael Marder suggests that plants express a form of "non-conscious intentionality" in how they grow, reproduce, and respond.[2]

What might paying attention to these aspects of plant life reveal about *Cannabis*? Consider, for example, how mature female plants respond when deprived of pollen: They enlarge and extend their sex organs. The whisker-like stigmas protruding from the flowers more than double in length and thickness, expanding the surface area available to receive pollen. This response increases the likelihood of fertilization.

The plants *tend* toward reproduction, in the literal sense of "moving in a particular direction."[3]

*Cannabis* is not alone in this regard; many other wind-pollinated species show similar responses when fertilization falls short. None of this comes as a surprise, given that plants generally allocate a large share of their energy to reproduction. Thousands of flowering species, including *Cannabis*, invest heavily in producing pollen, flowers, and protective compounds. These reproductive efforts promote genetic diversity in the next generation, thereby strengthening the resilience of plant populations.[4]

Some botanists describe free-living *Cannabis* as a "promiscuous" species. Male plants release huge amounts of pollen, and each female is typically fertilized by many different males. Most lineages interbreed freely, and domesticated varieties readily hybridize with wild populations. When left to its own devices, *Cannabis* thrives on the genetic diversity created by wind pollination between distant individuals.[5]

When humans prevent the pollination of female plants and propagate them through cuttings or clones, they impose an evolutionary constraint on *Cannabis*. A century of prohibition, eradication campaigns, and cultivation practices relying on asexual propagation of hybrid plants has resulted in a significant loss of genetic diversity within the species.[6]

Over time, *Cannabis* has shown remarkable adaptability in its relationship with humans, true to its nature as a flexible fiber plant. Few species have adjusted as seamlessly to domestication, selective breeding, and indoor cultivation. Yet, despite its responsiveness to human influence, *Cannabis* has shown no inclination to abandon sexual reproduction.[7]

* * *

Plants interact with other organisms through the chemicals they produce, some of which target specific groups of animals. For example, the tobacco plant generates colorful, fragrant flowers and sweet nectar to attract and reward insect and bird pollinators. At the same time, it synthesizes nicotine, an insect neurotoxin concentrated throughout the plant, particularly in the glandular hairs on its leaves. For most leaf-eating insects, consuming nicotine overstimulates the nervous system and can lead to paralysis or death. Because insects and humans share similar neurotransmitter systems, nicotine also affects human neurons, stimulating them at low doses and becoming toxic at higher ones. Despite its significant effects on humans, there is no evidence that nicotine evolved for our benefit; rather, it serves as a chemical defense against insect herbivores.[8]

Unlike tobacco, *Cannabis* is a wind-pollinated species with long, green flowers that effectively capture airborne pollen. Since the plant does not rely on animal pollinators, it has no need for bright colors or nectar. Instead, it can devote the bulk of its chemical output to defense, synthesizing a wide array of protective compounds, including terpenes and cannabinoids. Terpenes are volatile, aromatic substances that deter herbivores and pests with their pungent odors and also play a role in defending the plant against pathogens. Cannabinoids, on the other hand, serve a wide range of protective functions, only a small number of which focus on animals, particularly insects.[9]

The plant's main psychoactive compound, THC, has several clear protective roles, but targeting the neurons of insects does not seem to be among them—if only because insects do not have cannabinoid receptors. Like many other plant species, *Cannabis* produces hundreds of chemical compounds, some of which affect animal biology in ways unrelated to their original functions. All this suggests that THC's psychoactive effects on the brains of vertebrates and humans arise from shared, but incidental, biological features between plants and animals.[10]

Humans have bred *Cannabis* for its psychoactive properties, driving THC concentrations to levels unprecedented for any single compound in other plant species. However, the plant itself does not appear to prioritize THC production without human intervention. When *Cannabis* escapes cultivation and reverts to a wild state, it "turns weedy" and rapidly sheds traits shaped by human selection. Within just a few generations, THC levels typically fall below 3%, indicating that elevated concentrations in certain cultivated varieties are a human artifact rather than a natural priority for the plant.[11]

* * *

Throughout history, those who scorned or mistreated *Cannabis* typically did so in a context of limited knowledge. More recently, the plant's supporters, who bred it for higher THC levels, did not always realize they were transforming it into something different, and potentially riskier. All known health risks associated with cannabis intensify with higher THC intake. Consumers need to know that using milder cannabis instead of more potent varieties allows for better dose control and can result in a more pleasurable and enriching experience.[12]

In many parts of the world today, *Cannabis* is still prohibited from growing or reproducing freely outdoors. Yet when cultivated in sunlight and living soil, it reveals its hardy, high-yielding nature. Just a few bushy plants can supply an entire community of users. In a truly free market, cannabis would likely cost no more than other leafy crops like tea—yet today, it sells for hundreds of times more by weight. Prohibition and regulation have turned the grass of fakirs into an expensive herb.[13]

Cultivated, celebrated, criminalized, eradicated, driven indoors, deprived of sexual reproduction, re-medicalized, and even re-legalized—*Cannabis* has had a spectacular, up-and-down trajectory through human history. With scientists now discovering that THC works by modulating neurotransmitters that regulate human brain chemistry, *Cannabis* may earn new respect. No longer the "weed of insanity," it may come to be understood as a powerful psychoactive plant that some can use productively, under the right conditions and with proper knowledge.

As understanding broadens, humans may once again allow this singular plant to live a self-determined existence—under the open sky, its genes carried by the wind. Knowledge for people and freedom for plants go together.

# Epilogue

Researching this book transformed my relationship with *Cannabis*. I started out considering the plant as a friend, but the more I learned, the more I could see its complexity. Nearly every time I opened one door, two more appeared.

I intended to study *Cannabis* as long as necessary to get a grasp on the subject, but in the end, the plant took me for a ride. Writing this book took twice as long as anticipated. When it was over, *Cannabis* stood before me like a wild horse, its mystery intact. Now I view my friend as a magnificent enigma—a plant about which there is still so much to learn.

Along the way, I came to see *Cannabis* as resilient, flexible, intense, complex, and powerful—traits clearly expressed in its biology. This led me to reflect inward, drawn by its ways. It would be pretentious to claim that I learned from the plant, but I certainly regard it now with more respect than I did a few years ago.

Studying the plant's remarkable capacities has also prompted me to reconsider several practical matters. For instance, I've become cautious about *Cannabis* as a hyperaccumulator. Its tendency to absorb toxins from the surrounding environment has made me more aware of the origins of the plants I consume, as each can carry risks depending on where it was grown.

Now, when I find seeds nestled in *Cannabis* flowers, I feel a sense of satisfaction. Although seeded flowers may contain less THC and offer less aroma than their seedless counterparts, they show that the plant has fulfilled its drive to reproduce.

In addition, I increasingly appreciate mild cannabis. Learning about the biphasic effects of cannabinoids helped me understand that lower doses can offer greater inspiration.

All told, I have come to know *Cannabis* as a generous plant, and I remain grateful for its support in shaping this telling of its story.

# Acknowledgments

I thank Corinne Petignat for her companionship, unfailing support, and intelligence.

I thank the following people for their critical feedback on different chapters: Barbara Moulton, Gaspar Narby, Loïk Narby, Arthur Narby, Beda Künzle, Jace Callaway, J. P. Harpignies, John Beauclerk, Rodolfo Augusto, Cosima Klinger-Paul, Solil Paul, and Jon Christensen.

I thank Joel Fotinos, Tigrane Hadengue, Michka Seeliger-Chatelain, Roger Liggenstorfer, Chris Heidrich, and Anna Dantes for their editorial advice and wisdom.

I thank Erin Barnett for her meticulous copyediting and Ysée Berdat for her timely IT support.

The following scholars and experts kindly responded to requests for information, providing precious insights: Chris Duvall, Ethan Russo, Isaac Campos, Guangpen Ren, Charles Tart, Stephanie Shubat, Tim Dobe, Iain Oswald, Gatot Supangkat Samidjo, Mitchell Westmoreland, George Fisher, and Marc Brüngger.

Django, the border collie, kept things real.

# Bibliography


Aamodt, Athelstane. "Staying Off the Grass?" *New Law Journal* 168, no. 7799 (2018): 22.

Abel, Ernest L. *Marihuana: The First Twelve Thousand Years.* New York: McGraw-Hill, 1982.

Academia Farmacéutica de la Capital de la República. *Farmacopea mexicana formada y publicada por la Academia farmacéutica de la capital de la República.* Mexico: Imprenta de Manuel de la Vega, 1846.

Acosta, Cristóval. *Tractado de las drogas, y medicinas de las Indias Orientales.* Burgos, Spain, 1578.

Adams, Axel J., et al. "'Zombie' Outbreak Caused by the Synthetic Cannabinoid AMB-FUBINACA in New York." *New England Journal of Medicine* 376, no. 3 (2017): 235–242.

Aghaei, Ardavan Mohammad, et al. "Sex Differences in the Acute Effects of Oral THC: A Randomized, Placebo-Controlled, Crossover Human Laboratory Study." *Psychopharmacology* 241 (2024): 2145–2155.

Ahmad, Rafiq, et al. "Phytoremediation Potential of Hemp (*Cannabis sativa* L.): Identification and Characterization of Heavy Metals Responsive Genes." *Clean Soil Air Water* 44, no. 2 (2016): 195–201.

Ahmed, Muhammad Z., et al. "Arthropod and Mollusk Pests of Hemp, *Cannabis sativa* (Rosales: Cannabaceae), and Their Indoor Management Plan in Florida." *Journal of Integrated Pest Management* 15, no. 1 (2024): 1–22.

Aina, Ademola, et al. "Genetic Diversity, Population Structure, and Cannabinoid Variation in Feral *Cannabis sativa* Germplasm from the United States." *Scientific Reports* 15 (2025): 20423.

Aïnouche, Linda. "Erased from Collective Memory: *Dreadlocks Story* Documentary Untangles the Hindu Legacy of Rastafari." In *Afro-Asian Connections in Latin America and the Caribbean*, edited by Luisa Marcela Ossa and Debbie Lee-DiStefano, 137–168. Lanham, MD: Lexington Books, 2019.

Alcohol.org.nz. "Alcohol poisoning." https://www.alcohol.org.nz/alcohol-its-effects/health-effects/alcohol-poisoning.

Al-Makrizi, Taky-Eddin. "De l'herbe des fakirs." In *Chrestomathie Arabe.* Tome 2, edited by Silvestre de Sacy, 115–155. Paris: Imprimerie Impériale, 1806.

André, Christelle Martine, et al. "Dietary Antioxidants and Oxidative Stress from a Human and Plant Perspective: A Review." *Current Nutrition & Food Science* 6, no. 1 (2010): 2–12.

Appendino, Giovanni. "The Early History of Cannabinoid Research." *Rendiconti Lincei—Scienze Fisiche e Naturali* 31 (2020): 919–929.

Ara, Katsutoshi, et al. "Foot Odor Due to Microbial Metabolism and Its Control." *Canadian Journal of Microbiology* 52, no. 4 (2006): 357–364.

Archer, Robert A., et al. "Nabilone." In *Cannabinoids as Therapeutic Agents*, edited by Raphael Mechoulam, 86–103. London: CRC Press, 1986.

Arkell, Thomas R., et al. "Effect of Cannabidiol and Delta9-Tetrahydrocannabinol on Driving Performance: A Randomized Clinical Trial." *JAMA* 324, no. 21 (2020): 2177–2186.

Arkell, Thomas R., et al. "The Failings of *Per se* Limits to Detect Cannabis-Induced Driving Impairment: Results from a Simulated Driving Study." *Traffic Injury Prevention* 22, no. 2 (2021): 102–107.

Armentano, Paul. "FDA Halts the Anti-Pot Pill." *High Times*, November 2007, 18.

Arseneault, Louise, et al. "Causal Association Between Cannabis and Psychosis: Examination of the Evidence." *British Journal of Psychiatry* 184 (2004): 110–117.

Arveiller, Jacques. "Le Cannabis en France au XIX^e^ Siècle: Une Histoire Médicale." *L'Evolution Psychiatrique* 78, no. 3 (2013): 451–484.

Ashton, C. Heather. "Pharmacology and Effects of Cannabis: A Brief Review." *British Journal of Psychiatry* 178 (2001): 101–106.

Asimov, Eric. "In Defense of Wine." *The New York Times*, June 26, 2024, D2.

Asuni, Tolani. "Socio-Psychiatric Problems of Cannabis in Nigeria." *UNODC Bulletin on Narcotics* 16, no. 2 (1964): 17–28.

Atalay, Sinemyiz, et al. "Antioxidative and Anti-Inflammatory Properties of Cannabidiol." *Antioxidants* 9, no. 1 (2020): 1–21.

Aubert-Roche, Louis-Rémy. *De la peste, ou typhus d'Orient, documents et observations, suivis d'un essai sur le hachisch et son emploi dans le traitement de la peste*. Paris: Librairie des Sciences Médicales, 1840.

Auwärter, Volker, et al. "'Spice' and Other Herbal Blends: Harmless Incense or Cannabinoid Designer Drugs?" *Journal of Mass Spectrometry* 44, no. 5 (2009): 832–837.

Auwärter, Volker, et al. "Stellungnahme der Grenzwertkommission zur Frage einer Änderung des Grenzwertes für D9-Tetrahydrocannabinol (THC) im Blutserum zur Feststellung des Vorliegens der Voraussetzungen des § 24a (2) StVG." *Blutalkohol* 59 (2022): 331–339.

Backmund, Markus, et al. "Empfehlungen der interdisziplinären Expertengruppe für die Festlegung eines THC-Grenzwertes im Strassenverkehr ( 24a Strassenverkehrsgesetz)—Langfassung." March 2024. https://bmdv.bund.de/SharedDocs/DE/Anlage/K/cannabis-expertengruppe-langfassung.pdf?_blob=publicationFile.

Baker, Katie J. M., et al. "The Race for All-Powerful Pot." *The New York Times*, January 26, 2025, A1.

Baldy, Marian W. *The University Wine Course*. 3rd ed. South San Francisco, CA: Wine Appreciation Guild, 1997.

Ballotta, Danilo, et al. "Cannabis Control in Europe." In *A Cannabis Reader: Global Issues and Local Experiences*, edited by Sharon Rödner Sznitman, et. al., Monograph series 8, vol. 1, 99–117. Lisbon: European Monitoring Centre for Drugs and Drug Addiction, 2008.

Bancroft, Hubert Howe. *The Native Races of the Pacific States of North America*. Vol. 1, *Wild Tribes*. New York: D. Appleton and Company, 1874.

Banister, Samuel D., et al. "Dark Classics in Chemical Neuroscience: Delta-9-Tetrahydrocannabinol." *ASC Chemical Neuroscience* 10, no. 5 (2019): 2160–2175.

Banister, Samuel D., and Mark Connor. "The Chemistry and Pharmacology of Synthetic Cannabinoid Receptor Agonists as New Psychoactive Substances: Origins." In *New Psychoactive Substances: Pharmacology, Clinical, Forensic and Analytical Toxicology. Handbook of Experimental Pharmacology,* Vol. 252, edited by Hans H. Maurer and Simon D. Brandt, 165–190. Cham, Switzerland: Springer, 2018.

Banister, Samuel D., et al. "Pharmacology of Valinate and *tert*-Leucinate Synthetic Cannabinoids 5F-AMBICA, 5F-AMB, 5F-ADB, AMB-FUBINACA, MDMB-FUBINACA, MDMB-CHMICA, and Their Analogues." *ACS Chemical Neuroscience* 7, no. 9 (2016): 1241–1254.

Barcaccia, Gianni, et al. "Potentials and Challenges of Genomics for Breeding Cannabis Cultivars." *Frontiers in Plant Science* 11 (2020): 1–19.

Bar-On, Yinon M., et al. "The Biomass Distribution on Earth." *Proceedings of the National Academy of Sciences* 115, no. 25 (2018): 6506–6511.

Barrett, Marilyn L., et al. "Isolation from *Cannabis sativa* L. of Cannflavin—A Novel Inhibitor of Prostaglandin Production." *Biochemical Pharmacology* 34, no. 11 (1985): 2019–2024.

Bassir Nia, Anahita, et al. "Sex Differences in the Acute Effects of Intravenous (IV) Delta-9-Tetrahydrocannabinol (THC)." *Psychopharmacology* 239, no. 5 (2022): 1621–1628.

Baudelaire, Charles. *Du vin et du haschisch, comparés comme moyens de multiplication de l'individualité.* Paris: Mille et une nuits, 1997. Originally published in 1851.

Baudelaire, Charles. *Les paradis artificiels: opium et haschisch.* Paris: Poulet-Malassis et de Broise Libraires-Editeurs, 1860. https://gallica.bnf.fr/ark:/12148/bpt6k5832770f.texteImage.

Bautista, Johanna L., et al. "Flavonoids in *Cannabis sativa*: Biosynthesis, Bioactivities, and Biotechnology." *ACS Omega* 6, no. 8 (2021): 5119–5123.

Beaty, Roger E., et al. "Creative Cognition and Brain Network Dynamics." *Trends in Cognitive Sciences* 20, no. 2 (2016): 87–95.

Behzad, Danial, et al. "Association of Driving with Blood THC: A Systematic Review." 2024. https://ssrn.com/abstract=4990489.

Belenko, Steven R., ed. *Drugs and Drug Policy in America: A Documentary History.* Westport, CT: Greenwood Press, 2000.

Benjamin, Walter. "Haschich à Marseille." *Les Cahiers du Sud* (January 1935): 26–33.

Benjamin, Walter. *On Hashish.* Cambridge, MA: The Belknap Press, 2006.

Bennett, Chris. "Early/Ancient History." In *The Pot Book*, edited by Julie Holland, 17–26. Rochester, VT: Park Street Press, 2010.

Bergamaschi, Mateus M., et al. "Impact of Prolonged Cannabinoid Excretion in Chronic Daily Cannabis Smokers' Blood on Per se Drugged Driving Laws." *Clinical Chemistry* 59, no. 3 (2013): 519–526.

Bergreen, Laurence. *Louis Armstrong: An Extravagant Life.* New York: Broadway Books, 1997.

Berman, Paula, et al. "A New ESI-LC/MS Approach for Comprehensive Metabolic Profiling of Phytocannabinoids in *Cannabis.*" *Scientific Reports* 8 (2018): 1–15.

Bernstein, Nirit, et al. "Interplay Between Chemistry and Morphology in Medical Cannabis (*Cannabis sativa* L.)." *Industrial Crops & Products* 129 (2019): 185–194.

Bertrand, A. "Armagnac, Brandy, and Cognac and Their Manufacture." In *Encyclopedia of Food Sciences and Nutrition.* 2nd ed., edited by Benjamin Caballero, 584–601. Cambridge, MA: Academic Press, 2003.

Bewley-Taylor, David, and Martin Jelsma. "Fifty Years of the 1961 Single Convention on Narcotic Drugs: A Reinterpretation." *Transnational Institute Series on Legislative Reform of Drug Policies* 12 (2011): 1–20, www.tni.org/files/download/dlr12.pdf.

Bewley-Taylor, David, et al. *The Rise and Decline of Cannabis Prohibition: The History of Cannabis in the UN Drug Control System and Options for Reform.* Amsterdam: Transnational Institute, 2014.

Bidwell, L. Cinnamon, et al. "Association of Naturalistic Administration of Cannabis Flower and Concentrates with Intoxication and Impairment." *JAMA Psychiatry* 77, no. 8 (2020): 787–796.

Bidwell, L. Cinnamon, et al. "A Naturalistic Study of Orally Administered vs. Inhaled Legal Market Cannabis: Cannabinoids Exposure, Intoxication, and Impairment." *Psychopharmacology* 239, no. 2 (2022): 385–397.

Bilalis, Dimitrios, et al. "*Cannabis sativa* L.: A New Promising Crop for Medical and Industrial Use." *BulletinUASVM Horticulture* 76, no. 2 (2019): 145–150.

Bilby, Kenneth. "The Holy Herb: Notes on the Background of Cannabis in Jamaica." *Caribbean Quarterly* (1985): 82–95.

Bilkei-Gorzo, Andras, et al. "A Chronic Low Dose of Delta9-Tetrahydrocannabinol (THC) Restores Cognitive Function in Old Mice." *Nature Medicine* 23 (2017): 782–787.

Birenboim, Matan, et al. "Quantitative and Qualitative Spectroscopic Parameters Determination of Major Cannabinoids." *Journal of Luminescence* 252 (2022): 119387.

Bisogno, Tiziana, et al. "Molecular Targets for Cannabidiol and Its Synthetic Analogues: Effect on Vanilloid VR1 Receptors and on the Cellular Uptake and Enzymatic Hydrolysis of Anandamide." *British Journal of Pharmacology* 134, no. 4 (2001): 845–852.

Black, Bobby. "Generation Dab." *High Times*, July 2013, 54–60.

Black, Sean. "Concentrated Cannabis, Part 1: Extractions 101." *High Times*, April 2017, 92–98.

Block, Robert I. "Does Heavy Marijuana Use Impair Human Cognition and Brain Function?" *JAMA* 275, no. 7 (1996): 560–561.

Block, Robert I., et al. "Acute Effects of Marijuana on Cognition: Relationships to Chronic Effects and Smoking Techniques." *Pharmacology Biochemistry and Behavior* 43, no. 3 (1992): 907–917.

Block, Robert I., and Mohammed M. Ghoneim. "Effects of Chronic Marijuana Use on Human Cognition." *Psychopharmacology* 110, nos. 1–2 (1993): 219–228.

Block, Walter. "Drug Prohibition: A Legal and Economic Analysis." *Journal of Business Ethics* 12, no. 9 (1993): 689–700.

Bockris, Victor. *Keith Richards: The Biography*. New York: Poseidon Press, 1992.

Boire, Richard Glen, and Kevin Feeney. *Medical Marijuana Law*. Berkeley, CA: Ronin Publishing, 2007.

Bonnie, Richard J., and Charles H. Whitebread II. "The Forbidden Fruit and the Tree of Knowledge: An Inquiry into the Legal History of American Marijuana Prohibition." *Virginia Law Review* 56, no. 6 (1970): 971–1169.

Bonnie, Richard J., and Charles H. Whitebread II. *The Marijuana Conviction: A History of Marijuana Prohibition in the United States*. Charlottesville: University of Virginia Press, 1974.

Booth, Judith, and Jörg Bohlmann. "Terpenes in *Cannabis sativa*—From Plant Genome to Humans." *Plant Science* 284 (2019): 67–72.

Booth, Martin. *Cannabis*. London: Bantam Books, 2003.

Borougerdi, Bradley J. *Commodifying Cannabis: A Cultural History of a Complex Plant in the Atlantic World*. Lanham, MD: Lexington Books, 2018.

Bosnyak, Dan, et al. "Use of a Novel EEG-Based Objective Test, the Cognalyzer, in Quantifying the Strength and Determining the Action Time of Cannabis Psychoactive Effects and Factors that May Influence Them Within an Observational Study Framework." *Neurology and Therapy* 11, no. 1 (2022): 51–72.

Bourassa, Maurice, and Pierre Vaugeois. "Effects of Marijuana Use on Divergent Thinking." *Creativity Research Journal* 13, nos. 3–4 (2001): 411–416.

Bowman, Marilyn, and Robert O. Pihl. "Cannabis: Psychological Effects of Chronic Heavy Use." *Psychopharmacologia* 29, no. 2 (1973): 159–170.

Bowrey, Thomas. *A Geographical Account of Countries Round the Bay of Bengal, 1669 to 1679*. Cambridge: The Hakluyt Society, 1905.

Braff, David L., et al. "Impaired Speed of Visual Processing in Marijuana Intoxication." *American Journal of Psychiatry* 138 (1981): 613–617.

Brands, Bruna, et al. "Cannabis, Impaired Driving, and Road Safety: An Overview of Key Questions and Issues." *Frontiers in Psychiatry* 12 (2021): 641549.

Braudel, Fernand. *The Structures of Everyday Life: The Limits of the Possible*. University of California Press: Berkeley, 1992.

Brellenthin, Angelique G., and Kelli F. Koltyn. "Exercise as an Adjunctive Treatment for Cannabis Use Disorder." *American Journal of Drug and Alcohol Abuse* 42, no. 5 (2016): 481–489.

Brierre de Boismont, Alexandre. "Au Rédacteur." *Journal des débats politiques et littéraires* (November 17, 1837): 3a.

Brierre de Boismont, Alexandre. *Des hallucinations : ou, Histoire raisonnée des apparitions, des visions, des songes, de l'extase, du magnétisme et du somnambulisme*. Paris: Germer Baillière, 1845.

Brierre de Boismont, Alexandre. "Expériences toxicologiques sur une substance inconnue." *Gazette médicale de Paris* 8, no. 18 (1840): 278–279.

British Parliamentary Papers. *Papers Relating to the Consumption of Ganja and Other Drugs in India*. London: Hansard, 1891.

Brown, Steven, and Eunseon Kim. "The Neural Basis of Creative Production: A Cross-Modal ALE Meta-Analysis." *Open Psychology* 3, no. 1 (2021): 103–132.

Broyd, Samantha J., et al. "Acute and Chronic Effects of Cannabinoids on Human Cognition." *Biological Psychiatry* 79, no. 7 (2016): 557–567.

Bruni, Natascia, et al. "Cannabinoid Delivery Systems for Pain and Inflammation Treatment." *Molecules* 23, no. 10 (2018): 2478.

Bundesministerium für Digitales und Verkehr. "Unabhängige Expertengruppe legt Ergebnis zu THC-Grenzwert im Strassenverkehr vor." March 28, 2024. https://bmdv.bund.de/SharedDocs/DE/Pressemitteilungen/2024/018-expertengruppe-thc-grenzwert-im-strassenverkehr.html.

Burdon, Roland. D., and Jacqui Aimers-Halliday. "Risk Management for Clonal Forestry with Pinus Radiata—Analysis and Review. 1: Strategic Issues and Risk Spread." *New Zealand Journal of Forestry Science* 33, no. 2 (2003): 156–180.

Burke, Sean V., et al. "Phylogenomics and Plastome Evolution of Tropical Forest Grasses (*Leptaspis*, *Streptochaeta*: Poaceae)." *Frontiers in Plant Science* 7 (2016): 1993.

Burroughs, William. "Letter from a Master Addict to Dangerous Drugs." *British Journal of Addiction* 53, no. 2 (1956): 119–131.

Burroughs, William. *Naked Lunch*. New York: Grove Press, 1966. Originally published in 1959.

Burroughs, William. "Points of Distinction Between Sedative and Consciousness-Expanding Drugs." In *The Marijuana Papers*, edited by David Solomon, 440–446. New York: New American Library, 1968.

Burroughs, William, and Allen Ginsberg. *The Yage Letters*. San Francisco: City Lights Books, 1963.

Burstein, Sumner. "Cannabidiol (CBD) and Its Analogs: A Review of Their Effects on Inflammation." *Bioorganic & Medicinal Chemistry* 23, no. 7 (2015): 1377–1385.

Ćaćić, Marija, et al. "Evaluation of Heavy Metals Accumulation Potential of Hemp (*Cannabis sativa* L.)." *Journal of Central European Agriculture* 20, no. 2 (2019): 700–711.

Calabrese, Edward J., and Alberto Rubio-Casillas. "Biphasic Effects of THC in Memory and Cognition." *European Journal of Clinical Investigation* 48, no. 5 (2018): 12920.

Calakos, Katina C., et al. "Mechanisms Underlying Sex Differences in Cannabis Use." *Current Addiction Reports* 4, no. 4 (2017): 439–453.

Calhoun, S. R., et al. "Abuse Potential of Dronabinol (Marinol)." *Journal of Psychoactive Drugs* 30, no. 2 (1998): 187–196.

Callaway, J. C. "Hempseed as a Nutritional Resource: An Overview." *Euphytica* 140 (2004): 65–72.

Callaway, J. C., and David W. Pate. "Hempseed Oil." In *Gourmet and Health-Promoting Specialty Oils*, edited by Robert A. Moreau and Afaf Kamal-Eldin, 185–213. Urbana, IL: AOCS Press, 2009.

Campbell, Iain. "Grain Whisky Distillation." In *Whisky: Technology, Production and Marketing*, edited by Inge Russell, 181–208. London: Academic Press, 2003.

Campos, Isaac. *Home Grown: Marijuana and the Origins of Mexico's War on Drugs*. Chapel Hill: University of North Carolina Press, 2012.

Caprioglio, Diego, et al. "Cannabinoquinones: Synthesis and Biological Profile." *Biomolecules* 11, no. 7 (2021): 991.

Cardeña, Etzel. "A Festschrift for a Consciousness Hummingbird: Charles T. Tart." *Journal of Anomalous Experience and Cognition* 3, no. 2 (2023): 222–227.

Carlin, Albert S., et al. "Social Facilitation of Marijuana Intoxication: Impact of Social Set and Pharmacological Activity." *Journal of Abnormal Psychology* 80, no. 2 (1972): 132–140.

Caron, Christina. "Psychosis, Addiction, Chronic Vomiting: As Weed Becomes More Potent, Teens are Getting Sick." *The New York Times*, June 23, 2022: A12.

Carrier, Neil, and Gernot Klantschnig. *Africa and the War on Drugs*. London: Zed, 2012.

Carrier, Neil, and Gernot Klantschnig. "Free the Weed: A Short History of Marijuana." *The Elephant*, August 22, 2019. https://www.theelephant.info/analysis/2019/08/22/free-the-weed-a-short-history-of-marijuana/.

Carter, Gregory T., et al. "Medicinal Cannabis: Rational Guidelines for Dosing." *IDrugs* 7, no. 5 (2004): 464–470.

Castaneto, Marisol S., et al. "Synthetic Cannabinoids Pharmacokinetics and Detection Methods in Biological Matrices." *Drug Metabolism Reviews* 47, no. 2 (2015): 1–52.

Castelli, Maria Paola, et al. "Male and Female Rats Differ in Brain Cannabinoid CB1 Receptor Density and Function and in Behavioural Traits Predisposing to Drug Addiction: Effect of Ovarian Hormones." *Current Pharmaceutical Design* 20, no. 13 (2014): 2100–2113.

Casto, Don. "Marijuana and the Assassins, an Etymological Investigation." *British Journal of Addiction* 65, no. 3 (1979): 219–225.

Celik, Zeynep Dilan, et al. "Effects of Terroir on the Terpene Compounds of Muscat of Bornova Native White Grape Variety Grown in Turkey." *BIO Web of Conferences* 5 (2015): 01004.

Cervantes, Jorge. *Indoor Marijuana Horticulture*. Vancouver, WA: Van Patten Publishing, 1983.

Chamberlin, J. Edward, and Barry Chevannes. "Ganja in Jamaica." In *Smoke: A Global History of Smoking*, edited by Sander L. Gilman and Zhou Xun, 144–153. London: Reaktion Books, 2004.

Chamovitz, Daniel. *What a Plant Knows: A Field Guide to the Senses of Your Garden and Beyond*. London: Oneworld Publications, 2012.

Chan, Gary C. K., et al. "User Characteristics and Effect Profile of Butane Hash Oil: An Extremely High-Potency Cannabis Concentrate." *Drug and Alcohol Dependence* 178 (2017): 32–38.

Chandra, Suman, et al. "Cannabis Cultivation: Methodological Issues for Obtaining Medical-Grade Product." *Epilepsy & Behavior* 70, part B (2017): 302–312.

Chandra, Suman, et al. "Light Dependence of Photosynthesis and Water Vapor Exchange Characteristics in Different High Delta-9-THC Yielding Varieties of *Cannabis sativa* L." *Journal of Applied Research on Medicinal and Aromatic Plants* 2, no. 2 (2015): 39–47.

Changeux, Jean-Pierre. "Discovery of the First Neurotransmitter Receptor: The Acetylcholine Nicotinic Receptor." *Biomolecules* 10, no. 4 (2020): 1–15.

Charlet, Katrin, et al. "The Dopamine System in Mediating Alcohol Effects in Humans." In *Behavioral Neurobiology of Alcohol Addiction*, edited by Wolfgang H. Sommer and Rainer Spanagel, 461–488. Berlin: Springer, 2013.

Charters, Steve, and Simone Pettigrew. "The Dimensions of Wine Quality." *Food Quality and Preference* 18, no. 7 (2007): 997–1007.

Chasteen, John Charles. *Getting High: Marijuana Through the Ages*. New York: Rowman and Littlefield, 2016.

Chen, Angela. "Why It Can Be Okay to Call It 'Marijuana' Instead of 'Cannabis.'" *The Verge*, April 19, 2018. https://www.theverge.com/2018/4/19/17253446/marijuana-cannabis-drugs-racist-language-history.

Childs, Emma, et al. "Dose-Related Effects of Delta-9-THC on Emotional Responses to Acute Psychosocial Stress." *Drug and Alcohol Dependence* 177 (2017): 136–144.

Chin, Grace S., et al. "The Pharmacodynamics, Pharmacokinetics, and Potential Drug Interactions of Cannabinoids." In *Cannabis in Medicine: An Evidence-Based Approach*, edited by Kenneth Finn, 49–61. Chem, Switzerland: Springer, 2020.

Chopra, I. C., and R. N. Chopra. "The Use of Cannabis Drugs in India." *UNODC Bulletin on Narcotics* 11, no. 1 (1957): 4–29.

Chouvy, Pierre-Arnaud. "Why the Concept of Terroir Matters for Drug Cannabis Production." *GeoJournal* 88, no. 1 (2022): 89–106.

Chouvy, Pierre-Arnaud, and Jennifer Macfarlane. "Agricultural Innovations in Morocco's Cannabis Industry." *International Journal of Drug Policy* 58 (2018): 85–91.

Cisternas-Fuentes, Anita, and Matthew H. Koski. "Effective Population Size Mediates the Impact of Pollination Services on Pollen Limitation." *Proceedings of the Royal Society B* 291, no. 2014 (2024): 20231519.

Citti, Cinzia, et al. "A Novel Phytocannabinoid Isolated from *Cannabis sativa* L. with an In Vivo Cannabimimetic Activity Higher Than Delta9-Tetrahydrocannabinol: Delta9-Tetrahydrocannabiphorol." *Scientific Reports* 9, no. 1 (2019): 20335.

Clarke, Robert Connell. *Hashish!* Los Angeles: Red Eye, 1998.

Clarke, Robert Connell. "Sinsemilla Heritage: What's in a Name?" In *The Cannabible*, edited by Jason King, 1–24. Berkeley, CA: Ten Speed Press, 2001.

Clarke, Robert Connell, and Mark D. Merlin. *Cannabis: Evolution and Ethnobotany*. Berkeley: University of California Press, 2013.

Clarke, Robert Connell, and Mark D. Merlin. "*Cannabis* Domestication, Breeding History, Present-day Genetic Diversity, and Future Prospects." *Critical Reviews in Plant Sciences* 35, no. 5–6 (2016): 293–327.

Clarke, Robert C., and David P. Watson. "*Cannabis* and Natural *Cannabis* Medicines." In *Marijuana and the Cannabinoids*, edited by Mahmoud A. ElSohly, 1–17. Totowa, NJ: Humana Press, 2007.

Clarke, Tristyn L., et al. "The Endocannabinoid System and Invertebrate Neurodevelopment and Regeneration." *International Journal of Molecular Sciences* 22, no. 4 (2021): 1–24.

Clements, David R., and Vanessa L. Jones. "Ten Ways That Weed Evolution Defies Human Management Efforts Amidst a Changing Climate." *Agronomy* 11, no. 2 (2021): 1–20.

Clouston, Thomas S. "The Cairo Asylum: Dr Warnock on Hasheesh Insanity." *Journal of Mental Science* 42, no. 179 (1896): 790–795.

Cockburn, R. *Annual Report on the Insane Asylums in Bengal for the Year 1874*. Calcutta: Bengal Secretariat Press, 1875. https://digital.nls.uk/indiapapers/browse/archive/83380367.

Cogan, Peter S. "The 'Entourage Effect' or 'Hodge-Podge Hashish': The Questionable Rebranding, Marketing, and Expectations of Cannabis Polypharmacy." *Expert Review of Clinical Pharmacology* 13, no. 8 (2020): 835–845.

Cole, Juan. *Napoleon's Egypt: Invading the Middle East*. New York: St. Martin's Griffin, 2007.

Colizzi, Marco, and Robin Murray. "Cannabis and Psychosis: What Do We Know and What Should We Do?" *British Journal of Psychiatry* 212, no. 4 (2018): 195–196.

Colizzi, Marco, et al. "Descriptive Psychopathology of the Acute Effects of Intravenous Delta-9-Tetrahydrocannabinol Administration in Humans." *Brain Sciences* 9, no. 4 (2019): 93.

Collison, Robert F., et al. "Light, Not Age, Underlies the Maladaptation of Maize and Miscanthus Photosynthesis to Self-Shading." *Frontiers in Plant Science* 11 (2020): 1–10.

Commission of Inquiry into the Non-Medical Use of Drugs. *Cannabis: A Report of the Commission of Inquiry into the Non-Medical Use of Drugs*. Ottawa: Information Canada, 1972.

Confédération Helvétique. *Ordonnance sur les essais pilotes au sens de la loi sur les stupéfiants du 31 mars 2021*. 812. 121.5. Berne: Confédération suisse.

Constitutional Court of South Africa. "Minister of Justice and Constitutional Development v Prince," 2018. https://collections.concourt.org.za/handle/20.500.12144/34547.

Coon, Caroline. "We Were the Welfare Branch of the Alternative Society." In *The Unsung Sixties: Memoirs of Social Innovation*, edited by Helene Curtis and Mimi Sanderson, 183–197. London: Whiting & Birch, 2004.

Coon, Caroline, and Rufus Harris. *The Release Report on Drug Offenders and the Law*. London: Sphere Books, 1969.

Courtwright, David T. *Forces of Habit: Drugs and the Making of the Modern World*. Cambridge, MA: Harvard University Press, 2001.

Crocq, Marc-Antoine. "History of Cannabis and the Endocannabinoid System." *Dialogues in Clinical Neuroscience* 22, no. 3 (2020): 223–228.

Curran, H. Valerie, et al. "Cognitive and Subjective Dose-Response Effects of Acute Oral Delta9-Tetrahydrocannabinol (THC) in Infrequent Cannabis Users." *Psychopharmacology* 164, no. 1 (2002): 61–70.

Curran, H. Valerie, et al. "Keep off the Grass? Cannabis, Cognition and Addiction." *Nature Reviews Neuroscience* 17 (2016): 293–306.

Cuypers, Eva, et al. "The Use of Pesticides in Belgian Illicit Indoor Cannabis Plantations." *Forensic Science International* 277 (2017): 59–65.

Daftary, Farhad. *The Assassin Legends: Myths of the Isma'ilis.* London: Tauris, 1994.

Daftary, Farhad. *The Isma'ilis: Their History and Doctrines.* 2nd ed. Cambridge: Cambridge University Press, 2007.

Dalton, W. S., et al. "Influence of Cannabidiol on Delta-9-Tetrahydrocannabinol Effects." *Clinical Pharmacology & Therapeutics* 19, no. 3 (1976): 300–309.

Darley, Charles F., et al. "Influence of Marihuana on Storage and Retrieval Processes in Memory." *Memory & Cognition* 1, no. 2 (1973): 196–200.

Darnton, John. "Nigeria's Dissident Superstar." *The New York Times Magazine*, July 24, 1977, 10–12, 22–26.

Dash, Vaidya Bhagwan. *Fundamentals of Ayurvedic Medicine.* Delhi: Konark Publishers, 1978.

Dawson, Warren D. "Studies in the Egyptian Medical Texts: III." *The Journal of Egyptian Archaeology* 20, nos. 1–2 (1934): 41–46.

DEA (Drug Enforcement Administration). "1989 Domestic Cannabis Eradication/Suppression Program." Washington, DC: United States Department of Justice, 1990.

DEA (Drug Enforcement Administration). "1993 Domestic Cannabis Eradication/Suppression Program." Washington, DC: United States Department of Justice, 1994.

De Backer, Benjamin, et al. "Evolution of the Content of THC and Other Major Cannabinoids in Drug-Type Cannabis Cuttings and Seedlings During Growth of Plants." *Journal of Forensic Sciences* 57, no. 4 (2012): 918–922.

De Dreu, Carsten K. W., et al. "Human Creativity: Functions, Mechanisms, and Social Conditioning." *Advances in Experimental Social Psychology* 69 (2024): 203–262.

Dekker, Jack. *Evolutionary Ecology of Weeds.* Ames, IA: Weeds-R-Us Press, 2011.

De Las Casas, Bartolomé de. *Historia de las Indias.* Vol. 1. Madrid: Imprenta de Miguel Ginesta, 1875.

De Meijer, Etienne. "The Chemical Phenotypes (Chemotypes) of *Cannabis.*" In *Handbook of Cannabis*, edited by Roger G. Pertwee, 88–110. Oxford: Oxford University Press, 2014.

De Sacy, Silvestre. "Mémoire sur la dynastie des Assassins, et sur l'étymologie de leur nom." *Mémoires de l'Institut Royal de France* 4 (1818): 1–84.

De Sacy, Silvestre. "Mémoire sur la dynastie des Assassins, et sur l'origine de leur nom." *Moniteur* 120 (1809): 1–13.

De Saulcy, Félicien. *Narrative of a Journey Round the Dead Sea and in the Bible Lands; in 1850 and 1851.* Vol. 1. 2nd ed. London: Richard Bentley, 1854.

Desaulniers Brousseau, Vincent, et al. "Cannabinoids and Terpenes: How Production of Photo-Protectants Can Be Manipulated to Enhance *Cannabis sativa* L. Phytochemistry." *Frontiers in Plant Science* 12 (2021): 1–13.

Desrosiers, Nathalie A., et al. "Phase I and II Cannabinoid Disposition in Blood and Plasma of Occasional and Frequent Smokers Following Controlled Smoked Cannabis." *Clinical Chemistry* 60, no. 4 (2014): 631–643.

Desrosiers, Nathalie A., et al. "Smoked Cannabis' Psychomotor and Neurocognitive Effects in Occasional and Frequent Smokers." *Journal of Analytical Toxicology* 39, no. 4 (2015): 251–261.

Devane, William A., et al. "A Novel Probe for the Cannabinoid Receptor." *Journal of Medicinal Chemistry* 35, no. 11 (1992): 2065–2069.

Devane, William A., et al. "Determination and Characterization of a Cannabinoid Receptor in Rat Brain." *Molecular Pharmacology* 34, no. 5 (1988): 605–613.

Devane, William A., et al. "Isolation and Structure of a Brain Constituent That Binds to the Cannabinoid Receptor." *Science* 258, no. 5090 (1992): 1946–1949.

DeVuono, Marieka V., and Linda A. Parker. "Cannabinoid Hyperemesis Syndrome: A Review of Potential Mechanisms." *Cannabis and Cannabinoid Research* 5, no. 2 (2020): 132–144.

De Wet, J. M.J. "The Origin of Weediness in Plants." *Proceedings of the Oklahoma Academy of Sciences* (1966): 14–17.

Di Ciano, Patricia, et al. "Cannabis and Driving in Older Adults." *JAMA Network Open* 7, no. 1 (2024): e2352233.

Di Forti, Marta, et al. "Proportion of Patients in South London with First-Episode Psychosis Attributable to Use of High Potency Cannabis: A Case-Control Study." *The Lancet Psychiatry* 2, no. 3 (2015): 233–238.

Di Forti, Marta, et al. "High-Potency Cannabis and the Risk of Psychosis." *British Journal of Psychiatry* 195, no. 6 (2009): 488–491.

Di Marzo, Vincenzo, et al. "The Endocannabinoid System and Its Therapeutic Exploitation." *Nature Reviews Drug Discovery* 3, no. 9 (2004): 771–784.

Di Marzo, Vincenzo, and Angelo Fontana. "Anandamide, an Endogenous Cannabinomimetic Eicosanoid: 'Killing Two Birds with One Stone.'" *Prostaglandins, Leukotrienes & Essential Fatty Acids* 53, no. 1 (1995): 1–11.

Di Marzo, Vincenzo, et al. "Leptin-Regulated Endocannabinoids Are Involved in Maintaining Food Intake." *Nature* 410, no. 6830 (2001): 822–825.

Di Marzo, Vincenzo, et al. "Endocannabinoids: Endogenous Cannabinoid Receptor Ligands with Neuromodulatory Action." *Trends in Neurosciences* 21, no. 12 (1998): 521–528.

Dobe, Timothy S. *Hindu Christian Faqir: Modern Monks, Global Christianity, and Indian Sainthood.* Oxford: Oxford University Press, 2015.

Dolphin, William. "Measuring Marijuana's Potency." In *The Big Book of Buds.* Vol. 2, edited by Ed Rosenthal, 92–97. Oakland, CA: Quick American Archives, 2004.

Donnan, Jennifer, et al. "Characteristics That Influence Purchase Choice for Cannabis Products: A Systematic Review." *Journal of Cannabis Research* 4, no. 1 (2022): 9.

Dos Santos, Naraya Araujo, and Wanderson Romão. "*Cannabis*—A State of the Art about the Millenary Plant: Part I." *Forensic Chemistry* 32, no. 8 (2023): 100470.

Dreyer, Benard P. "Sustained Animus Toward Latino Immigrants—Deadly Consequences for Children and Families." *New England Journal of Medicine* 381, no. 13 (2019): 1196–1198.

Driscoll, Lawrence. *Reconsidering Drugs: Mapping Victorian and Modern Drug Discourses.* London: Palgrave Macmillan, 2001.

Dronkers, Ben. "A History of Cannabis in Holland." In *The Big Book of Buds*, edited by Ed Rosenthal, 41–44. Oakland, CA: Quick American Archives, 2001.

Dryburgh, Laura M., et al. "Cannabis Contaminants: Sources, Distribution, Human Toxicity and Pharmacological Effects." *British Journal of Clinical Pharmacology* 84, no. 11 (2018): 2468–2476.

D'Souza, Deepak Cyril, et al. "The Psychotomimetic Effects of Intravenous Delta-9-Tetrahydrocannabinol in Healthy Individuals: Implications for Psychosis." *Neuropsychopharmacology* 29, no. 8 (2004): 1558–1572.

Duvall, Chris S. *The African Roots of Marijuana.* Durham, NC: Duke University Press, 2019.

Duvall, Chris S. "A Brief Agricultural History of Cannabis in Africa, from Prehistory to Canna-Colony." *EchoGéo* 48 (2019): 1–27.

Duvall, Chris S. *Cannabis*. London: Reaktion Books, 2015.

Duvall, Chris S. "Cannabis and Tobacco in Precolonial and Colonial Africa." In *Oxford Research Encyclopedia of African History*, edited by Thomas Spear. New York: Oxford University Press, 2017.

Duvall, Chris S. "Drug Laws, Bioprospecting and the Agricultural Heritage of *Cannabis* in Africa." *Space and Polity* 20, no. 1 (2016): 10–25.

Earlenbaugh, Emily. "Why Less Is Often More with Medical Cannabis." *The Cannigma*, November 10, 2019. https://cannigma.com/research/why-less-is-often-more-with-medical-cannabis/

Earleywine, Mitch. *Understanding Marijuana: A New Look at the Scientific Evidence*. Oxford: Oxford University Press, 2002.

Eichhorn Bilodeau, Samuel, et al. "An Update on Plant Photobiology and Implications for Cannabis Production." *Frontiers in Plant Science* 10 (2019): 1–15.

Eliade, Mircea. *Shamanism: Archaic Techniques of Ecstasy*. London: Arkana, 1964.

ElSohly, Mahmoud A., et al. "Potency Trends of Delta-9-THC and Other Cannabinoids in Confiscated Marijuana from 1980–1997." *Journal of Forensic Sciences* 45, no. 1 (2000): 24–30.

EMCDDA. *Cannabis and Driving: Questions and Answers for Policymaking*. Lisbon, Portugal: European Monitoring Centre for Drugs and Drug Addiction, 2018.

EMCDDA. *EU Drug Market: Cannabis—In-depth Analysis*. Lisbon, Portugal: European Monitoring Centre for Drugs and Drug Addiction, 2023.

EMCDDA. *Fentanils and Synthetic Cannabinoids: Driving Greater Complexity into the Drug Situation—An Update from the EU Early Warning System*. Lisbon, Portugal: European Monitoring Centre for Drugs and Drug Addiction, 2018.

EMCDDA. *Synthetic Cannabinoids in Europe—A Review*. Lisbon, Portugal: European Monitoring Centre for Drugs and Drug Addiction, 2021.

Emmanuel, François. "Dans la cristallerie (Retour sur l'expérience mescalinienne de Henri Michaux)." *Midis de la poésie*, November 17, 2015. https://www.francoisemmanuel.be/wp-content/uploads/2022/09/Dans-la-cristallerie.pdf

Englund, Amir, et al. "Can We Make Cannabis Safer?" *The Lancet Psychiatry* 4, no. 8 (2017): 643–648.

Englund, Amir, et al. "Cannabidiol Inhibits THC-Elicited Paranoid Symptoms and Hippocampal-Dependent Memory Impairment." *Journal of Psychopharmacology* 27, no. 1 (2013): 19–27.

Englund, Amir, et al. "Does Cannabidiol Make Cannabis Safer? A Randomised, Double-blind, Cross-over Trial of Cannabis with Four Different CBD:THC Ratios." *Neuropsychopharmacology* 48, no. 6 (2023): 869–876.

Englund, Amir, et al. "Cannabis in the Arm: What Can We Learn from Intravenous Cannabinoid Studies?" *Current Pharmaceutical Design* 18, no. 32 (2012): 4906–4914.

Escondido, Nico. "The Strongest Concentrates on Earth." *High Times*, September 2018, 58–64.

European Food Safety Authority. "Modification of the Existing Maximum Residue Levels and Setting of Import Tolerances for Fluopyram in Various Crops." *EFSA Journal* 21, no. 6 (2023): 1–58.

Evangelista, Dominic A., et al. "Fossil Calibrations for the Cockroach Phylogeny (Insecta, Dictyoptera, Blattodea), Comments on the Use of Wings for Their Identification, and a Redescription of the Oldest Blaberidae." *Palaeontologia Electronica* 20, no. 3 (2017): 1–23.

Fachner, Joerg. "An Ethno-Methodological Approach to Cannabis and Music Perception, with EEG Brain Mapping in a Naturalistic Setting." *Anthropology of Consciousness* 17, no. 2 (2006): 78–103.

Fachner, Joerg. "Out of Time? Music, Consciousness States and Neuropharmacological Mechanisms of an Altered Temporality." *Proceedings of the 7th Triennial Conference of European Society for the Cognitive Sciences of Music (ESCOM 2009), Jyväsklä, Finland*, edited by Jukka Louhivuori et al., 103–109.

Farquhar, J. N. "The Fighting Ascetics of India." *Bulletin of the John Rylands Library* 9, no. 2 (1925): 431–452.

Faust-Socher, Achinoam, et al. "Enhanced Creative Thinking under Dopaminergic Therapy in Parkinson Disease." *Annals of Neurology* 75, no. 6 (2014): 935–942.

Felder, Christian C., and Michelle Glass. "Cannabinoid Receptors and Their Endogenous Agonists." *Annual Review of Pharmacology and Toxicology* 38 (1998): 179–200.

Ferber, Sari Goldstein, et al. "The 'Entourage Effect': Terpenes Coupled with Cannabinoids for the Treatment of Mood Disorders and Anxiety Disorders." *Current Neuropharmacology* 18, no. 2 (2020): 87–96.

Fernández Olmos, Margarite, and Lizabeth Paravisini-Gebert. *Creole Religions of the Caribbean: An Introduction from Vodou and Santería to Obeah and Espiritismo*. New York: New York University Press, 2011.

Fernández-Ruiz, Javier, et al. "Cannabinoid-Dopamine Interaction in the Pathophysiology and Treatment of CNS Disorders." *CNS Neuroscience & Therapeutics* 16, no. 3 (2010): e72–91.

Filer, Crist N. "Acidic Cannabinoid Decarboxylation." *Cannabis and Cannabinoid Research* 7, no. 3 (2022): 262–273.

Finlay, David B., et al. "Do Toxic Synthetic Cannabinoid Receptor Agonists Have Signature *In vitro* Activity Profiles? A Case Study of AMB-FUBINCA." *ACS Chemical Neuroscience* 10, no. 10 (2019): 4350–4360.

Finlay, David B., et al. "Terpenoids from Cannabis Do Not Mediate an Entourage Effect by Acting at Cannabinoid Receptors." *Frontiers in Pharmacology* 11 (2020): 359.

Finn, Kenneth, ed. *Cannabis in Medicine: An Evidence-Based Approach*. Cham, Switzerland: Springer, 2020.

Firman, James W., et al. "Chemoinformatic Consideration of Novel Psychoactive Substances: Compilation and Preliminary Analysis of a Categorised Dataset." *Molecular Informatics* 38, nos. 8–9 (2019): e1800142.

Fischer-Tiné, Harald. "Britain's Other Civilising Mission: Class Prejudice, European 'Loaferism' and the Workhouse-System in Colonial India." *The Indian Economic and Social History Review* 42, no. 3 (2005): 295–338.

Fisher, George. "Racial Myths of the Cannabis War." *Boston University Law Review* 101 (2021): 933–977.

Fitzgerald, Robert L., et al. "Driving Under the Influence of Cannabis: Impact of Combining Toxicology Testing with Field Sobriety Tests." *Clinical Chemistry* 69, no. 7 (2023): 724–733.

Flemming, T., et al. "Chemistry and Biological Activity of Tetrahydrocannabinol and its Derivatives." In *Bioactive Heterocycles IV. Topics in Heterocyclic Chemistry*, vol. 10, edited by M. T. H. Khan, 1–42. Berlin: Springer, 2007.

Fletcher, Jack M., et al. "Cognitive Correlates of Long-Term Cannabis Use in Costa Rican Men." *Archives of General Psychiatry* 53, no. 11 (1996): 1051–1057.

Flores-Sanchez, Isvett Josefina, and Robert Verpoorte. "PKS Activities and Biosynthesis of Cannabinoids and Flavonoids in *Cannabis sativa* L. Plants." *Plant Cell Physiology* 49, no. 12 (2008): 1767–1782.

Fordjour, Eric, et al. "*Cannabis*: A Multifaceted Plant with Endless Potentials." *Frontiers in Pharmacology* 14 (2023): 1–36.

Fossier, Albert E. "The Mariahuana Menace." *New Orleans Medical and Surgical Journal* 84, no. 4 (1931): 247–252.

Frank, Mel. *Marijuana Grower's Insider's Guide*. Los Angeles: Red Eye Press, 1988.

Frank, Mel, and Ed Rosenthal. *Marijuana Grower's Guide*. Berkeley, CA: And/Or Press, 1978.

Franz, Chlodwig, and Johannes Novak. "Sources of Essential Oils." In *Handbook of Essential Oils: Science, Technology, and Applications*. 3rd ed., edited by K. Hüsnü Can Baser and Gerhard Buchbauer, 41–83. New York: CRC Press, 2020.

Freeman, Abigail M., et al. "How Does Cannabidiol (CBD) Influence the Acute Effects of Delta-9-Tetrahydrocannabinol (THC) in Humans? A Systematic Review." *Neuroscience & Biobehavioral Reviews* 107 (2019): 696–712.

Freeman, Tom P., and Valentina Lorenzetti. "'Standard THC Units': A Proposal to Standardize Dose Across All Cannabis Products and Methods of Administration." *Addiction* 115, no. 7 (2020): 1207–1216.

Freeman, Tom P., and Adam R. Winstock. "Examining the Profile of High-Potency Cannabis and Its Association with Severity of Cannabis Dependence." *Psychological Medicine* 45, no. 15 (2015): 3181–3189.

French Republic. *Pièces diverses et correspondance relatives aux opérations de l'armée d'Orient en Égypte*. Paris: Baudoin, 1801. https://gallica.bnf.fr/ark:/12148/bpt6k364155/f2.item.

Fride, E., et al. "Critical Role of the Endogenous Cannabinoid System in Mouse Pup Suckling and Growth." *European Journal of Pharmacology* 419, nos. 2–3 (2001): 207–214.

Frongia, Francesca, et al. "Sound Perception and Its Effects in Plants and Algae." *Plant Signaling & Behavior* 15, no. 12 (2020): e1828674.

Frost, Ram, et al. "What Can the Brain Teach Us about Winemaking? An fMRI Study of Alcohol Level Preferences." *PLoS One* 10, no. 3 (2015): e0119220.

Gadde, Kishore M., and David B. Allison. "Cannabinoid-1 Receptor Antagonist, Rimonabant, for Management of Obesity and Related Risks." *Circulation* 114, no. 9 (2006): 974–984.

Gagliano, Monica. "Breaking the Silence: Green Mudras and the Faculty of Language in Plants." In *The Language of Plants: Science, Philosophy, Literature*, edited by Monica Gagliano et al., 84–100. Minneapolis: University of Minnesota Press, 2017.

Gagnon, Mathieu, et al. "High Levels of Pesticides Found in Illicit Cannabis Inflorescence Compared to Licensed Samples in Canadian Study Using Expanded 327 Pesticides Multiresidue Method." *Journal of Cannabis Research* 5, no. 1 (2023): 34.

Gaoni, Yechiel, and Raphael Mechoulam. "Isolation, Structure, and Partial Synthesis of an Active Constituent of Hashish." *Journal of the American Chemical Society* 86, no. 8 (1964): 1646–1647.

Gately, Iain. *Tobacco: A Cultural History of How an Exotic Plant Seduced Civilization*. New York: Grove Press, 2001.

Gautier, Théophile. "Charles Baudelaire." In *Les Fleurs du mal, précédées d'une Notice par Théophile Gautier*, by Charles Baudelaire, 1–75. Paris: Calmann-Lévy Editeurs, 1908. Originally printed in 1860. https://gallica.bnf.fr/ark:/12148/bpt6k8571036.texteImage.

Gautier, Théophile. "Le club des Hachichins," *Revue des deux mondes*, February 1846, 520–535. http://revuedesdeuxmondes.fr/article-revue/le-club-des-hachichins/.

Gautier, Théophile. "Le hachisch," *La Presse*, July 10, 1843. Reprinted in *Psychotropes* 20, vol. 4 (2014): 111–118. https://shs.cairn.info/revue-psychotropes-2014-4-page-111?lang=fr&ref=doi.

Gervasi, María Gracia, et al. "Anandamide Capacitates Bull Spermatozoa Through CB1 and TRPV1 Activation." *PLoS One* 6, no. 2 (2011): e16993.

Ghosh, Abhishek. "Prisoners With Drug Use Disorders During Covid-19 Pandemic: Caught Between a Rock and a Hard Place." *Asian Journal of Psychiatry* 54 (2020): 102332.

Ghosh, Jamini Mohan. *Sanyasi and Fakir Raiders in Bengal.* Calcutta: Bengal Secretariat Book Depot, 1930.

Gidley, Jennifer M. *The Future: A Very Short Introduction*. Oxford: Oxford University Press, 2017.

Gieringer, Dale H. "Economics of Cannabis Legalization." Paper submitted for the Drug Policy Foundation Conference, November 1993. https://www.mpp.org/issues/economics/economics-cannabis-legalization/.

Gieringer, Dale H. "The Forgotten Origins of Cannabis Prohibition in California." *Contemporary Drug Problems* 26, no. 2 (1999): 237–288.

Gieringer, Dale H. "The Origins of Cannabis Prohibition in California." https://www.druglibrary.org/schaffer/history/California_Marijuana_Law_History/California_Marijuana_Law_History_TOC.html. (This paper is an updated version of the previous entry.)

Giffen, P. J., et al. *Panic and Indifference: The Politics of Canada's Drug Laws*. Ottawa: Canadian Centre on Substance Abuse, 1991.

Gingrich, Jeremy, et al. "Review of the Oral Toxicity of Cannabidiol (CBD)." *Food and Chemical Toxicology* 176 (2023): 1–9.

Ginsberg, Allen. "First Manifesto to End the Bringdown." In *The Marijuana Papers*, edited by David Solomon, 230–248. New York: New American Library, 1968.

Ginsberg, Allen. "Henri Michaux." In *Deliberate Prose: Selected Essays, 1952–1995*, by Allen Ginsberg, 444–448. New York: HarperCollins, 2000.

Ginsberg, Allen. *Howl and Other Poems*. San Francisco: City Lights Books, 1959. Originally published in 1956.

Girdhar, Madhuri, et al. "Comparative Assessment for Hyperaccumulatory and Phytoremediation Capability of Three Wild Weeds." *3 Biotech* 4, no. 6 (2014): 579–589.

Global Commission on Drug Policy. *War on Drugs*. Geneva: GCDP, 2011.

Gobbi, Gabriella, et al. "Association of Cannabis Use in Adolescence and Risk of Depression, Anxiety, and Suicidality in Young Adulthood: A Systematic Review and Meta-analysis." *JAMA Psychiatry* 76, no. 4 (2019): 426–434.

Gokhale, Balkrishna Govind. "Tobacco in Seventeenth-Century India." *Agricultural History* 48, no. 4 (Oct. 1974): 484–492.

Gonzalez, Adolfo. "Sinsemilla: Origins of Modern Cannabis." *Cannabis Sommelier*, May 16, 2019. https://cannabissommelier.com/sinsemilla-origins-of-modern-cannabis/.

Goodden, Joe. *Riding So High: The Beatles and Drugs*. London: Pepper & Pearl, 2017.

Goodman, Jordan. *Tobacco in History: The Cultures of Dependence*. London: Routledge, 1993.

Gorelick, Jonathan, and Nirit Bernstein. "Chemical and Physical Elicitation for Enhanced Cannabinoid Production in Cannabis." In Cannabis sativa *L.—Botany and Biotechnology*, edited by Suman Chandra et al., 439–456. Cham, Switzerland: Springer, 2017.

Gowin, Joshua L., et al. "Brain Function Outcomes of Recent and Lifetime Cannabis Use." *JAMA Network Open* 8, no. 1 (2025): e2457069.

Grassa, Christopher J., et al. "A New *Cannabis* Genome Assembly Associates Elevated Cannabidiol (CBD) with Hemp Introgressed into Marijuana." *New Phytologist* 230, no. 4 (2021): 1665–1679.

Gray, Paul, et al. "The Use of Synthetic Cannabinoid Receptor Agonists (SCRAs) Within the Homeless Population: Motivations, Harms, and the Implications for Developing an Appropriate Response." *Addiction Research & Theory* 29, no. 1 (2021): 1–10.

Green, Bob, et al. "Being Stoned: A Review of Self-Reported Cannabis Effects." *Drug and Alcohol Review* 22, no. 4 (2003): 453–460.

Green, Nile. "Breaking the Begging Bowl: Morals, Drugs, and Madness in the Fate of the Muslim *Faqir*." *South Asian History and Culture* 5, no. 2 (2014): 226–245.

Green, Nile. *Islam and the Army in Colonial India: Sepoy Religion in the Service of Empire*. Cambridge: Cambridge University Press, 2009.

Green, Nile. *Sufism: A Global History*. Oxford: Wiley-Blackwell, 2012.

Grinspoon, Peter. *Seeing Through the Smoke: A Cannabis Specialist Untangles the Truth About Marijuana.* Essex, CT: Prometheus Books, 2023.

Gross, Madeleine E., et al. "Why Creatives Don't Find the Oddball Odd: Neural and Psychological Evidence for Atypical Salience Processing." *Brain and Cognition* 178 (2024): 106178.

Gross, Madeleine E., and Jonathan W. Schooler. "Standing Out: An Atypical Salience Account of Creativity." *Trends in Cognitive Sciences* 28, no. 7 (2024): 597–599.

Gross, Robert Lewis. *The Sadhus of India: A Study of Hindu Asceticism*. Jaipur, India: Rawat Publications, 1992.

Grotenhermen, Franjo. "Pharmacokinetics and Pharmacodynamics of Cannabinoids." *Clinical Pharmacokinetics* 42, no. 4 (2003): 327–360.

Guba, David A. *Taming Cannabis: Drugs and Empire in Nineteenth-Century France*. Montreal and Kingston: McGill-Queen's University Press, 2020.

Gülck, Thies, and Birger Lindberg Møller. "Phytocannabinoids: Origins and Biosynthesis." *Trends in Plant Science* 25, no. 10 (2020): 985–1004.

Gunther, Robert T. *The Greek Herbal of Dioscorides*. Oxford: Oxford University Press, 1934.

Hadengue, Tigrane, et al. *Le livre du cannabis: le XXI[e] sera-t-il psychédélique?* Geneva: Georg Editeur, 1999.

Hager, Steven. "Inside Cannabis Castle." *High Times*, March 1987, 37–48, 93.

Hahner, Astrid. "Zweitgutachten erwartet im Schweizer Pilotprojekt 'Weed Care.'" *Krautinvest*, September 27, 2022. https://krautinvest.de/zweitgutachten-erwartet-im-schweizer-pilotprojekt-weed-care/.

Hall, Wayne. "Is Cannabis Use Psychotogenic?" *The Lancet* 367, no. 9506 (2006): 193–195.

Hallam, Christopher. "Reaching Out from the Sixties." *Druglink*, January/February (2007): 10–12.

Hamdan, Marwan A. "Jazz Aesthetics Speak Loud in Allen Ginsberg's *Howl*: A Thematic Cultural Sketch." *International Journal of English Literature and Social Sciences* 4, no. 1 (2019): 81–88.

Hamilton, Alexander. *A New Account of the East Indies, Being the Observations and Remarks of Capt. Alexander Hamilton, Who Spent Time There from the Year 1688 to 1723*. Vol. 1. Edinburgh: John Mosman, 1727.

Hammond, David. "Communicating THC Levels and 'Dose' to Consumers: Implications for Product Labelling and Packaging of Cannabis Products in Regulated Markets." *International Journal of Drug Policy* 91 (2019): 102509.

Hammond, David, et al. "Trends in the Use of Cannabis Products in Canada and the USA, 2018–2020: Findings from the International Cannabis Policy Study." *International Journal of Drug Policy* 105 (2022): 103716.

Hammond, John. *John Hammond on Record: An Autobiography*. New York: Penguin, 1981.

Hamowy, Ronald. *Dealing with Drugs: Consequences of Government Control*. San Francisco: Pacific Research Institute for Public Policy, 1987.

Hampson, A. J., et al. "Cannabidiol and Delta-9-Tetrahydrocannabinol Are Neuroprotective Antioxidants." *Proceedings of the National Academy of Sciences* 95 (1998): 8268–8273.

Hansen, Jolene. "Growing Under High Light Intensities." *Cannabis Business Times*, October 2021. https://www.cannabisbusinesstimes.com/article/growing-under-high-light-intensities-lighting-report/.

Hanuš, Lumír Ondřej, and Yotam Hod. "Terpenes/Terpenoids in *Cannabis*: Are They Important?" *Medical Cannabis and Cannabinoids* 3, no. 1 (2020): 25–60.

Hanuš, Lumír Ondřej, et al. "Phytocannabinoids: A Unified Critical Inventory." *Natural Product Reports* 33, no. 12 (2016): 1357–1392.

Harari, Yuval Noah. *Homo Deus: A Brief History of Tomorrow*. New York: Signal, 2016.

Hardwick, Sheila, and Leslie King. *Home Office Cannabis Potency Study 2008*. St Albans, UK: Home Office Scientific Development Branch, 2008.

Harpignies, J. P., ed. *Visionary Plant Consciousness: The Shamanic Teachings of the Plant World*. Rochester, VT: Park Street Press, 2007.

Hart, Carl, et al. "Effects of Acute Smoked Marijuana on Complex Cognitive Performance." *Neuropsychopharmacology* 25, no. 5 (2001): 757–765.

Hasan, Alkomiet, et al. "Cannabis Use and Psychosis: A Review of Reviews." *European Archives of Psychiatry and Clinical Neuroscience* 270, no. 4 (2020): 403–412.

Hazekamp, Arno. "Letter: The Herbal Way—A Response to Ethan Russo." *Cannabinoids* 2, no. 3 (2007): 20–21.

Hazekamp, Arno, et al. "Chemistry of Cannabis." In *Comprehensive Natural Products II: Chemistry and Biology*, Vol. 3. Edited by Lew Mander and Hung-Wen Liu, 1033–1084. Oxford: Elsevier, 2010.

Hazekamp, Arno, et al. "Evaluation of a Vaporizing Device (Volcano®) for the Pulmonary Delivery of Tetrahydrocannabinol." *Journal of Pharmaceutical Sciences* 96, no. 6 (2005): 1308–1317.

Health Canada. *Information for Health Care Professionals: Cannabis (Marihuana, Marijuana) and the Cannabinoids*. Ottawa: Health Canada, 2018.

Helmer, John. *Drugs and Minority Oppression*. New York: The Seabury Press, 1975.

Henman, Anthony. "War on Drugs Is War on People." *The Ecologist* 10, nos. 8/9 (1980): 282–289.

Henman, Anthony, and Osvaldo Pessoa Jr. *Diamba Sarabamba*. São Paulo: Ground, 1986.

Herer, Jack. *The Emperor Wears No Clothes*. Van Nuys, CA: Ah Ah Publishing, 1985.

Herkenham, Miles, et al. "Characterization and Localization of Cannabinoid Receptors in Rat Brain: A Quantitative In Vitro Autoradiographic Study." *The Journal of Neuroscience* 11, no. 2 (1991): 563–583.

Herodotus. *The Histories*. Translated by G. C. Macaulay. Wrocław, Poland: Amazon Fulfillment, 2020.

Hesami, Mohsen, et al. "Recent Advances in Cannabis Biotechnology." *Industrial Crops and Products* 158 (2020): 1–20.

Hill, Matthew. "Perspective: Be Clear About the Real Risks." *Nature* 525, no. 7570 (2015): S14.

Hillig, Karl W., and Paul G. Mahlberg. "A Chemotaxonomic Analysis of Cannabinoid Variation in *Cannabis* (Cannabaceae)." *American Journal of Botany* 91, no. 6 (2004): 966–975.

Hindocha, Chandni, et al. "Cannabidiol Reverses Attentional Bias to Cigarette Cues in a Human Experimental Model of Tobacco Withdrawal." *Addiction* 113, no. 9 (2018): 1696–1705.

Hines, Lindsey A., et al. "Association of High-Potency Cannabis Use with Mental Health and Substance Use in Adolescence." *JAMA Psychiatry* 77, no. 10 (2020): 1044–1051.

Hoare, Stephen. "How Soho's Early Jazz Clubs Paved the Way for a Multiracial Britain." *Londonist*, Updated April 6, 2023. https://londonist.com/london/music/soho-jazz-clubs-history#.

Hobbs, Jack M., et al. "Evaluation of Pharmacokinetics and Acute Anti-Inflammatory Potential of Two Oral Cannabidiol Preparations in Healthy Adults." *Phytotherapy Research* 34, no. 7 (2020): 1696–1703.

Holdsworth, Elizabeth A., et al. "Human Milk Cannabinoid Concentrations and Associations with Maternal Factors: The Lactation and Cannabis (LAC) Study." *Breastfeeding Medicine* 19, no. 7 (2024): 515–524.

Holland, Ellen. *Weed: A Connoisseur's Guide to Cannabis*. New York: Quarto Publishing, 2021.

Home Office. *Cannabis: Report by the Advisory Committee on Drug Dependence*. London: Her Majesty's Stationery Office, 1968.

Houghton, E. M., and H. C. Hamilton. "Pharmacological Study of Cannabis Americana (Cannabis Sativa)." *American Journal of Pharmacy* 80 (1908): 16–20.

Howlett, Allyn C., et al. "International Union of Pharmacology. XXVII. Classification of Cannabinoid Receptors." *Pharmacological Reviews* 54, no. 2 (2002): 161–202.

Howlett, Allyn C., et al. "Stereochemical Effects of 11-OH-Delta 8-Tetrahydrocannabinol-Dimethylheptyl to Inhibit Adenylate Cyclase and Bind to the Cannabinoid Receptor." *Neuropharmacology* 29, no. 2 (1990): 161–165.

Howlett, Allyn C., et al. "The Spicy Story of Cannabimimetic Indoles." *Molecules* 26, no. 20 (2021): 1–30.

Hudak, John. *Marijuana: A Short History*. Washington, DC: Brookings Institution Press, 2020.

Hudak, John, et al. *Uruguay's Cannabis Law: Pioneering a New Paradigm*. Washington, DC: Washington Office on Latin America, 2018.

Huestis, Marilyn A. "Human Cannabinoid Pharmacokinetics." *Chemistry & Biodiversity* 4, no. 8 (2007): 1770–1804.

Huestis, Marilyn A., and Michael L. Smith. "Cannabinoid Pharmacokinetics and Disposition in Alternative Matrices." In *Handbook of Cannabis*, edited by Roger G. Pertwee, 300–316. Oxford: Oxford University Press, 2014.

Huffman, John W., et al. "Design, Synthesis and Pharmacology of Cannabimimetic Indoles." *Bioorganic & Medicinal Chemistry Letters* 4, no. 4 (1994): 563–566.

Hughes, Anne Randall, et al. "Ecological Consequences of Genetic Diversity." *Ecology Letters* 11, no. 6 (2008): 609–623.

Hurd, Yasmin L., et al. "Cannabidiol for the Reduction of Cue-Induced Craving and Anxiety in Drug-Abstinent Individuals with Heroin Use Disorder: A Double-Blind Randomized Placebo-Controlled Trial." *American Journal of Psychiatry* 176, no. 11 (2019): 911–922.

Hussain, Tajammul, et al. "*Cannabis sativa* Research Trends, Challenges, and New-Age Perspectives." *iScience* 24, no. 12 (2021): 103391.

Hutchinson, Harry William. "Patterns of Marihuana Use in Brazil." In *Cannabis and Culture*, edited by Vera Rubin, 173–184. The Hague: Mouton, 1975.

Huxley, Aldous. *The Doors of Perception*. New York: Harper & Row, 1954.

Hybertson, Brooks M., et al. "Oxidative Stress in Health and Disease: The Therapeutic Potential of Nrf2 Activation." *Molecular Aspects of Medicine* 32, no. 4–6 (2011): 234–246.

Ibn-al-Baytar, Abd Allah ibn Ahmad. *Traité des simples.* Vol. 3. Paris: Imprimerie Nationale, 1883.

Ibn Iyas, Muhammad. *An Account of the Ottoman Conquest of Egypt.* London: The Royal Asiatic Society, 1921.

Indian Hemp Drugs Commission. *Report of the Indian Hemp Drugs Commission.* 8 vols. Simla: Government Central Printing Office, 1894. https://digital.nls.uk/indiapapers/browse/archive/74574104.

Institut Technique du Chanvre. *Le chanvre industriel: Guide technique.* Troyes, France: Editions Institut Technique du Chanvre, 2007.

International Council on Alcohol, Drugs & Traffic Safety. "Cannabis & Driving: 2: Recent Experimental Evidence." 2020. https://www.icadtsinternational.com/Fact-Sheets.

Ishrat, Saba, et al. "Association Between Cannabis Use and Brain Structure and Function: An Observational and Mendelian Randomisation Study." *BMJ Mental Health* 27, no. 1 (2024): 1–8.

Iversen, Leslie. *The Science of Marijuana.* 3rd. ed. Oxford: Oxford University Press, 2018.

Jackson, Joy. "Prohibition in New Orleans: The Unlikeliest Crusade." *Louisiana History: The Journal of the Louisiana Historical Society* 19, no. 3 (1978): 261–284.

Jacobson, Matthew Frye. *Whiteness of a Different Color: European Immigrants and the Alchemy of Race.* Cambridge, MA: Harvard University Press, 1999.

Jansen, A. C. M. "The Economics of Cannabis-Cultivation in Europe." Paper presented at the 2nd European Conference on Drug Trafficking and Law Enforcement. Paris, September 26–27, 2002. https://www.cedro-uva.org/lib/jansen.economics.html.

Jeffers, Abra M., et al. "Association of Cannabis Use with Cardiovascular Outcomes Among US Adults." *Journal of the American Heart Association* 13, no. 5 (2024): e030178.

Jiang, Hong-En, et al. "A New Insight into *Cannabis sativa* (Cannabaceae) Utilization from 2500-year-old Yanghai Tombs, Xinjiang, China." *Journal of Ethnopharmacology* 108, no. 3 (2006): 414–422.

Jiloha, R. C. "Lunatic Asylums: A Business of Profit During the Colonial Empire in India." *Indian Journal of Psychiatry* 63, no. 1 (2021): 84–87.

Jin, Dan, et al. "Secondary Metabolites Profiled in Cannabis Inflorescences, Leaves, Stem Barks, and Roots for Medicinal Purposes." *Scientific Reports* 10, no. 1 (2020): 3309.

Jin, Dan, et al. "Cannabis Indoor Growing Conditions, Management Practices, and Post-Harvest Treatment: A Review." *American Journal of Plant Sciences* 10, no. 6 (2019): 925–946.

Johnson, Cameron S., et al. "Controlling New Psychoactive Substances in New Zealand." *Australian Journal of Forensic Sciences* 55, no. 5 (2023): 670–688.

Johnson, M. Ross, and Lawrence S. Melvin. "The Discovery of Nonclassical Cannabinoid Analgetics." In *Cannabinoids as Therapeutic Agents*, edited by Raphael Mechoulam, 121–145. London: CRC Press, 1986.

Johnson, Nick. "American Weed: A History of Cannabis Cultivation in the United States." *EchoGéo* 48 (2019): 1–22.

Johnson, Nick. *Grass Roots: A History of Cannabis in the American West.* Corvallis: Oregon State University Press, 2017.

Johnston, James F. W. *The Chemistry of Common Life.* Vol. 2. New York: Appleton, 1855.

Jones, Alana. "Isovaleric Acid." In *The Oxford Companion to Beer*, edited by Garrett Oliver, 498. Oxford: Oxford University Press, 2011.

Jones, Katy A., et al. "Cannabis and Ecstasy/MDMA: Empirical Measures of Creativity in Recreational Users." *Journal of Psychoactive Drugs* 41, no. 4 (2009): 323–329.

Jones, Max, and John Chilton. *Louis: The Louis Armstrong Story, 1900–1971*. Boston: Little, Brown and Co., 1971.

Jones, Reese T. "Mental Illness and Drugs: Pre-Existing Psychopathology and Response to Psychoactive Drugs." In *Drug Use in America: Patterns and Consequences of Drug Use*. Vol. 1 of *Drug Use in America: Problem in Perspective: Appendix*, edited by National Commission on Marihuana and Drug Abuse, 373–397. Washington, DC: Superintendent of Documents, U.S. Government Print Office, 1973.

Joy, Janet E., et al. *Marijuana and Medicine*. Washington, DC: National Academy Press, 1999.

Joyce, Tommy. "The THC Dosage Guide: Flower, Edibles, Concentrates and More." *Key to Cannabis*, December 4, 2024. https://keytocannabis.com/the-thc-dosage-guide-flower-edibles-concentrates-and-more/.

Jugl, Sebastian, et al. "Much Ado about Dosing: The Needs and Challenges of Defining a Standardized Cannabis Unit." *Medical Cannabis and Cannabinoids* 4, no. 2 (2021): 121–124.

Justice Canada. Cannabis Act (S.C. 2018, c.16). https://laws-lois.justice.gc.ca/eng/acts/C-24.5/fulltext.html.

Kalant, Harold. "Medicinal Use of Cannabis: History and Current Status." *Pain Research & Management* 6, no. 2 (2001): 80–91.

Kalinowski, Agnieszka, and Keith Humphreys. "Governmental Standard Drink Definitions and Low-Risk Alcohol Consumption Guidelines in 37 Countries." *Addiction* 111, no. 7 (2016): 1293–1298.

Kamel, Ibrahim, et al. "Myocardial Infarction and Cardiovascular Risks Associated with Cannabis Use: A Multicenter Retrospective Study." *JACC Advances* 4, no. 5 (2025): 101698.

Karniol, I. G., et al. "Cannabidiol Interferes with the Effects of Delta9-Tetrahydrocannabinol in Man." *European Journal of Pharmacology* 28, no. 1 (1974): 172–177.

Karschner, Erin L., et al. "Do Delta9-Tetrahydrocannabinol Concentrations Indicate Recent Use in Chronic Cannabis Users?" *Addiction* 104, no. 12 (2009): 2041–2048.

Kayo. *The Sinsemilla Technique: An Insight into a Cultivation Production Technique*. San Francisco: Last Gasp Publications, 1982.

Kayo. "Some Big Well-Lighted Places." *High Times*, September 1986, 46.

Keck, François, et al. "The Global Human Impact on Biodiversity." *Nature* 641 (2025): 395–400.

Kelleher, Lucia M., et al. "The Effects of Cannabis on Information-Processing Speed." *Addictive Behaviors* 29, no. 6 (2004): 1213–1219.

Kendell, Robert. "Cannabis Condemned: The Proscription of Indian Hemp." *Addiction* 98, no. 2 (2003): 143–151.

Kerr, Hem Chundra. "Report on the Cultivation of and Trade in Ganja in Bengal." In *Papers Relating to the Consumption of Ganja and Other Drugs in India*. British Parliamentary Papers 66 (1891): 94–154. London: Hansard.

Kessler, Danny, et al. "Changing Pollinators as a Means of Escaping Herbivores." *Current Biology* 20, no. 3 (2010): 237–242.

Khalifa, Ahmad M. "Traditional Patterns of Hashish Use in Egypt." In *Cannabis and Culture*, edited by Vera Rubin, 195–205. The Hague: Mouton & Co., 1975.

Kiernan, Victor G. *Tobacco: A History*. London: Hutchinson Radius, 1991.

Kimmerer, Robin Wall. *Braiding Sweetgrass: Indigenous Wisdom, Scientific Knowledge, and the Teachings of Plants*. Minneapolis: Milkweed Editions, 2013.

King, Leslie A., et al. *An Overview of Cannabis Potency in Europe*. Lisbon, Portugal: European Monitoring Centre for Drugs and Drug Addiction, 2004.

Kitdumrongthum, Sarunya, and Dunyaporn Trachootham. "An Individuality of Response to Cannabinoids: Challenges in Safety and Efficacy of Cannabis Products." *Molecules* 28, no. 6 (2023): 2791.

Klantschnig, Gernot. "The Origins of Cannabis Prohibition in Nigeria and the Indian Hemp Decree of 1966." In *Cannabis: Global Histories*, edited by Lucas Richert and James H. Mills, 225–246. Cambridge, MA: MIT Press, 2021.

Klein, Axel. "Nigeria and the Drugs War." *Review of African Political Economy* 26, no. 79 (1999): 51–73.

Klonoff, Harry, et al. "Neuropsychological Effects of Marijuana." *Canadian Medical Association Journal* 108, no. 2 (1973): 150–156.

Knodt, Michael. "44% THC—Does Berlin Have the Strongest Weed in the World?" *Sensi Seeds*, April 23, 2020. www.sensiseeds.com/en/blog/44-thc-does-berlin-have-the-strongest-weed-in-the-world/.

Koeppel, Dan. *Banana: The Fate of the Fruit That Changed the World*. New York: Hudson Press, 2008.

Komenda, Michael, and Ralf Koppmann. "Monoterpene Emissions from Scots Pine (*Pinus sylvestris*): Field Studies of Emission Rate Variabilities." *Journal of Geophysical Research* 107, no. D13 (2002): 4161.

Komp, Ellen. "Cannabis & 'Muggles': An Etymology." *Leafly*, Updated January 12, 2022. https://www.leafly.com/news/lifestyle/etymology-of-muggle.marijuana.

Kovalchuk, Igor, et al. "The Genomics of *Cannabis* and Its Close Relatives." *Annual Review of Plant Biology* 71 (2020): 713–739.

Kozak, Theresa, et al. "Reimagining Research with Pregnant Women and Parents Who Consume Cannabis in the Era of Legalization: The Value of Integrating Intersectional Feminist and Participatory Approaches." *Cannabis and Cannabinoid Research* 7, no. 1 (2022): 11–15.

Kozma, Liat. "Cannabis Prohibition in Egypt, 1880–1939: From Local Ban to League of Nations Diplomacy." *Middle Eastern Studies* 47, no. 3 (2011): 443–460.

Krassner, Paul. "The Secret History of Sinsemilla." *High Times*, February 1999, 36–38.

Kumar, Kaavya Krishna, et al. "Structure of a Signaling Cannabinoid Receptor 1-G Protein Complex." *Cell* 176, no. 3 (2019): 448–458.

Lamarck, Jean-Baptiste. *Encyclopédie méthodique: Botanique*. Tome 1. Paris: Panckoucke, 1783.

Lambert, Didier M., and Christopher J. Fowler. "The Endocannabinoid System: Drug Targets, Lead Compounds, and Potential Therapeutic Applications." *Journal of Medicinal Chemistry* 48, no. 16 (2005): 5059–5087.

Lambo, Thomas Adeoye. "Medical and Social Problems of Drug Addiction in West Africa." *UNODC Bulletin on Narcotics* 1 (1965): 3–13.

Lane, Véronique. *The French Genealogy of the Beat Generation: Burroughs, Ginsberg and Kerouac's Appropriations of Modern Literature, from Rimbaud to Michaux*. New York: Bloomsbury Academic, 2017.

Lapham, Sandra C. "The Limits of Tolerance: Convicted Alcohol-Impaired Drivers Share Experiences Driving Under the Influence." *The Permanente Journal* 14, no. 2 (2010): 26–30.

Lapoint, Jeff, and Lewis S. Nelson. "Synthetic Cannabinoids: The Newest, Almost Illicit Drug of Abuse." *Emergency Medicine* (February 2011): 26–28. https://cdn.mdedge.com/files/s3fs-public/Document/September-2017/043020026.pdf.

Laprairie, R. B., et al. "Cannabidiol is a Negative Allosteric Modulator of the Cannabinoid CB1 Receptor." *British Journal of Pharmacology* 172, no. 20 (2015): 4790–4805.

Lastreto, Nikki, and Swami Chaitanya. "Why Choose Sungrown Cannabis, and What Makes Outdoor Cultivation Superior?" *Merry Jane*, June 13, 2019. https://merryjane.com/culture/why-choose-sungrown-cannabis-and-what-makes-outdoor-cultivation-superior/#:~:text=So%20what%20is%20it%20that,farmers%20who%20grow%20the%20plant.

Laursen, Lucas. "Botany: The Cultivation of Weed." *Nature* 525 (2015): S4–S5.

Laverty, Kaitlin U., et al. "A Physical and Genetic Map of *Cannabis sativa* Identifies Extensive Rearrangements at the *THC/CBD Acid Synthase* Loci." *Genome Research* 29, no. 1 (2019): 146–156.

Lawn, Will, et al. "The Acute Effects of Cannabis with and without Cannabidiol in Adults and Adolescents: A Randomised, Double-Blind, Placebo-Controlled, Crossover Experiment." *Addiction* 118, no. 7 (2023): 1282–1294.

League of Nations. *International Opium Convention.* Treaty Series 81 (1929): 319–358.

League of Nations. *Records of the Second Opium Conference.* Vol. I, *Plenary Meetings, Texts of the Debates.* Geneva: League of Nations, 1925.

Lee, Jiwon. "Tetrahydrocannabinol and Dopamine D1 Receptor." *Frontiers in Neuroscience* 18 (2024): 1360205.

Lee, Martin A. *Smoke Signals: A Social History of Marijuana—Medical, Recreational, and Scientific.* New York: Scribner, 2012.

Leggett, Ted. "A Review of the World Cannabis Situation." *UNODC Bulletin on Narcotics*, 58, nos. 1–2 (2006): 1–155.

Leggett, Ted, and Thomas Pietschmann. "Global Cannabis Cultivation and Trafficking." In *A Cannabis Reader: Global Issues and Local Experiences*, edited by Sharon Rödner Sznitman et al., 187–212. Lisbon, Portugal: EMCDDA, 2008.

Lemberger, Louis. "Nabilone: A Synthetic Cannabinoid of Medicinal Utility." In *Marihuana and Medicine*, edited by Gabriel G. Nahas et al., 561–566. Totowa, NJ: Humana Press, 1999.

Leos-Toro, Cesar, et al. "Cannabis Labelling and Consumer Understanding of THC Levels and Serving Sizes." *Drug and Alcohol Dependence* 208 (2020): 107843.

Leung, Janni, et al. "Do Cannabis Users Reduce Their THC Dosages When Using More Potent Cannabis Products? A Review." *Frontiers in Psychiatry* 12 (2021): 630602.

Levine, Michael. *Deep Cover: The Inside Story of How DEA Infighting, Incompetence and Subterfuge Lost Us the Biggest Battle of the Drug War.* New York: Delacorte Press, 1990.

Lewis, Bernard. *The Assassins: A Radical Sect in Islam.* New York: Oxford University Press, 1967.

Li, Hui-Lin. "An Archaeological and Historical Account of Cannabis in China." *Economic Botany* 28, no. 4 (1974): 437–448.

Lichtman, Aron H., and Billy R. Martin. "Analgesic Properties of THC and Its Synthetic Derivatives." In *Marihuana and Medicine*, edited by Gabriel G. Nahas et al., 511–526. Totowa, NJ: Humana Press, 1999.

Liktor-Busa, Erika, et al. "Analgesic Potential of Terpenes Derived from *Cannabis sativa*." *Pharmacological Reviews* 73, no. 4 (2021): 98–126.

Lindigkeit, Rainer, et al. "Spice: A Never Ending Story?" *Forensic Science International* 191, nos. 1–3 (2009): 58–63.

Linnaeus, Caroli A. *Species Plantarum*, vol. II. Holmiae, Sweden: Laurentii Salvii, 1772.

Liu, Icy. "The Science of Tasting, Drinking and Thinking Wine." *Littlewine*, March 15, 2024. www.littlewine.io/learn-about-wine/wine-neuroscience.

Livingston, Samuel J., et al. "Cannabis Glandular Trichomes Alter Morphology and Metabolite Content During Flower Maturation." *The Plant Journal* 101, no. 1 (2020): 37–56.

Llewellyn, David, et al. "Indoor Grown Cannabis Yield Increased Proportionally with Light Intensity, but Ultraviolet Radiation Did Not Affect Yield or Cannabinoid Content." *Frontiers in Plant Science* 13 (2022): 1–12.

Lo, Jamie O., et al. "Prenatal Cannabis Use and Neonatal Outcomes: A Systematic Review and Meta-Analysis." *JAMA Pediatrics* 179, no. 7 (2025): 738–746. Advance online publication. https://doi.org/10.1001/jamapediatrics.2025.0689.

Lobato-Freitas, Carolina, et al. "Overview of Synthetic Cannabinoids ADB-FUBINACA and AMB-FUBINACA: Clinical, Analytical, and Forensic Implications." *Pharmaceuticals* 14, no. 3 (2021): 186.

Loflin, Mallory, and Mitch Earleywine. "A New Method of Cannabis Ingestion: The Dangers of Dabs?" *Addictive Behaviors* 39, no. 10 (2014): 1430–1433.

Logan, Barry, et al. *An Evaluation of Data from Drivers Arrested for Driving Under the Influence in Relation to Per Se Limits for Cannabis.*" Washington, DC: American Automobile Association Foundation for Traffic Safety, 2016.

López Dávila, Edelbis, et al. "Pesticides Residues in Tobacco Smoke: Risk Assessment Study." *Environmental Monitoring and Assessment* 192, no. 9 (2020): 615.

Lorenzen, David N. "Warrior Ascetics in Indian History." *Journal of the American Oriental Society* 98, no. 1 (1978): 61–75.

Louw, Leanie, and Marius G. Lambrechts. "Grape-Based Brandies: Production, Sensory Properties and Sensory Evaluation." In *Alcoholic Beverages*, edited by John Piggott, 281–298. Sawston, UK: Woodhead Publishing, 2012.

Lowe, Henry, et al. "Non-Cannabinoid Metabolites of *Cannabis sativa* L. with Therapeutic Potential." *Plants* 10, no. 2 (2021): 400.

Lowe, Henry, et al. "The Endocannabinoid System: A Potential Target for the Treatment of Various Diseases." *International Journal of Molecular Sciences* 22, no. 17 (2021): 9472.

Lubman, Dan I., et al. "Cannabis and Adolescent Brain Development." *Pharmacology & Therapeutics* 148 (2015): 1–16.

Lubotsky, Alexander. "Scythian Elements in Old Iranian." In *Indo-Iranian Languages and Peoples (Proceedings of the British Academy)*, edited by Nicholas Sims-Williams, 189–202. Oxford: Oxford University Press, 2002.

Lundqvist, Thomas. *Cognitive Dysfunctions in Chronic Cannabis Users Observed During Treatment: An Integrative Approach.* Stockholm: Almqvist & Wiksell, 1995.

Lydon, John, et al. "UV-B Radiation Effects on Photosynthesis, Growth and Cannabinoid Production of Two *Cannabis sativa* Chemotypes." *Photochemistry and Photobiology* 46, no. 2 (1987): 201–206.

Lynch, Ryan C., et al. "Genomic and Chemical Diversity in *Cannabis*." *Critical Reviews in Plant Sciences* 35, nos. 5–6 (2016): 349–363.

Maalouf, Amin. *Samarkand: A Novel.* New York: Interlink Books, 1998.

MacCallum, Caroline A., and Ethan B. Russo. "Practical Considerations in Medical Cannabis Administration and Dosing." *European Journal of Internal Medicine* 49 (2018): 12–19.

Maccoun, Robert J. "Commentary on Niesink et al. (2015): Interpreting Trends in Tetrahydrocannabinol Potency—Three Stories, One of Which May Be True." *Addiction* 110, no. 12 (2015): 1951–1952.

Macher, Rayna B., and Mitchell Earleywine. "Enhancing Neuropsychological Performance in Chronic Cannabis Users: The Role of Motivation." *Journal of Clinical and Experimental Neuropsychology* 34, no. 4 (2012): 405–415.

Mack, Alison, and Janet Joy. *Marijuana as Medicine? The Science Beyond the Controversy.* Washington, DC: National Academy Press, 2000.

Madras, Bertha K. "Cannabinoid and Marijuana Neurobiology." In *Cannabis and Medicine: An Evidence-Based Approach*, edited by Kenneth Finn, 25–47. Cham, Switzerland: Springer, 2020.

Magagnini, Gianmaria, et al. "The Effect of Light Spectrum on the Morphology and Cannabinoid Content of *Cannabis sativa* L." *Medical Cannabis and Cannabinoids* 1, no. 1 (2018): 19–27.

Malfeito-Ferreira, Manuel. "Fine Wine Flavour Perception and Appreciation: Blending Neuronal Processes, Tasting Methods and Expertise." *Trends in Food Science & Technology* 115 (2021): 332–346.

Malfeito-Ferreira, Manuel. "Fine Wine Recognition and Appreciation: It Is Time to Change the Paradigm of Wine Tasting." *Food Research International* 174, no. 2 (2023): 113668.

Mallatt, Jon, et al. "Debunking a Myth: Plant Consciousness." *Protoplasma* 258, no. 3 (2021): 459–476.

Manthey, Jakob, et al. "Germany's Cannabis Act: A Catalyst for European Drug Policy Reform?" *The Lancet Regional Health—Europe* 42 (2024): 100929.

Marcial, Guillermo, et al. "Intraspecific Variation in Essential Oil Composition of the Medicinal Plant *Lippia integrifolia* (Verbenaceae). Evidence for Five Chemotypes." *Phytochemistry* 122 (2016): 203–212.

Marcotte, Thomas D., et al. "Driving Performance and Cannabis Users' Perception of Safety: A Randomized Clinical Trial." *JAMA Psychiatry* 79, no. 3 (2022): 201–209.

Marder, Michael. *The Philosopher's Plant: An Intellectual Herbarium.* New York: Columbia University Press, 2014.

Marder, Michael. "What Is Plant-Thinking?" *Klesis Revue Philosophique* 25 (2013): 124–143.

Marincolo, Sebastián. *What Hashish Did to Walter Benjamin: Mind-Altering Essays on Marijuana.* Stuttgart, Germany: Khargala Press, 2015.

Marron, Tali R., et al. "Chain Free Association, Creativity, and the Default Mode Network." *Neuropsychologia* 118, Part A (2018): 40–58.

Marshall, Jules, and Floris Leeuwenberg, in collaboration with Ben Dronkers and Gerbrand Korevaar. *Weed of Wonder: Explore the World of Cannabis Through the Collection of Ben Dronkers and the Hash Marihuana & Hemp Museum.* Amsterdam: Hash Marihuana & Hemp Museum, 2021.

Matzneller, Philipp, et al. "Canopy Management." In *Handbook of Cannabis Production in Controlled Environments*, edited by Youbin Zheng, 189–212. London: CRC Press, 2022.

Mayseless, Naama, et al. "Generating Original Ideas: The Neural Underpinning of Originality." *NeuroImage* 116 (2015): 232–239.

McCartney, Danielle, et al. "Are Blood and Oral Fluid Delta9-Tetrahydrocannabinol (THC) and Metabolite Concentrations Related to Impairment? A Meta-Regression Analysis." *Neuroscience and Biobehavioral Reviews* 134 (2022): 104433.

McCartney, Danielle, et al. "How Long Does a Single Oral Dose of Cannabidiol Persist in Plasma? Findings from Three Clinical Trials." *Drug Testing and Analysis* 15, no. 3 (2023): 334–344.

McDaniel, Patricia A., et al. "The Tobacco Industry and Pesticide Regulations: Case Studies from Tobacco Industry Archives." *Environmental Health Perspectives* 113, no. 12 (2005): 1659–1665.

McKeon, Caroline M., et al. "Human Land Use Is Comparable to Climate as a Driver of Global Plant Occurrence and Abundance Across Life Forms." *Global Ecology and Biogeography* 32, no. 9 (2023): 1618–1631.

McKernan, Kevin J., et al. "Sequence and Annotation of 42 Cannabis Genomes Reveals Extensive Copy Number Variation in Cannabinoid Synthesis and Pathogen Resistance Genes." *bioRxiv* (2020).

McPartland, John M. "*Cannabis* Systematics at the Levels of Family, Genus, and Species." *Cannabis and Cannabinoid Research* 3, no. 1 (2018): 203–212.

McPartland, John M. "A Survey of Hemp Diseases and Pests." In *Advances in Hemp Research*, edited by Paolo Ranalli, 109–131. London: CRC Press, 1999.

McPartland, John M., et al. *Hemp Diseases and Pests: Management and Biological Control.* New York: CABI Publishing, 2000.

McPartland, John M., William Hegman, and Tengwen Long. "*Cannabis* in Asia: Its Center of Origin and Early Cultivation, Based on a Synthesis of Subfossil Pollen and Archaeobotanical Studies." *Vegetation History and Archaeobotany* 28 (2019): 671–702.

McPartland, John M., et al. "Evolutionary Origins of the Endocannabinoid System." *Gene* 370 (2006): 64–74.

McPartland, John M., and Kevin J. McKernan. "Contaminants of Concern in Cannabis: Microbes, Heavy Metals and Pesticides." In Cannabis sativa *L.—Botany and Biotechnology*, edited by Suman Chandra et al., 457–474. Cham, Switzerland: Springer, 2017.

McPartland, John M., and Ethan B. Russo. "Cannabis and Cannabis Extracts: Greater Than the Sum of Their Parts?" *Journal of Cannabis Therapeutics* 1, nos. 3–4 (2001): 103–132.

McPartland, John M., and Ethan B. Russo. "Non-Phytocannabinoid Constituents of Cannabis and Herbal Synergy." In *Handbook of Cannabis*, edited by Roger G. Pertwee, 280–295. Oxford: Oxford University Press, 2014.

McPartland, John M., and Zahra Sheikh. "A Review of *Cannabis sativa*-based Insecticides, Miticides, and Repellents." *Journal of Entomology and Zoology Studies* 6, no. 6 (2018): 1288–1299.

Mead, Alice P. "International Control of Cannabis." In *Handbook of Cannabis*, edited by Roger G. Pertwee, 44–64. Oxford: Oxford University Press, 2014.

Mechoulam, Raphael, and Shimon Ben-Shabat. "From *Gan-zi-gun-nu* to Anandamide and 2-Arachidonoylglycerol: The Ongoing Story of Cannabis." *Natural Products Reports* 16, no. 2 (1999): 131–143.

Mechoulam, Raphael, and Lumir Hanuš. "A Historical Overview of Chemical Research on Cannabinoids." *Chemistry and Physics of Lipids* 108, nos. 1–2 (2000): 1–13.

Mechoulam, Raphael, et al. "Early Phytocannabinoid Chemistry to Endocannabinoids and Beyond." *Nature Reviews Neuroscience* 15 (2014): 757–764.

Mechoulam, Raphael, et al. "Stereochemical Requirements for Cannabimimetic Activity." In *Structure-Activity Relationships of the Cannabinoids*, edited by Rao S. Rapaka and Alexandros Makriyannis, 15–30. Washington, DC: U.S. Government Printing Office, 1987.

Mechoulam, Raphael, and Yuval Shvo. "Hashish. I. The Structure of Cannabidiol." *Tetrahedron* 19, no. 12 (1963): 2073–2078.

Mehmedic, Zlatko, et al. "Potency Trends of Delta-9-THC and Other Cannabinoids in Confiscated Cannabis Preparations from 1993 to 2008." *Journal of Forensic Sciences* 55, no. 5 (2010): 1209–1217.

Melamede, Robert. "Harm Reduction—The Cannabis Paradox." *Harm Reduction Journal* 2 (2005): 17.

Melvin, Lawrence S., and M. Ross Johnson. "Structure-Activity Relationships of Tricyclic and Nonclassical Bicyclic Cannabinoids." *NIDA Research Monograph* 79 (1987): 31–47.

Merlin, Mark D. "Ancient Use of *Ephedra* in Eurasia and the Western Hemisphere." In *Ancient Psychoactive Substances*, edited by Scott M. Fitzpatrick, 71–111. Gainesville: University Press of Florida, 2018.

Merlin, Mark D. "Archaeological Evidence for the Tradition of Psychoactive Plant Use in the Old World." *Economic Botany* 57, no. 3 (2003): 295–323.

Merlin, Mark D., and Robert C. Clarke. "Cannabis in Ancient Central Eurasian Burials." In *Ancient Psychoactive Substances*, edited by Scott M. Fitzpatrick, 20–42. Gainesville: University Press of Florida, 2018.

Merrill, Frederick T. *Marihuana: The New Dangerous Drug*. Washington, DC: Foreign Policy Association, 1938.

Merrill, Thomas F. *Allen Ginsberg*. Boston: Twayne Publishers, 1988.

Meyer, Heidi C., et al. "The Role of the Endocannabinoid System and Genetic Variation in Adolescent Brain Development." *Neuropsychopharmacology Reviews* 43 (2018): 21–33.

Meyer, Jerrold S., and Linda F. Quenzer. *Psychopharmacology: Drugs, the Brain, and Behaviour*. Oxford: Oxford University Press, 2018.

Mezzrow, Milton "Mezz," and Bernard Wolfe. *Really the Blues*. New York: Citadel Press, 1990. Originally published in 1946.

Michaux, Henri. *Connaissance par les gouffres*. Paris: Gallimard, 1961.

Michaux, Henri. *Œuvres complètes*. Vol. 2. Paris: Gallimard, 2001.

Michaux, Henri. *Œuvres complètes*. Vol. 3. Paris: Gallimard, 2004.

Mikos, Robert A., and Cindy D. Kam. "Has the 'M' Word Been Framed? Marijuana, Cannabis, and Public Opinion." *PLoS One* 14, no. 10 (2019): e0224289.

Mikuriya, Tod H. *Marijuana: Medical Papers 1839–1972*. Oakland, CA: Medi-Comp, 1973.

Miles, Barry. *The Beat Hotel: Ginsberg, Burroughs, and Corso in Paris, 1958–1963*. New York: Grove Press, 2000.

Miles, Barry. *Call Me Burroughs: A Life*. New York: Twelve, 2013.

Miller, Bryan Lee, et al. "Exploring Butane Hash Oil Use: A Research Note." *Journal of Psychoactive Drugs* 48, no. 1 (2016): 44–49.

Miller, Christine L., et al. "Marijuana and Suicide: Case-Control Studies, Population Data and Potential Neurochemical Mechanisms." In *Cannabis in Medicine: An Evidence-Based Approach,* edited by Kenneth Finn, 90–105. Chem, Switzerland: Springer, 2020.

Miller, Naseem S. "Driving Under the Influence of Marijuana: Explainer and Research Roundup." *The Mandarin*, August 13, 2024. https://themandarin.com.au/252371-driving-under-the-influence-of-marijuana-explainer-and-research-roundup/.

Mills, Evan. "The Carbon Footprint of Indoor *Cannabis* Production." *Energy Policy* 46 (2012): 58–67.

Mills, James H. *Cannabis Britannica: Empire, Trade, and Prohibition*. Oxford: Oxford University Press, 2003.

Mills, James H. *Cannabis Nation: Control and Consumption in Britain, 1928–2008*. Oxford: Oxford University Press, 2013.

Mills, James H. *Madness, Cannabis and Colonialism: The "Native-Only" Lunatic Asylums of British India, 1857–1900*. New York: Palgrave Macmillan, 2000.

Milman, Garry, et al. "Plasma Cannabinoid Concentrations During Dronabinol Pharmacotherapy for Cannabis Dependence." *Therapeutic Drug Monitoring* 36, no. 2 (2014): 218–224.

Miner, Grace L., and William L. Bauerle. "Seasonal Responses of Photosynthetic Parameters in Maize and Sunflower and Their Relationship with Leaf Functional Traits." *Plant, Cell & Environment* 42, no. 5 (2019): 1561–1574.

Ministère de la Santé. *Cultivation of Cannabis at Home and Reduction of Penalties for Small Quantities*. Luxembourg: Le Gouvernement du Grand-Duché de Luxembourg. https://cannabis-information.lu/wp-content/uploads/2023/07/USAGE_RECREATIF_EN.pdf.

Minorsky, Peter V. "The 'Plant Neurobiology' Revolution." *Plant Signaling & Behavior* 19, no. 1 (2024): 2345413.

Misra, Prathit. "Rastafari: The Indian Connection." *Jamaican Gleaner*, March 26, 2021.

Mizumoto, Nobuaki, and Thomas Bourguignon. "The Evolution of Body Size in Termites." *Proceedings of the Royal Society B* 288, no. 1963 (2021): 20211458.

Moher, Melissa, et al. "Light Intensity Can Be Used to Modify the Growth and Morphological Characteristics of Cannabis During the Vegetative Stage of Indoor Production." *Industrial Crops and Products* 183 (2022): 114909.

Mongan, Deirdre, and Jean Long. *Standard Drink Measures Throughout Europe; People's Understanding of Standard Drinks and Their Use in Drinking Guidelines, Alcohol Survey and Labelling*. Lisbon, Portugal: RARHA, 2015.

Moore, Brian L., and Michele A. Johnson. *"They Do as They Please": The Jamaican Struggle for Cultural Freedom After Morant Bay*. Kingston, Jamaica: University of the West Indies Press, 2011.

Moosmann, Bjoern, Nadine Roth and Volker [Auwärter]. "Hair Analysis for Delta9-Tetrahydrocannabinolic Acid A (THCA-A) and Delta9-Tetrahydrocannabinol (THC) After Handling Cannabis Plant Material." *Drug Testing and Analysis* 8, nos. 1–2 (2016): 128–132.

Moreau, Jacques-Joseph. *Du hachisch et de l'aliénation mentale*. Études Psychologiques. Paris: Librairie de Fortin, Masson et Cie, 1845.

Moreau, Jacques-Joseph. *Mémoire sur le traitement des hallucinations par le Datura stramonium*. Paris: J. Rouvier et E. Le Bouvier, 1841.

Moreau, Jacques-Joseph. "Recherches sur les aliénés, en Orient: Notes sur les établissements qui leur sont consacrés à Malte, au Caire, à Smyrne, à Constantinople." *Annales Médico-Psychologiques* (1843): 1–30. https://gallica.bnf.fr/ark:/12148/bpt6k5620432b/f2.item.texteImage.

Moreno-Sanz, Guillermo. "Can You Pass the Acid Test? Critical Review and Novel Therapeutic Perspectives of Delta-9-Tetrahydrocannabinolic Acid A." *Cannabis and Cannabinoid Research* 1, no. 1 (2016): 124–130.

Morgan, Bill, ed. *I Greet You at the Beginning of a Great Career: The Selected Correspondence of Lawrence Ferlinghetti and Allen Ginsberg, 1955–1997*. San Francisco: City Lights Books, 2015.

Morgan, Bill, and Nancy J. Peters. *Howl on Trial: The Battle for Free Expression*. San Francisco: City Light Books, 2006.

Morrot, Gil, et al. "The Colors of Odors." *Brain and Language* 79, no. 2 (2001): 309–320.

Morrow, Paul L., et al. "An Outbreak of Deaths Associated with AMB-FUBINACA in Auckland NZ." *EClinicalMedicine* 25 (2020): 100460.

Moskowitz, Herbert, and D. Florentino. *A Review of the Literature on the Effects of Low Doses of Alcohol on Driving-Related Skills*. Washington, DC: National Highway Traffic Safety Administration, 2000.

Mott, Luiz. "A Maconha na História do Brasil." In *Diamba Sarabamba*, edited by Anthony Henman and Osvaldo Pessoa Jr., 117–135. São Paulo: Ground, 1986.

Mountain Girl. *The Primo Plant: Growing Sinsemilla Marijuana*. Berkeley, CA: Leaves of Grass/Wingbow Press, 1977.

Musto, David F. "The 1937 Marihuana Tax Act." In *Marijuana: Medical Papers 1839–1972*, edited by Tod H. Mikuriya, 419–440. Oakland, CA: Medi-Comp, 1973.

Musto, David F. *The American Disease: Origins of Narcotic Control*. 3rd ed. Oxford: Oxford University Press, 1999.

Naim-Feil, Erez, et al. "The Cannabis Plant as a Complex System: Interrelationships Between Cannabinoid Compositions, Morphological, Physiological and Phenological Traits." *Plants* 12, no. 3 (2023): 493.

Narby, Jeremy, and Francis Huxley. *Shamans Through Time: 500 Years on the Path to Knowledge*. New York: Jeremy P. Tarcher/Penguin, 2001.

Narby, Jeremy, with Rafael Chanchari Pizuri. *Plant Teachers: Ayahuasca, Tobacco, and the Pursuit of Knowledge*. Novato, CA: New World Library, 2021.

National Commission on Marihuana and Drug Abuse. *Marihuana: A Signal of Misunderstanding: The Official Report of the National Commission on Marihuana and Drug Abuse*. New York: The New American Library, 1972.

National Institute of Standards and Technology. "NIST Researchers to Test New Approach for Detecting Cannabis in Breath." August 15, 2024. www.nist.gov/news-events/news/2024/08/nist-researchers-test-new-approach-detecting-cannabis-breath.

National Library of Medicine. "Bifenthrin." 2016. https://toxnet.nlm.nih.gov.

National Medical Convention. *The Pharmacopoeia of the United States of America*. Philadelphia: Lippincott, Grambo, & Co., 1851.

Nazaryan, Alexander. "Your Weed Habit May Be Messing with Your Sperm." *The New York Times*, April 8, 2025, D6.

Neocision. "Grow Light Intensity Affects Yield." September 27, 2023. https://neocisiongrowlights.com/led-grow-lights-light-intensity/.

Newmeyer, Matthew N., et al. "Free and Glucuronide Whole Blood Cannabinoids' Pharmacokinetics after Controlled Smoked, Vaporized, and Oral Cannabis Administration in Frequent and Occasional Cannabis Users: Identification of Recent Cannabis Intake." *Clinical Chemistry* 62, no. 12 (2016): 1579–92.

Niebuhr, Carsten. *Beschreibung von Arabien. Aus eigenen Beobachtungen und im Lande selbst gesammelten Nachrichten*. Copenhagen, Denmark: Nicolaus Möller, 1772. https://gdz.sub.uni-goettingen.de/id/PPN504575031.

Niebuhr, Carsten. *Description de l'Arabie, faite sur des observations propres et des avis recueillis dans les lieux mêmes*. 2 vols. Amsterdam: S. J. Baalde, 1774. https://wellcomecollection.org/works/z7spa2qu/items?canvas=7.

Niebuhr, Carsten. *Travels Through Arabia, and Other Countries in the East*. Vol. 2. Edinburgh: Morison and Son, 1792. https://archive.org/details/39020024848031-travels-through-arabia-_niebuhr-carsten_1792_2.

Nitzan, Keren, et al. "An Ultra-Low Dose of Delta9-Tetrahydrocannabinol Alleviates Alzheimer's

Disease-Related Cognitive Impairments and Modulates TrkB Receptor Expression in a 5XFAD Mouse Model." *International Journal of Molecular Sciences* 23, no. 16 (2022): 9449.

Nobles, Melissa, et al. "Ending Racism Is Key to Better Science: A Message from *Nature*'s Guest Editors." *Nature* 610 (2022): 419–420.

Nordstrom, Benjamin R., and Carl L. Hart. "Assessing Cognitive Functioning in Cannabis Users: Cannabis Use History an Important Consideration." *Neuropsychopharmacology* 31 (2006): 2798–2799.

Norman, Philip. *The Stones*. London: Elm Tree Books, 1984.

Nour, Samar Gamal. "'Claiming the Mad': Implications of the Introduction of the Mental Asylum in Colonial Egypt." PhD diss., American University in Cairo, 2014.

Nutt, David. *Drugs without the Hot Air: Making Sense of Legal and Illegal Drugs*. Cambridge: UIT Cambridge, 2020.

Olaniyan, Tejumola. *Arrest the Music! Fela and His Rebel Art and Politics*. Bloomington: Indiana University Press, 2004.

Oleson, Erik B., et al. "Cannabinoid Receptor Activation Shifts Temporally Engendered Patterns of Dopamine Release." *Neuropsychopharmacology* 39 (2014): 1441–1452.

Oleson, Erik B., and Joseph F. Cheer. "A Brain on Cannabinoids: The Role of Dopamine Release in Reward Seeking." *Cold Spring Harbor Perspectives in Medicine* 2, no. 8 (2012): 012229.

Oleson, Erik B., et al. "Cannabinoid Modulation of Dopamine Release During Motivation, Periodic Reinforcement, Exploratory Behavior, Habit Formation, and Attention." *Frontiers in Synaptic Neuroscience* 13 (2021): 660218.

Olson, Steve, with Dean R. Gerstein. *Alcohol in America: Taking Action to Prevent Abuse*. Washington, DC: National Academy Press, 1985.

O'Shaughnessy, William B. *On the Preparations of the Indian Hemp, or Gunjah (Cannabis Indica), Their Effects on the Animal System in Health, and Their Utility in the Treatment of Tetanus and Other Convulsive Disorders*. Calcutta: Bishop's College Press, 1839.

O'Sullivan, Saoirse. "Phytocannabinoids and the Cardiovascular System." In *Handbook of Cannabis*, edited by Roger G. Pertwee, 208–226. Oxford: Oxford University Press, 2014.

Oswald, Iain W. H., et al. "Identification of a New Family of Prenylated Volatile Sulfur Compounds in Cannabis Revealed by Comprehensive Two-Dimensional Gas Chromatography." *ACS Omega* 6, no. 47 (2021): 31667–31676.

Oswald, Iain W. H., et al. "Minor, Nonterpenoid Volatile Compounds Drive the Aroma Differences of Exotic *Cannabis*." *ACS Omega* 8, no. 42 (2023): 39203–39216.

Panlilio, Leigh V., et al. "Cannabinoid Abuse and Addiction: Clinical and Preclinical Findings." *Clinical Pharmacology & Therapeutics* 97, no. 6 (2015): 616–627.

Papaseit, Esther, et al. "Cannabinoids: From Pot to Lab." *International Journal of Medical Sciences* 15, no. 12 (2018): 1286–1295.

Pardal, Mafalda, et al. *Alternatives to Profit-Maximising Commercial Models of Cannabis Supply for Non-Medical Use*. Cambridge: RAND Corporation, 2023.

Pardal, Mafalda, and Elle Wadsworth. "Strictly Regulated Cannabis Retail Models with State Control Can Provide Lessons in How Jurisdictions Can Regulate THC." *Addiction* 118, no. 6 (2023): 1005–1007.

Park, Sang-Hyuck, et al. "Contrasting Roles of Cannabidiol as an Insecticide and Rescuing Agent for Ethanol-Induced Death in the Tobacco Hornworm *Manduca sexta*." *Scientific Reports* 9, no. 1 (2019): 10481.

Parker, Linda A. *Cannabinoids and the Brain.* Cambridge, MA: MIT Press, 2017.

Parker, Robert. *Parker's Wine Buyer's Guide: The Complete, Easy-to-Use Reference on Recent Vintages, Prices, and Ratings for More than 8,000 Wines from All Around the Major Wine Regions.* 7th ed. New York: Simon & Schuster, 2008.

Parliament of Uruguay. Ley no. 19.172. https://archivo.presidencia.gub.uy/sci/leyes/2013/12/cons_min_803.pdf.

Parloff, Roger. "How Pot Became Legal." *Fortune* 160, no. 6 (2009): 140–162.

Parr, Wendy V. "Demystifying Wine Tasting: Cognitive Psychology's Contribution." *Food Research International* 124 (2019): 230–233.

Pasman, Joëlle A., et al. "GWAS of Lifetime Cannabis Use Reveals New Risk Loci, Genetic Overlap with Psychiatric Traits, and a Causal Effect of Schizophrenia Liability." *Nature Neuroscience* 21 (2018): 1161–1170.

Pasternak, Gavril W., and Ying-Xian Pan. "Mu Opioids and Their Receptors: Evolution of a Concept." *Pharmacological Reviews* 65, no. 4 (2013): 1257–1317.

Pate, David W. "Chemical Ecology of *Cannabis.*" *Journal of the International Hemp Association* 1, no. 2 (1994): 29, 32–37.

Patel, Jason, and Raman Marwaha. *Cannabis Use Disorder.* Treasure Island, FL: StatPearls Publishing, 2024.

Paulus, Victoria, et al. "Cannabidiol in the Context of Substance Use Disorder Treatment: A Systematic Review." *Addictive Behaviors* 132 (2022): 107360.

Payne, Kelly S., et al. "Cannabis and Male Fertility: A Systematic Review." *Journal of Urology* 202, no. 4 (2019): 674–681.

Pearlson, Godfrey. *Weed Science: Cannabis Controversies and Challenges.* London: Academic Press, 2020.

Pearlson, Godfrey D., et al. "Cannabis and Driving." *Frontiers in Psychiatry* 12 (2021): 689444.

Pellati, Federica, et al. "*Cannabis sativa* L. and Nonpsychoactive Cannabinoids: Their Chemistry and Role Against Oxidative Stress, Inflammation, and Cancer." *BioMed Research International* 12 (2018): 1691428.

Pennypacker, Sarah D., et al. "Potency and Therapeutic THC and CBD Ratios: U.S. Cannabis Markets Overshoot." *Frontiers in Pharmacology* 13 (2022): 921493.

Perisetti, Abhilash, et al. "Cannabis Hyperemesis Syndrome: An Update on the Pathophysiology and Management." *Annals of Gastroenterology* 33, no. 6 (2020): 571–578.

Pertwee, Roger G. "Cannabinoid Pharmacology: The First 66 Years." *British Journal of Pharmacology* 147, suppl. 1 (2006): S163–S171.

Pertwee, Roger G. "Pharmacological, Physiological and Clinical Implications of the Discovery of Cannabinoid Receptors: An Overview." In *Cannabinoid Receptors*, edited by Roger G. Pertwee, 1–34. London: Academic Press, 1995.

Pertwee, Roger G. "Sites and Mechanisms of Action." In *Cannabis and Cannabinoids: Pharmacology, Toxicology, and Therapeutic Potential*, edited by Franjo Grotenhermen and Ethan Russo, 73–87. London: Routledge, 2002.

Pertwee, Roger G., et al. "International Union of Basic and Clinical Pharmacology. LXXIX. Cannabinoid Receptors and Their Ligands: Beyond CB1 and CB2." *Pharmacological Reviews* 62, no. 4 (2010): 588–631.

Pessoa de Castro, Yeda. "Towards a Comparative Approach of Bantuisms in Iberoamerica." In *AfricAmericas: Itineraries, Dialogues, and Sounds*, edited by Ineke Phaf-Rheinberger and Tiago de Oliveira Pinto, 79–90. Madrid, Spain: Iberoamericana, 2008.

Peterman, David Randall. *Marijuana Use and Highway Safety*. Washington, DC: Congressional Research Service, 2019.

Peynaud, Emile. *The Taste of Wine: The Art and Science of Wine Appreciation*. London: Macdonald & Co, 1987.

Pic, Muriel. "Par la voie des nerfs. Henri Michaux et les psychotropes." *Mouvements* 2, no. 86 (2016): 142–150.

Pinch, William R. *Peasants and Monks in British India*. Berkeley: University of California Press, 1996.

Piomelli, Daniele, and Ethan B. Russo. "The *Cannabis sativa* Versus *Cannabis indica* Debate: An Interview with Ethan Russo, MD." *Cannabis and Cannabinoid Research* 1, no. 1 (2016): 44–46.

Piper, Alan. "The Mysterious Origins of the Word 'Marijuana.'" *Sino-Platonic Papers* 153 (2005): 1–17.

Pisanti, Simona, and Maurizio Bifulco. "Modern History of Medical *Cannabis*: From Widespread Use to Prohibitionism and Back." *Trends in Pharmacological Sciences* 38, no. 3 (2017): 195–198.

Plumb, Jeremy, et al. "The Nose Knows: Aroma, but Not THC Mediates the Subjective Effects of Smoked and Vaporized Cannabis Flower." *Psychoactives* 1, no. 2 (2020): 70–86.

Pocuca, Nina, et al. "The Effects of Cannabis Use on Cognitive Function in Healthy Aging: A Systematic Scoping Review." *Archives of Clinical Neuropsychology* 36, no. 5 (2021): 673–685.

Pollan, Michael. *The Botany of Desire: A Plant's-Eye View of the World*. New York: Random House, 2001.

Pollan, Michael. *How to Change Your Mind*. New York: Penguin Press, 2018.

Pollard, John K., et al. *Review of Technology to Prevent Alcohol-and Drug-Impaired Crashes: Update*. (Report No. DOT HS 813 542). Washington, DC: National Highway Traffic Safety Administration, 2024.

Pollan, Michael. *Second Nature: A Gardner's Education*. New York: Dell, 1991.

Pollastro, Federica, et al. "Cannabis Phenolics and Their Bioactivities." *Current Medical Chemistry* 25, no. 10 (2018): 1160–1185.

Poorter, Hendrik, et al. "A Meta-Analysis of Plant Responses to Light Intensity for 70 Traits Ranging from Molecules to Whole Plant Performance." *New Phytologist* 223, no. 3 (2019): 1073–1105.

Pope, Harrison G., and Deborah Yurgelun-Todd. "The Residual Cognitive Effects of Heavy Marijuana Use in College Students." *JAMA* 275, no. 7 (1996): 521–527.

Potter, David. "The Propagation, Characterisation and Optimisation of *Cannabis sativa* L. as a Phytopharmaceutical." PhD diss., King's College London, 2009.

Potter, David J. "A Review of the Cultivation and Processing of Cannabis (*Cannabis sativa* L.) for Production of Prescription Medicines in the UK." *Drug Testing and Analysis* 6, nos. 1–2 (2014): 31–38.

Potter, David John. "Cannabis Horticulture." In *Handbook of Cannabis*, edited by Roger G. Pertwee, 65–88. Oxford: Oxford University Press, 2014.

Potter, David J. et al. "Potency of Delta9-THC and Other Cannabinoids in Cannabis in England in 2005: Implications for Psychoactivity and Pharmacology." *Journal of Forensic Sciences* 53, no. 1 (2008): 90–94.

Potter, David J., and Paul Duncombe. "The Effect of Electrical Lighting Power and Irradiance on Indoor-Grown Cannabis Potency and Yield." *Journal of Forensic Sciences* 57, no. 3 (2012): 618–622.

Potter, Gary, et al. "Domestic Cannabis Cultivation Around the World: The GCCRC's ICCQ 2.0." 2021. Paper presented at the conference of the International Society for the Study of Drug Policy, Lisbon, Portugal, November 22, 2022.

Poulsen, H. A., and G. J. Sutherland. "The Potency of Cannabis in New Zealand from 1976 to 1996." *Science & Justice* 40, no. 3 (2000): 171–176.

Pow, Joni Lee, et al. "Cannabis Use and Psychotic Disorders in Diverse Settings in the Global South: Findings from INTREPID II." *Psychological Medicine* 53, no. 15 (2023): 7062–7069.

Power, Robert A., et al. "Genetic Predisposition to Schizophrenia Associated with Increased Use of Cannabis." *Molecular Psychiatry* 19, no. 11 (2014): 1201–1204.

Preuss, Ulrich W., et al. "Cannabis Use and Car Crashes: A Review." *Frontiers in Psychiatry* 12 (2021): 643315.

Prieto-Arenas, Laura, et al. "Gender Differences in Dual Diagnoses Associated with Cannabis Use: A Review." *Brain Sciences* 12, no. 3 (2022): 388.

Prince, Mark A., et al. "Quantifying Cannabis: A Field Study of Marijuana Quantity Estimation." *Psychology of Addictive Behaviors* 32, no. 4 (2018): 426–433.

Proctor, Robert N., and Londa Schiebinger, eds. *Agnotology: The Making and Unmaking of Ignorance*. Stanford, CA: Stanford University Press, 2008.

Pulver, Benedikt, et al. "EMCDDA Framework and Practical Guidance for Naming Synthetic Cannabinoids." *Drug Testing and Analysis* 15, no. 3 (2022): 255–276.

Punja, Zamir K. "Emerging Diseases of *Cannabis sativa* and Sustainable Management." *Pest Management Science* 77, no. 9 (2021): 3857–3870.

Purdy, Cassandra. "Master of Hash: Frenchy Cannoli Has a Plan to Change the World of Hashmaking." *High Times*, September 2019, 44–52.

Quammen, David. "Planet of Weeds: Tallying the Losses of Earth's Animals and Plants." *Harper's Magazine*, October 1998, 57–70.

Raber, Jeffrey C., et al. "Understanding Dabs: Contamination Concerns of Cannabis Concentrates and Cannabinoid Transfer During the Act of Dabbing." *The Journal of Toxicological Sciences* 40, no. 6 (2015): 797–803.

Raguso, Robert A., and André Kessler. "Speaking in Chemical Tongues: Decoding the Language of Plant Volatiles." In *The Language of Plants: Science, Philosophy, Literature*, edited by Monica Gagliano et al., 27–61. Minneapolis: University of Minnesota Press, 2017.

Raimondi, Giorgia, et al. "Phytomanagement of Chromium-Contaminated Soils Using *Cannabis sativa* (L.)." *Agronomy* 10, no. 9 (2020): 1223.

Rami, Esha, et al. "An Overview of Plant Secondary Metabolites, Their Biochemistry and Generic Applications." *The Journal of Phytopharmacology* 10, no. 5 (2021): 421–428.

Randall, Robert C., ed. *Cancer Treatment and Marijuana Therapy*. Washington, DC: Galen Press, 1990.

Rathge, Adam R. "Cannabis Cures: American Medicine, Mexican Marijuana, and the Origins of the War on Weed, 1840–1937." PhD diss. Boston College, 2017.

Rathge, Adam R. "Mapping the Muggleheads: New Orleans and the Marijuana Menace, 1920–1930." *Southern Spaces*, October 23, 2018. https://southernspaces.org/2018/mapping-muggleheads-new-orleans-and-marijuana-menace-1920-1930/.

Rattansi, Ali. *Racism: A Very Short Introduction*. 2nd ed. Oxford: Oxford University Press, 2020.

Ray, Cecily S. "The Hookah—the Indian Waterpipe." *Current Science* 96, no. 10 (2009): 1319–1323.

Raz, Noa, et al. "Selected Cannabis Terpenes Synergize with THC to Produce Increased CB1 Receptor Activation." *Biochemical Pharmacology* 212 (2023): 115548.

Reber, Arthur S., et al. *The Penguin Dictionary of Psychology*. 4th ed. London: Penguin, 2017.

Ren, Guangpeng, et al. "Large-Scale Whole-Genome Resequencing Unravels the Domestication History of *Cannabis sativa*." *Science Advances* 7, no. 29 (2021): eabg2286.

Ren, Meng, et al. "The Origins of Cannabis Smoking: Chemical Residue Evidence from the First Millennium BCE in the Pamirs." *Science Advances* 5, no. 6 (2019): eaaw1391.

Rendon, Jim. *Super-Charged: How Outlaws, Hippies, and Scientists Reinvented Marijuana*. Portland, OR: Timber Press, 2012.

Resin, Harry. "Growing for Terpenes." *High Times*, May 2023, 36–44.

Rettew, David C., et al. "Blunted: The Effects of Cannabis on Cognition and Motivation." In *Cannabis in Medicine*, edited by Kenneth Finn, 66–74. Cham, Switzerland: Springer, 2020.

Riboulet-Zemouli, Kenzi. "'Cannabis' Ontologies I: Conceptual Issues with *Cannabis* and Cannabinoids Terminology." *Drug Science, Policy and Law* 6 (2020): 1–37.

Riboulet-Zemouli, Kenzi, et al. "Briefing on the International Scientific Assessment of Cannabis: Processes, Stakeholders, and History." *The Crimson Digest* 1 (2018): 1–145.

Richards, Keith, with James Fox. *Life*. New York: Little, Brown and Company, 2010.

Rinaldi-Carmona, Murielle, et al. "SR141716A, a Potent and Selective Antagonist of the Brain Cannabinoid Receptor." *Federation of European Biochemical Societies* 350, nos. 2–3 (1994): 240–244.

Ritchay, Megan M., et al. "Resting State Functional Connectivity in the Default Mode Network: Relationships Between Cannabis Use, Gender, and Cognition in Adolescents and Young Adults." *NeuroImage: Clinical* 30 (2021): 102664.

Roberts, Chris. "Emerald Cup: The Best Marijuana Doesn't Always Have the Most THC." *High Times*, December 13, 2016. www.hightimes/grow/emerald-cup-the-best-marijuana-doesn't-always-have-the-most-thc/.

Roberts, Chris. "Science Reveals the Cannabis Industry's Greatest Lie: You're Buying Weed Wrong (and So Is Everyone Else)." *Forbes*, June 16, 2020. https://www.forbes.com/sites/chrisroberts/2020/06/16/science-reveals-the-cannabis-industrys-greatest-lie-youre-buying-weed-wrong-and-so-is-everyone-else/.

Robinson, B. B. "Hemp." *Farmers' Bulletin*, no. 1935, United States Department of Agriculture (1943): 1–16.

Rodriguez-Morrison, Victoria, et al. "Cannabis Yield, Potency, and Leaf Photosynthesis Respond Differently to Increasing Light Levels in an Indoor Environment." *Frontiers in Plant Science* 12 (2021): 646020.

Rosenberg, Eli, and Nate Schweber. "33 Suspected of Overdosing on Synthetic Marijuana in Brooklyn." *The New York Times*, July 13, 2016, A17.

Rosenthal, Ed. *Closet Cultivator*. Oakland, CA: Quick American Archives, 1999.

Rosenthal, Ed. "Growing Dutch." *High Times*, June 1985, 70–71.

Rosenthal, Ed. "The Miniature Forest." *High Times*, July 1997, 82–85.

Rosenthal, Ed, with David Downs. *Beyond Buds: Marijuana Extracts—Hash, Vaping, Dabbing, Edibles & Medicines*. Piedmont, CA: Quick American, 2014.

Rosenthal, Franz. *The Herb: Hashish versus Medieval Muslim Society*. Leiden, The Netherlands: E. J. Brill, 1971.

Rosenthal, Franz. *Man versus Society in Medieval Islam*. Leiden, The Netherlands: Brill, 2015.

Rubin, Vera, and Lambros Comitas. *Ganja in Jamaica: A Medical Anthropological Study of Chronic Marihuana Use*. The Hague, The Netherlands: Mouton & Co, 1975.

Runco, Mark A., and Ivonne Chand. "Cognition and Creativity." *Educational Psychology Review* 7, no. 3 (1995): 243–267.

Russo, Ethan B. "Cannabis in India: Ancient Lore and Modern Medicine." In *Cannabinoids as Therapeutics*, edited by Raphael Mechoulam, 1–21. Basel, Switzerland: Birkhäuser Verlag, 2005.

Russo, Ethan B. "The Case for the Entourage Effect and Conventional Breeding of Clinical Cannabis: No 'Strain,' No Gain." *Frontiers in Plant Science* 9 (2019): 1969.

Russo, Ethan B. "Current Therapeutic Cannabis Controversies and Clinical Trial Design Issues." *Frontiers in Pharmacology* 7 (2016): 309.

Russo, Ethan B. "History of Cannabis and Its Preparations in Saga, Science, and Sobriquet." *Chemistry & Biodiversity* 4, no. 8 (2007): 1614–1648.

Russo, Ethan B. "Letter: Cannabinoid Medicines and the Need for the Scientific Method." *Cannabinoids* 2, no. 2 (2007): 16–19.

Russo, Ethan B. "The Pharmacological History of Cannabis." In *Handbook of Cannabis*, edited by Roger G. Pertwee, 23–43. Oxford: Oxford University Press, 2014.

Russo, Ethan B. "The Solution to the Medicinal Cannabis Problem." In *Ethical Issues in Chronic Pain Management*, edited by Michael E. Schatman, 165–194. New York: Informa Healthcare, 2007.

Russo, Ethan B. "Synthetic and Natural Cannabinoids: The Cardiovascular Risk." *The British Journal of Cardiology* 22, no. 1 (2015): 7–9.

Russo, Ethan B. "Taming THC: Potential Cannabis Synergy and Phytocannabinoid-Terpenoid Entourage Effects." *British Journal of Pharmacology* 163, no. 7 (2011): 1344–1364.

Russo, Ethan B., and Geoffrey W. Guy. "A Tale of Two Cannabinoids: The Therapeutic Rationale for Combining Tetrahydrocannabinol and Cannabidiol." *Medical Hypotheses* 66, no. 2 (2006): 234–246.

Russo, Ethan B., et al. "Phytochemical and Genetic Analyses of Ancient Cannabis from Central Asia." *Journal of Experimental Botany* 59, no. 15 (2008): 4171–4182.

Russo, Ethan B., and Jahan Marcu. "Cannabis Pharmacology: The Usual Suspects and a Few Promising Leads." *Advances in Pharmacology* 80 (2017): 67–134.

Russo, Ethan B., and John M. McPartland. "Cannabis Is More Than Simply Delta9-Tetrahydrocannabinol." *Psychopharmacology* 165 (2003): 431–432.

Russo, Ethan B., et al. "Cannabinoid Hyperemesis Syndrome Survey and Genomic Investigation." *Cannabis and Cannabinoid Research* 7, no. 3 (2022): 336–344.

Ryu, Byeong Ryeol, et al. "Conversion Characteristics of Some Major Cannabinoids from Hemp (*Cannabis sativa* L.) Raw Materials by New Rapid Simultaneous Analysis Method." *Molecules* 26, no. 14 (2021): 4113.

Ryz, Natasha R., et al. "Cannabis Roots: A Traditional Therapy with Future Potential for Treating Inflammation and Pain." *Cannabis and Cannabinoid Research* 2, no. 1 (2017): 210–216.

Sabaghi, Dario. "Thailand Has Become the First Asian Country to Decriminalize Recreational Cannabis." *Forbes*, January 27, 2022. https://www.forbes.com/sites/dariosabaghi/2022/01/27/thailand-has-become-the-first-asian-country-to-decriminalize-recreational-cannabis/?sh=b48989d4b210.

Salonen, Erkki. "Über einige Lehnwörter aus dem Nahen Osten im Griechischen und Lateinischen." *ARCTOS, Acta Philologica Fennica* 8 (1974): 139–144.

Sam, Amir H., et al. "Rimonabant: From RIO to Ban." *Journal of Obesity* (2011): 432607.

Samidjo, Gatot Supangkat. "Growth Pattern of Sunflower on Some Light Intensity in The Coastal Land." *IOP Conference Series: Earth and Environmental Science* 752 (2021): 012019.

Santiago, Marina, et al. "Absence of Entourage: Terpenoids Commonly Found in *Cannabis sativa* Do Not Modulate the Functional Activity of Delta-9-THC at Human CB1 and CB2 Receptors." *Cannabis and Cannabinoid Research* 4, no. 3 (2019): 165–176.

Santora, Marc. "Drug 85 Times as Potent as Marijuana Caused a 'Zombielike' State in Brooklyn." *The New York Times*, December 14, 2016, A24.

Sarne, Yosef. "THC For Age-Related Cognitive Decline?" *Aging* 10, no. 12 (2018): 3628–3629.

Sarne, Yosef, et al. "Reversal of Age-Related Cognitive Impairments in Mice by an Extremely Low Dose of Tetrahydrocannabinol." *Neurobiology of Aging* 61 (2018): 177–186.

Saunders, Richard, and Jacob Cannon. "Dab Myths Debunked." *High Times*, August 2021, 79–82.

Schafer, Gráinne, et al. "Investigating the Interaction Between Schizotypy, Divergent Thinking and Cannabis Use." *Consciousness and Cognition* 21, no. 1 (2012): 292–298.

Schievenini, José Domingo. "A Historical Approach to the Criminalization of Marijuana Use in Mexico." In *Cannabis: Global Histories*, edited by Lucas Richert and James H. Mills, 131–156. Cambridge, MA: MIT Press, 2021.

Schievenini, José Domingo, and Carlos A. Pérez Ricart. "Pasado y presente de los usos medicinales del cannabis en México." *Redes* 26, no. 50 (2020): 115–145.

Schofield, Michael. *The Strange Case of Pot*. London: Penguin, 1971.

Scholten, W. K. "Guidelines for Cultivating Cannabis for Medicinal Purposes." *Journal of Cannabis Therapeutics* 3, no. 2 (2003): 51–61.

Schonbeck, Mark. "An Ecological Understanding of Weeds." *eOrganic*, January 20, 2009. https://eorganic-org/node/2314.

Schuler, Monica. *"Alas, Alas, Kongo": A Social History of Indentured African Immigration into Jamaica, 1841–1865*. Baltimore: Johns Hopkins University Press, 1980.

Schultes, Richard Evans, and Albert Hofmann. *Plants of the Gods: Origins of Hallucinogenic Use*. New York: McGraw-Hill, 1979.

Schumacher, Michael. *Dharma Lion: A Biography of Allen Ginsberg*. New York: St. Martin's Press, 1992.

Schwabe, Anna L., et al. "Uncomfortably High: Testing Reveals Inflated THC Potency on Retail *Cannabis* Labels." *PLoS One* 18, no. 4 (2023): e0282396.

Schwope, David M., et al. "Identification of Recent Cannabis Use: Whole-Blood and Plasma Free and Glucuronidated Cannabinoid Pharmacokinetics Following Controlled Smoked Cannabis Administration." *Clinical Chemistry* 57, no. 10 (2011): 1406–1414.

Science Media Center Germany. "Neuer Grenzwert für THC im Strassenverkehr," June 6, 2024. www.sciencemediacenter.de/angebote/24086.

Scientific American. "Hasheesh and Its Smokers and Eaters." *Scientific American* 14, no. 7 (October 1858): 49.

Scott, J. Cobb, et al. "Association of Cannabis with Cognitive Functioning in Adolescents and Young Adults: A Systematic Review and Meta-analysis." *JAMA Psychiatry* 75, no. 6 (2018): 585–595.

Seely, Kathryn A., et al. "Spice Drugs Are More Than Harmless Herbal Blends: A Review of the Pharmacology and Toxicology of Synthetic Cannabinoids." *Progress in Neuro-Psychopharmacology and Biological Psychiatry* 39, no. 2 (2012): 234–243.

Segal, Mark. "Cannabinoids and Analgesia." In *Cannabinoids as Therapeutic Agents*, edited by Raphael Mechoulam, 105–120. London: CRC Press, 1986.

Seltenrich, Nate. "Into the Weeds: Regulating Pesticides in Cannabis." *Environmental Health Perspectives* 127, no. 4 (2019): 1–8.

Sewell, R. Andrew, et al. "The Effect of Cannabis Compared with Alcohol on Driving." *American Journal on Addictions* 18, no. 3 (2009): 185–193.

Shapira, Guy, et al. "Hippocampal Differential Expression Underlying the Neuroprotective Effect of Delta-9-Tetrahydrocannabinol Microdose on Old Mice." *Frontiers in Neuroscience* 17 (2023): 1182932.

Shapiro, Harry. *Waiting for the Man: The Story of Drugs and Popular Music*. Rev. ed. London: Helter Skelter, 1999.

Shea, David. *The Dabistan, or School of Manners*. Vol. II. Translated by Anthony Troyer. Paris: The Oriental Translation Fund of Great Britain and Ireland, 1843.

Shepherd, Gordon M. "Neuroenology: How the Brain Creates the Taste of Wine." *Flavour* 4 (2015): 19.

Shepherd, Gordon M. *Neuroenology: How the Brain Creates the Taste of Wine*. New York: Columbia University Press, 2017.

Shevyrin, Vadim, et al. "On a New Cannabinoid Classification System: A Sight on the Illegal Market of Novel Psychoactive Substances." *Cannabis and Cannabinoid Research* 1, no. 1 (2016): 186–194.

Shofty, Ben, et al. "The Default Network is Causally Linked to Creative Thinking." *Molecular Psychiatry* 27 (2022): 1848–1854.

Sholler, Dennis J., et al. "Sex Differences in the Acute Effects of Oral and Vaporized Cannabis Among Healthy Adults." *Addiction Biology* 26, no. 4 (2021): e12968.

Sigman, Zoe. "Decarboxylating Cannabis: Why Does Eating a Raw Cannabis Bud Not Get You High But Eating a THC Edible Does?" *Project CBD*, February 22, 2020. www.projectcbd.org/guidane/decarboxylating-cannabis.

Silver, Robert J. "The Endocannabinoid System of Animals." *Animals* 9, no. 9 (2019): 686.

Singer, Merrill, and Greg Mirhej. "High Notes: The Role of Drugs in the Making of Jazz." *Journal of Ethnicity in Substance Abuse* 5, no. 4 (2006): 1–38.

Singh, Nilay Vishal, and Vinay Kumar Singh. "Indigenous Plant *Cannabis sativa*: A Comprehensive Ethnobotanical and Pharmacological Review." *European Journal of Biological Research* 13, no. 2 (2023): 114–128.

Small, Ernest. *Cannabis: A Complete Guide*. London: CRC Press, 2017.

Small, Ernest. "Classification of *Cannabis sativa* L. in Relation to Agricultural, Biotechnological, Medical and Recreational Utilization." In Cannabis sativa *L.—Botany and Biotechnology*, edited by Suman Chandra et al., 1–62. Cham, Switzerland: Springer, 2017.

Small, Ernest, and H. D. Beckstead. "Cannabinoid Phenotypes in *Cannabis Sativa*." *Nature* 245 (1973): 147–148.

Small, Ernest, and Steve G. U. Naraine. "Expansion of Female Sex Organs in Response to Prolonged Virginity in *Cannabis sativa* (Marijuana)." *Genetic Resources and Crop Evolution* 63 (2016): 339–348.

Smallwood, Jonathan, et al. "Cooperation Between the Default Mode Network and the Frontal-Parietal Network in the Production of an Internal Train of Thought." *Brain Research* 1428 (2012): 60–70.

Smart, Rosanna, et al. "Variation in Cannabis Potency and Prices in a Newly Legal Market: Evidence From 30 Million Cannabis Sales in Washington State." *Addiction* 112, no. 12 (2017): 2167–2177.

Smith, Peter Andrey. "How Iron-Clad Is the Iron Law of Prohibition?" *Undark.org*, October 16, 2023. https://undark.org/2023/10/16/iron-law-drugs/.

Société des Nations. *Actes de la deuxième conférence de l'opium.* Vol. I, *Séances plénières, compte rendu des débats.* Geneva: Société des Nations, 1925.

Sohn, Emily. "Balancing Act." *Nature* 572, no. 7771 (2019): S16–S18.

Solomon, David. *The Marijuana Papers.* New York: New American Library, 1968.

Solowij, Nadia. *Cannabis and Cognitive Functioning.* Cambridge: Cambridge University Press, 1998.

Solowij, Nadia. "Do Cognitive Impairments Recover Following Cessation of Cannabis Use?" *Life Sciences* 56, nos. 23–24 (1995): 2119–2126.

Solowij, Nadia, et al. "A Randomised Controlled Trial of Vaporised Delta-9-Tetrahydrocannabinol and Cannabidiol Alone and in Combination in Frequent and Infrequent Cannabis Users: Acute Intoxication Effects." *European Archives of Psychiatry and Clinical Neuroscience* 269, no. 1 (2019): 17–35.

Solvyns, Frans Baltazar. *Les Hindoûs.* Vol. 3. Paris: Chez l'Auteur, 1811.

Sommano, Sarana Rose, et al. "The Cannabis Terpenes." *Molecules* 25, no. 24 (2020): 5792.

Sonnini, Charles-Nicolas-Sigisbert. *Voyage dans la haute et basse Égypte: fait par ordre de l'ancien gouvernement, et contenant des observations de tous genres.* Vol. 3. Paris: F. Buisson, 1798.

Spindle, Tory R., et al. "Acute Effects of Smoked and Vaporized Cannabis in Healthy Adults Who Infrequently Use Cannabis: A Crossover Trial." *JAMA Network Open* 1, no. 7 (2018): e184841.

Spindle, Tory R., et al. "Vaporized D-Limonene Selectively Mitigates the Acute Anxiogenic Effects of Delta9-Tetrahydrocannabinol in Healthy Adults Who Intermittently Use Cannabis." *Drug and Alcohol Dependence* 257 (2024): 111267.

Stack, George M., et al. "Cannabinoids Function in Defense Against Chewing Herbivores in *Cannabis sativa* L." *Horticulture Research* 10, no. 11 (2023): uhad207.

Stasilowicz-Krzemień, Anna, et al. "Determining Antioxidant Activity of Cannabis Leaves Extracts from Different Varieties—Unveiling Nature's Treasure Trove." *Antioxidants* 12, no. 7 (2023): 1390.

Stein, Sara B. *My Weeds: A Gardener's Botany.* New York: Harper and Row, 1990.

Stempfer, Miriam, et al. "Analysis of Cannabis Seizures by Non-Targeted Liquid Chromatography-Tandem Mass Spectrometry." *Journal of Pharmaceutical and Biomedical Analysis* 205 (2021): 114313.

Stevens, Murphy. *How to Grow the Finest Marijuana Indoors.* Seattle: Sun Magic Publishing Co., 1979.

Stewart, Dominique. "Early Encounters in Colonial Jamaica: Hindu and Rastafari Divine Metaphysics." *Journal of Interreligious Studies* 32, no. 1 (2021): 40–62.

Stoa, Ryan. *Craft Weed: Family Farming and the Future of the Marijuana Industry.* Cambridge, MA: MIT Press, 2024.

Stockberger, W. W. "Drug Plants Under Cultivation." *Farmers' Bulletin*, no. 663, United States Department of Agriculture (1915): 1–39.

Stockberger, W. W. "Production of Drug-Plant Crops in the United States." *United States Department of Agriculture Yearbook* (1917): 169–176.

Storck, Wilhelm, et al. "Cardiovascular Risk Associated with the Use of Cannabis and Cannabinoids: A Systematic Review and Meta-Analysis." *Heart*, June 17 (2025): heartjnl-2024–325429.

Struik, Dicky, et al. "The Modulating Role of Sex and Anabolic-Androgenic Steroid Hormones in Cannabinoid Sensitivity." *Frontiers in Behavioral Neuroscience* 12 (2018): 249.

Sullivan, Nicholas, et al. "Determination of Pesticide Residues in Cannabis Smoke." *Journal of Toxicology* (2013): 378168.

Summers, Hailey M., et al. "The Greenhouse Gas Emissions of Indoor Cannabis Production in the United States." *Nature Sustainability* 4 (2021): 644–650.

Swift, Wendy, et al. "Analysis of Cannabis Seizures in NSW, Australia: Cannabis Potency and Cannabinoid Profile." *PLoS One* 8, no. 7 (2013): e70052.

Szejko, Natalia, et al. "Medicinal Use of Different *Cannabis* Strains: Results from a Large Prospective Survey in Germany." *Pharmacopsychiatry* 57, no. 3 (2024): 133–140.

Taiz, Lincoln, et al. "Plants Neither Possess nor Require Consciousness." *Trends in Plant Science*, 24, no. 8 (2019): 677–687.

Taleb, Nassim Nicholas. *The Black Swan: The Impact of the Highly Improbable*. New York: Random House, 2010.

Tanney, Cailun A. S., et al. "Cannabis Glandular Trichomes: A Cellular Metabolite Factory." *Frontiers in Plant Science* 12 (2021): 721986.

Tart, Charles T. *On Being Stoned: A Psychological Study of Marijuana Intoxication*. Palo Alto, CA: Science and Behavior Books, 1971.

Taylor, Bayard. "The Vision of Hasheesh." *Putnam's Monthly Magazine of American Literature, Science, and Art* 3, no. 16 (April 1854): 402–408.

Taylor, Juliet M., et al. "Neuroinflammation and Oxidative Stress: Co-Conspirators in the Pathology of Parkinson's Disease." *Neurochemistry International* 62, no. 5 (2013): 803–819.

Thomas, Brian F., et al. "Cannabinoid Designer Drugs: Effects and Forensics." In *Handbook of Cannabis*, edited by Roger G. Pertwee, 710–729. Oxford: Oxford University Press, 2014.

Thomas, Sarah A., et al. "Cannabis Vaping Is Associated with Past 30-Day Suicide Attempts and Suicidal Ideation Among Adolescents in a Psychiatric Inpatient Setting." *JAACAP Open* 2, no. 4 (2024): 263–273.

Thomsen, Kristine Rømer, et al. "Cannabinoids for the Treatment of Cannabis Use Disorder: New Avenues for Reaching and Helping Youth?" *Neuroscience and Biobehavioral Reviews* 132 (2022): 169–180.

Thomsen, Lucy Rose. "Factors Influencing the Toxicity of AMB-FUBINACA, a Synthetic Cannabinoid." PhD diss., University of Otago, 2024. https://hdl.handle.net/10523/16665.

Thornton, Mark. "Alcohol Prohibition Was a Failure." *Cato Institute Policy Analysis* 157, July 17, 1991. https://www.cato.org/sites/cato.org/files/pubs/pdf/pa157.pdf.

Thornton, Mark. "The Potency of Illegal Drugs." *Journal of Drug Issues* 28, no. 3 (1998): 725–740.

Tomko, Andrea M., et al. "Anti-Cancer Potential of Cannabinoids, Terpenes, and Flavonoids Present in Cannabis." *Cancers* 12, no. 7 (2020): 1985.

Torkamaneh, Davoud, and Andrew M. P. Jones. "Cannabis, the Multibillion Dollar Plant That No Genebank Wanted." *Genome* 65, no. 1 (2021): 1–5.

Trac, Judy, et al. "Cannabidiol Oxidation Product HU-331 Is a Potential Anticancer Cannabinoid-Quinone: A Narrative Review." *Journal of Cannabis Research* 3, no. 1 (2021): 11.

Trecki, Jordan, et al. "Synthetic Cannabinoid-Related Illnesses and Death." *New England Journal of Medicine* 373, no. 2 (2015): 103–107.

Trewavas, Anthony. "The Foundations of Plant Intelligence." *Interface Focus* 7, no. 3 (2017): 20160098.

Twohey, Megan, et al. "As America's Marijuana Use Grows, So Do the Harms." *The New York Times*, October 5, 2024, A1.

Uchiyama, Nahoko, et al. "New Cannabimimetic Indazole Derivatives, *N*-(1-amino-3-methyl-1-oxobutan-2-yl)-1-pentyl-1*H*-indazole-3-carboxamide (AB-PINACA) and *N*-(1-amino-3-methyl-1-oxobutan-2-yl)-1-(4-fluorobenzyl)-1*H*-indazole-3-carboxamide (AB-FUBINACA) Identified as Designer Drugs in Illegal Products." *Forensic Toxicology* 31, no. 1 (2013): 93–100.

United Nations. *Commentary on the Single Convention on Narcotic Drugs, 1961*. United Nations: New York, 1973.

United Nations Office on Drugs and Crime (UNODC). *World Drug Report 2006*, Vol. 1, *Analysis*. Vienna: United Nations Office on Drugs and Crime, 2006.

UNODC (United Nations Office on Drugs and Crime). *World Drug Report 2022. Booklet 3, Drug Market Trends: Cannabis and Opioids*. Vienna: United Nations Office on Drugs and Crime, 2022.

United States Congress. *House Ways and Means Committee, Hearings on H.R. 6385: Taxation of Marihuana*, 75th Congress, 1st Session. Washington, DC: United States Government Printing Office, 1937.

United States General Accounting Office. "Drug Control: DEA's Strategies and Operations in the 1990s." Washington, DC: United States General Accounting Office, 1999.

Vacek, Jan, et al. "Antioxidant Function of Phytocannabinoids: Molecular Basis of Their Stability and Cytoprotective Properties Under UV-Irradiation." *Free Radical Biology Medicine* 164 (2021): 258–270.

Van der Kooy, Frank, et al. "Cannabis Smoke Condensate II: Influence of Tobacco on Tetrahydrocannabinol Levels." *Inhalation Toxicology* 21, no. 2 (2009): 87–90.

Van der Steur, Sanne J., et al. "Factors Moderating the Association between Cannabis Use and Psychosis Risk: A Systematic Review." *Brain Sciences* 10, no. 2 (2020): 97.

Vandrey, Ryan, et al. "Pharmacokinetic Profile of Oral Cannabis in Humans: Blood and Oral Fluid Disposition and Relation to Pharmacodynamic Outcomes." *Journal of Analytical Toxicology* 41, no. 2 (2017): 83–99.

Veal, Michael E. *Fela: The Life and Times of an African Musical Icon*. Philadelphia: Temple University Press, 2000.

Veit, Markus. "Quality Requirements for Medicinal Cannabis and Respective Products in the European Union—Status Quo." *Planta Medica* 89, no. 8 (2023): 808–823.

Vergara, Daniela, et al. "Gene Copy Number Is Associated with Phytochemistry in *Cannabis sativa*." *AoB Plants* 11, no. 6 (2019): plz074.

Voelker, Rodger, and Mowgli Holmes. "Pesticide Use on Cannabis." *Cannabis Safety Institute*, 2015. https://cdn.technologynetworks.com/tn/Resources/pdf/pesticide-use-on-cannabis.pdf.

Volk, David W., et al. "Reciprocal Alterations in Cortical Cannabinoid Receptor 1 Binding Relative to Protein Immunoreactivity and Transcript Levels in Schizophrenia." *Schizophrenia Research* 159, no. 1 (2014): 124–129.

Volkow, Nora D., et al. "Adverse Health Effects of Marijuana Use." *New England Journal of Medicine* 370, no. 23 (2014): 2219–2227.

Volkow, Nora D., et al. "Effects of Cannabis Use on Human Behavior, Including Cognition, Motivation, and Psychosis: A Review." *JAMA Psychiatry* 73, no. 3 (2016): 292–297.

Vuilleumier, Caroline, et al. "Cannabinoids in the Treatment of Cannabis Use Disorder: Systematic Review of Randomized Controlled Trials." *Frontiers in Psychiatry* 13 (2022): 867878.

Wagley, Charles, and Eduardo Galvão. *Os índios Tenetehara: Uma cultura em transição.* Rio de Janeiro: Ministério da Educação e Cultura, 1961.

Wagley, Charles, and Eduardo Galvão. "The Tenetehara." In *Handbook of South American Indians.* Vol 3, *The Tropical Forest Tribes*, edited by Julian H. Steward, 137–148. Washington, DC: Smithsonian Institution, Bureau of American Ethnology, U.S. Government Printing Office, 1948.

Walton, Robert P. *Marihuana: America's New Drug Problem.* Philadelphia: J. B. Lippincott Company, 1938.

Wang, Yan-Hong, et al. "Quantitative Determination of Delta9-THC, CBG, CBD, Their Acidic Precursors and Five Other Neutral Cannabinoids by UHPLC-UV-MS." *Planta Medica* 84, no. 4 (2018): 260–266.

Ware, Mark A. "Synthetic Psychoactive Cannabinoids Licensed as Medicines." In *Handbook of Cannabis*, edited by Roger G. Pertwee, 393–414. Oxford: Oxford University Press, 2014.

Warnick, Benjamin J., et al. "Head in the Clouds? Cannabis Users' Creativity in New Venture Ideation Depends on Their Entrepreneurial Passion and Experience." *Journal of Business Venturing* 36, no. 2 (2021): 106088.

Warnock, John. "Insanity from Hasheesh." *Journal of Mental Science* 49, no. 204 (1903): 96–110.

Warnock, John. "Twenty-Eight Years' Lunacy Experience in Egypt (1895–1923), Part III." *Journal of Mental Science* 70, no. 291 (1924): 579–612.

Wartenberg, Ariani C., et al. "Cannabis and the Environment: What Science Tells Us and What We Still Need to Know." *Environmental Science & Technology Letters* 8, no. 2 (2021): 98–107.

Webster, Peter. "Marijuana and Music: A Speculative Exploration." *Journal of Cannabis Therapeutics* 1, no. 2 (2001): 93–105.

Weckowicz, Thaddeus E., et al. "Effect of Marijuana on Divergent and Convergent Production Cognitive Tests." *Journal of Abnormal Psychology* 84, no. 4 (1975): 386–398.

Weiblen, George D., et al. "Gene Duplication and Divergence Affecting Drug Content in *Cannabis sativa.*" *New Phytologist* 208, no. 4 (2015): 1241–1250.

Weinberg, Gerhard L. *World War II: A Very Short Introduction.* Oxford: Oxford University Press, 2014.

Weissman, Albert, et al. "Cannabimimetic Activity from CP-47,497, a Derivative of 3-Phenylcyclohexanol." *Journal of Pharmacology and Experimental Therapeutics* 223, no. 2 (1982): 516–523.

Werz, Oliver, et al. "Cannflavins from Hemp Sprouts, a Novel Cannabinoid-Free Hemp Food Product, Target Microsomal Prostaglandin $E_2$ Synthase-1 and 5-Lipoxygenase." *PharmaNutrition* 2, no. 3 (2014): 53–60.

White, Michael A., and Nicholas R. Burns. "The Risk of Being Culpable for or Involved in a Road Crash After Using Cannabis: A Systematic Review and Meta-Analyses." *Drug Science, Policy and Law* 7 (2021): e20503245211055381.

Whiteside, Henry O. *Menace in the West: Colorado and the American Experience with Drugs, 1873—1963.* Denver: Colorado Historical Society, 1997.

Whiting, Frederick. "Monstrosity on Trial: The Case of *Naked Lunch*." *Twentieth Century Literature* 52, no. 2 (2006): 145–174.

World Health Organization (WHO), *Expert Committee on Drugs Liable to Produce Addiction, Third Report*. Technical Report Series no. 57. Geneva: World Health Organization, 1952.

Wikipedia. "Pioneer Species." https://en.wikipedia.org/wiki/Pioneers_species. Accessed July 19, 2025.

Wilbert, Johannes. *Tobacco and Shamanism in South America*. New Haven, CT: Yale University Press, 1987.

Wiley, Jenny L., et al. "Hijacking of Basic Research: The Case of Synthetic Cannabinoids." *RTI Press Occasional Paper* No. OP-0007–11. Research Triangle Park, NC: RTI Press, 2011. https://www.rti.org/rti-press-publication/hijacking-basic-research-case-synthetic-cannabinoids.

Wilson, James T., et al. "HU-331 and Oxidized Cannabidiol Act as Inhibitors of Human Topoisomerase IIα and β." *Chemical Research in Toxicology* 31 (2018): 137–144.

Wilson, Jason. *Curious About Cannabis: A Scientific Introduction to a Controversial Plant*. 2nd. ed. Poland: New Learning Enterprises, LLC, 2020.

Wise, Thomas A. "Practical Remarks on Insanity as it Occurs Among the Inhabitants of Bengal." *Monthly Journal of Medical Science* 5, no. 30 (1852): 500–515.

Wolff, Mariana. "What Makes Cannabis Smell Like Cheese, Skunk, Diesel" *The Oz.*, July 7, 2023. https://theounce.ca/features/what-makes-cannabis-smell-like-cheese-skunk-and-diesel/#.

Wong, Kristin, et al. "Establishing Legal Limits for Driving Under the Influence of Marijuana." *Injury Epidemiology* 1, no. 1 (2014): 26.

Wood, George B., and Franklin Bache. *The Dispensatory of the United States of America*. 10th ed. Philadelphia: Lippincott, Grambo, and Co., 1854.

Wood, Shea, et al. "Canada's THC Unit: Applications for the Legal Cannabis Market." *International Journal of Drug Policy* 128 (2024): 104457.

World Wildlife Fund. *Living Planet Report 2024: A System in Peril*. Gland, Switzerland: WWF, 2024.

Wujastyk, Dominik. "Cannabis in Traditional Indian Herbal Medicine." In *Ayurveda at the Crossroads of Care and Cure*, edited by A. Salema, 45–73. Lisbon, Portugal: Centro de História de Além-Mar, 2002.

Wymore, Erica M., et al. "Persistence of Delta-9-Tetrahydrocannabinol in Human Breast Milk." *JAMA Pediatrics* 175, no. 6 (2021): 623–634.

Yadav, Shubh Pravat Singh, et al. "A Comprehensive Review of the Production Technology of *Cannabis sativa* L. with its Current Legal Status and Botanical Features." *Fundamental and Applied Agriculture* 8, nos. 1–2 (2023): 458–474.

Yawney, Carole. "Lions in Babylon: The Rastafarians of Jamaica as a Visionary Movement." PhD diss., McGill University, 1978.

Yeager, Ashley. "How K2 and Other Synthetic Cannabinoids Got Their Start in the Lab." *The Scientist*, November 27, 2018. https://www.the-scientist.com/how-k2-and-other-synthetic-cannabinoids-got-their-start-in-the-lab-65145.

Yule, Henry, and Arthur Coke Burnell. *Hobson-Jobson: Being a Glossary of Anglo-Indian Colloquial Words and Phrases, and of Kindred Terms*. London: John Murray, 1886.

Zabelina, Darya L., et al. "Dopamine and the Creative Mind: Individual Differences in Creativity Are Predicted by Interactions Between Dopamine Genes DAT and COMT." *PLoS One* 11, no. 1 (2016): e0146768.

Zamarripa, C. Austin, et al. "Assessment of Orally Administered Delta-9-Tetrahydrocannabinol When Coadministered with Cannabidiol on Delta-9-Tetrahydrocannabinol Pharmacokinetics and Pharmacodynamics in Healthy Adults: A Randomized Clinical Trial." *JAMA Network Open* 6, no. 2 (2023): e2254752.

Zamberletti, Erica, and Tiziana Rubino. "Dos(e)Age: Role of Dose and Age in the Long-Term Effect of Cannabinoids on Cognition." *Molecules* 27, no. 4 (2022): 1411.

Zandkarimi, Fereshteh, et al. "Comparison of the Cannabinoid and Terpene Profiles in Commercial Cannabis from Natural and Artificial Cultivation." *Molecules* 28, no. 2 (2023): 833.

Zhao, Rui, et al. "Volatile Terpenes and Terpenoids from Workers and Queens of *Monomorium chinense* (Hymenoptera: Formicidae)." *Molecules* 23, no. 11 (2018): 2838.

Zuardi, Antonio Waldo. "Cannabidiol: From an Inactive Cannabinoid to a Drug with Wide Spectrum of Action." *Brazilian Journal of Psychiatry* 30, no. 3 (2008): 271–280.

Zuardi, Antonio Waldo, et al. "Action of Cannabidiol on the Anxiety and Other Effects Produced by Delta9-THC in Normal Subjects." *Psychopharmacology* 76 (1982): 245–250.

Zuardi, Antonio Waldo, et al. "Cannabidiol, a *Cannabis sativa* Constituent, as an Antipsychotic Drug." *Brazilian Journal of Medical and Biological Research* 39, no. 4 (2006): 421–429.

# Notes

## INTRODUCTION

1. All the quotes in this introduction appear in the main text, where they are fully referenced.
2. The term "drug"—defined by the *Oxford English Dictionary* as "a medicine or other substance which has a marked physiological effect when taken into the body"—likely originates from the medieval Dutch word *droge*, meaning "dry," in reference to dried medicinal plants. In this sense, the term aptly describes cannabis, a dried herbal substance. See Riboulet-Zemouli, "Cannabis Ontologies I," 6, for an insightful discussion of the double-edged meaning of the word "cannabis."

## 1: USEFUL WEED, ENIGMATIC PLANT

1. Regarding biomass production, see Marshall and Leeuwenberg, *Weed of Wonder*, 29. Although many people refer to *Cannabis* as an herb, Small, *Cannabis*, 1, notes that herbs are defined as lacking significant woody tissues, while mature *Cannabis* plants possess woody stems. Similarly, a *Cannabis* "seed" is more accurately termed an "achene," meaning a fruit that contains a single seed.
2. Small, *Cannabis*, 5–6, notes that female plants cultivated in greenhouses or winterless climates can survive for several years but gradually lose vigor.
3. See Small, *Cannabis*, 27, 89. Frank and Rosenthal, *Marijuana Grower's Guide*, 135: "*Cannabis* is a nitrophile, a lover of nitrogen. Given ample N, *Cannabis* will outgrow practically any plant." Plants require nitrogen for growth because it is an essential component of amino acids, the building blocks of living cells.
4. See Schonbeck, "An Ecological Understanding of Weeds."
5. See Small, *Cannabis*, 407, 42, 411–412; Wikipedia, "Pioneer Species"; Ahmad et al., "Phytoremediation Potential of Hemp"; Girdhar et al., "Comparative Assessment"; and Raimondi et al., "Phytomanagement of Chromium-Contaminated Soils."
6. See Clarke and Merlin, *Cannabis: Evolution and Ethnobotany*, 24–27, 31, 74; McPartland and Long, "Cannabis in Asia"; Russo, "History of Cannabis," 1619; and Small, *Cannabis*, 31.

7. See Small, *Cannabis*, 27–29, and Clarke and Merlin, *Cannabis*, 31.
8. See Dekker, *Evolutionary Ecology of Weeds*, 14; Clements and Jones, "Weed Evolution Defies Human Management"; and Small, *Cannabis*, 35–36. The definition of *weed* is from *The Concise Oxford Dictionary* (10th ed., 1999), 1623. Stein, *My Weeds*, 14: "A weed is a plant that is not only in the wrong place but intends to stay." Stein addresses the challenges of defining a "weed" objectively and indisputably. Pollan, *Second Nature*, 124, comparing weeds to garden plants, describes them as "truly a different order of being, more versatile, better equipped, swifter, craftier—simply more adroit at the work of being a plant."
9. De Wet, "The Origin of Weediness in Plants," 16: "All weeds are endowed with an amazing phenotypic plasticity." See also Clarke and Merlin, *Cannabis*, 29, 17, 37, 345; Frank and Rosenthal, *Marijuana Grower's Guide*, 15–17, 176; Small, *Cannabis*, 48, 59–60; Marshall and Leeuwenberg, *Weed of Wonder*, 23–24; and Lynch et al., "Genomic and Chemical Diversity in *Cannabis*," 350.
10. Dekker, *Evolutionary Ecology of Weeds*, 20–22: "It was these vigorous colonizing species that humans gathered and selected among for the very earliest crops. Both weeds and crops often begin with a common progenitor. . . . Some crops were derived from weeds, and some weeds from crops, but the most common pattern is crop-weed complexes in which both are derived from a progenitor species (the wild-crop-weed complex)." See also Clarke and Merlin, *Cannabis*, 57, 24, 27, 74; Russo, "History of Cannabis," 1619; and Ren et al., "Large-Scale Whole-Genome Resequencing," 3.
11. Clarke and Merlin, *Cannabis*, 1, refer to *Cannabis* as "the multipurpose plant."
12. Small and Beckstead, "Cannabinoid Phenotypes," 148, write that THC-rich varieties "usually originated from countries south of latitude 30° N." See also Small, *Cannabis*, 31, 73; Campos, *Home Grown*, 62; Marshall and Leeuwenberg, *Weed of Wonder*, 29; and Frank and Rosenthal, *Marijuana Grower's Guide*, 33. Regarding the increase of cannabinoid production according to light energy, see Russo, "Taming THC"; Lydon et al., "UV-B Radiation Effects"; Pate, "Chemical Ecology of *Cannabis*"; Hillig and Mahlberg, "Chemotaxonomic Analysis"; Magagnini et al., "Effect of Light Spectrum"; Desaulniers Brousseau et al., "Cannabinoids and Terpenes"; and chapter 10 of this book.
13. See Clarke and Merlin, *Cannabis*, 31; Small, *Cannabis*, 45, 49, 298; and Moosmann and Volker, "Hair Analysis," who show that THC can be absorbed through the fingers of those who handle the plant.
14. The plant produces THC (tetrahydrocannabinol) in its acidic form, tetrahydrocannabinolic acid (THCA). THCA sheds a small cluster of atoms (carbon-oxygen-oxygen-hydrogen) and becomes THC when heated. This process, known as decarboxylation, transforms nonpsychoactive THCA into psychoactive THC—see chapter 10; and Sigman, "Decarboxylating Cannabis." Ren et al., "Large-Scale Whole-Genome Resequencing," 4, propose that "strong divergent selection for increased fiber or drug production" began approximately 4,000 years ago. Duvall, *African Roots of Marijuana*, 41, referring to the discovery of 4,000-year-old *Cannabis* seeds in Pakistan's Indus Valley, writes: "Psychoactive cannabis drug use originated in South Asia."
15. The quotes are from Ren et al., "The Origins of Cannabis Smoking," 1–3.
16. See Jiang et al., "A New Insight," 421; Merlin and Clarke, "Cannabis in Ancient Central Eurasian Burials," 35; and Russo et al., "Phytochemical and Genetic Analyses." See Narby and Huxley, *Shamans Through Time*, 75–78, for a historical consideration of the term *shaman*.

17. Regarding the Egyptian and Indian texts, see Russo, "History of Cannabis," 1622, 1631; Russo, "Cannabis in India," 3; Duval, "Cannabis and Tobacco," 43; Dawson, "Studies in the Egyptian Medical Texts," 44–45; and Booth, *Cannabis*, 45. Regarding the Chinese text, see Bennett, "Early/Ancient History," 18; Russo, "History of Cannabis," 1636; and Li, "Cannabis in China," 445–446. Duvall, *Cannabis*, 39, 46, contrasts the Sanskrit words *sana* and *bhanga*, stating that the former "was probably the one that initially meant 'cannabis,'" whereas "the original meaning of *bhanga* was probably something like 'psychoactive drug plant.' Ancient uses of *bhanga* mainly suggest a use, not necessarily any specific plant. . . . written accounts of bhang do not certainly refer to *Cannabis* until about 1200 CE. *Bhang* sometimes meant other plants, and other common names could mean *Cannabis* or various unrelated species." See also Merlin, "Ancient Use of *Ephedra*," 97. Chasteen, *Getting High*, 106, writes concerning the use of psychoactive cannabis in India before the 1800s that "we have breathtakingly little evidence to represent centuries of a supposedly common practice." He attributes part of this absence to the fact that many ancient Indian texts were written on dried palm leaves, which the country's hot and humid climate eventually caused to deteriorate. Several early texts from India and Iran have been interpreted as referring to psychoactive *Cannabis*, but they, too, are clouded by uncertainty and contradictory explanations. For more than a century, scholars have debated the identity of *soma/haoma*, a plant frequently mentioned in ancient Indian and Iranian texts. Some suggest it was *Cannabis*, while others propose *Ephedra*, Syrian rue, henbane, or even the fly-agaric mushroom. Despite extensive debate, the mystery remains unsolved. The *Rig Veda* (India) and the *Avesta* (Iran/Afghanistan) provide no clear evidence of psychoactive cannabis use, yet speculation persists. Historian of religion Mircea Eliade argued that ancient Iranians used cannabis smoke to induce "shamanic ecstasy," citing references to haoma in Iranian texts and Herodotus's account of the Scythians. Eliade claimed cannabis intoxication was "the most elementary technique of ecstasy." He also stated that this use of "narcotics" was "a decadence in shamanic technique"—see Eliade, *Shamanism*, 399–401. However, there is no empirical evidence to support his theory. Burning cannabis seeds alone does not induce altered states, and shamanic practices—often involving singing, drumming, or ingesting psychoactive plants—leave few concrete traces. While such practices may have existed in ancient Iran, proving them remains elusive.
18. The quote is from Bennett, "Early/Ancient History," 20. See also Russo, "History of Cannabis," 1630; and Schultes and Hofmann, *Plants of the Gods*, 95.
19. Herodotus, *The Histories*, 161. I have slightly modified the translation for clarity, replacing "hemp" with "cannabis" and incorporating insights from the adaptation in Chasteen, *Getting High*, 116. See also Clarke and Merlin, *Cannabis*, 216.
20. Duvall, *Cannabis*, 51, writes that inferring psychoactivity from Herodotus's Scythian story is "incorrect and unnecessary, and an example of the historian's fallacy of presentism, or anachronistically applying modern ideas to past events. The presumption is that any smoking *kánnabis* must be psychoactive because marijuana smoking is nowadays common. . . . Scythians burned hempseeds for unknown reasons."
21. Lubotsky, "Scythian Elements," 190, writes: "Unfortunately, we know next to nothing about the Scythian of that period—we have only a couple of personal and tribal names in Greek and Persian sources at our disposal—and cannot even determine with any degree of certainty whether it was a single language." Adding to this uncertainty, Salonen, "Über einige Lehnwörter,"140, writes: "It is reasonable to assume that the Greek word

κάνναβις (*kánnabis*), as well as the Latin word *cannabis,* originate from the same source as the Akkadian expression *qunnabu* meaning 'hemp.'" This suggests two possible origins of the Greek word κάνναβις: 1) it could be a loanword from Scythian, though the corresponding Scythian word is unknown; or 2) it could be a loanword from Akkadian, which had long served as the lingua franca of much of the ancient Near East and still had scholarly practitioners at the time when Herodotus was writing. In brief, the etymology of *cannabis* remains patchy (Beda Künzle, personal communication). For a third possible origin, see Duvall, *Cannabis*, 15, who favors the Proto-Indo-European word *śaṇa* as the root word for *kannabis*, German *Hanf*, English *hemp*, and Dutch *hennep*.

22. Russo, "History of Cannabis," 1638. See also Booth, *Cannabis*, 48, 39; and Gunther, *The Greek Herbal of Dioscorides*, 390.
23. For the first quote, see Rosenthal, *The Herb*, 36. For the other quotes, see Russo, "Cannabis in India," 3, 5; and Wujastyk, "Cannabis in Traditional Indian Herbal Medicine," 56–57. Wujastyk comments: "A careful examination of the references to *cannabis* in pre-modern India turns up some surprises, the first of which is that reliable references to *cannabis* do not date from before about AD 1000" (p. 55).
24. See Merlin, "Archaeological Evidence," 314–315; Duvall, "Cannabis and Tobacco," 7; and Duvall, *African Roots of Marijuana*, 16.
25. See Duvall, *Cannabis*, 46; and Clarke and Merlin, *Cannabis*, 238. Duvall, *African Roots of Marijuana*, 72, writes that African smoking technologies "transformed the plant drug, changing it from a slow-acting edible drug into a fast-acting, easily dosed pharmacological agent."
26. See Duvall, *Cannabis*, 94. Duvall, *African Roots of Marijuana*, 91–92: "Cannabis probably began arriving in Angola in the mid-1700s. Its documented history in western Central Africa began in 1803 when Portuguese administrators lamented that Africans in Angola cultivated cannabis for smoking rather than fiber."
27. Duvall, *Cannabis*, 91–92.
28. See Duvall, *African Roots of Marijuana*, 57; and Clarke and Merlin, *Cannabis*, 128–129. Chasteen, *Getting High*, 126: "Knowledge of the plant's drug potential, and how to cultivate it, seems not to have reached Western Europe before the nineteenth century."
29. According to Merlin, "Archaeological Evidence," 313, the first specific identification of *Cannabis* in Europe dates back 2,500 years. Numerous Europeans mentioned the medicinal virtues of (nonpsychoactive) cannabis, including Pliny the Elder in the first century, Hildegard von Bingen in 1158, Leonhart Fuchs in 1542, François Rabelais in 1546, and Nicholas Culpeper in 1653—see Russo, "The Pharmacological History." See also Ryz et al., "Cannabis Roots"; Acosta, *Tractado de las drogas*, 360–361; Duvall, "Cannabis and Tobacco," 7; and Chopra and Chopra, " Use of the Cannabis," 4.
30. See Russo, "The Case for the Entourage Effect," 2; Linnaeus, *Species Plantarum*, 2:1027; and Lamarck, *Encyclopédie méthodique*, 695. In 1855, Scottish agricultural chemist James F. W. Johnston argued that *Cannabis sativa* and *Cannabis indica* were the same plants, writing: "In the sap of this plant—probably in all countries—there exists a peculiar resinous substance, in which the esteemed narcotic virtue resides. In northern climates, the proportion of this resin in the several parts of the plant is so small as to have escaped general observation. . . . But in the warmer regions of the East, the resinous substance is so abundant as to exude naturally, and in sensible quantity, from the flowers, from the leaves, and from the young twigs of the hemp plant"—see Johnston, *Chemistry of Common Life*, 2:100–101.

31. See Duvall, *African Roots of Marijuana*, 47; Pate, "Chemical Ecology of Cannabis"; Small, *Cannabis*, 219; and Tanney et al., "Cannabis Glandular Trichomes," 4. See also chapter 10 on cannabinoids and UV radiation.
32. Ren et al., "Large-Scale Whole-Genome Resequencing," consider the different kinds of *Cannabis* as varieties within a single species. In contrast, Barcaccia et al., "Potentials and Challenges," confirm *Cannabis* as a single species but distinguish two subspecies, which they call "sativa" and "indica," while Hillig and Mahlberg, "Chemotaxic Analysis," support a two-species concept with six subspecies, and Clarke and Merlin, *Cannabis*, 17, argue for three species (*C. sativa, C. indica,* and *C. ruderalis*), divided into a total of seven subspecies. Meanwhile, Pearlson, *Weed Science*, 55, advises against using the word *strain* to refer to varieties of *Cannabis*, writing: "'Strain' is the term used correctly by microbiologists to describe bacteria, and incorrectly by some cannabis breeders and dispensaries to describe varieties of the plant. . . . This term is not employed in botany but is used casually to refer to variations within cultivars and their offspring. So we can cross 'strain' off our list of terms to ID cannabis varieties." Russo, "The Case for the Entourage Effect," 2, writes that he "has chosen to eschew the irreconcilable taxonomic debate as an unnecessary distraction, and rather emphasize that only biochemical and pharmacological distinctions between *Cannabis* accessions are relevant."
33. See Piomelli and Russo, "*Cannabis sativa* Versus *Cannabis indica*," 46.
34. McPartland, "*Cannabis* Systematics," 210.
35. See Piomelli and Russo, "*Cannabis sativa* Versus *Cannabis indica*," 45. Duvall, "Cannabis and Tobacco," 2, refers to the plant's "invisible character of psychoactive potential."
36. Kovalchuk et al., "The Genomics of *Cannabis*," 5, attribute the existence of "key gaps in our evolutionary knowledge" of *Cannabis* to its "long-standing prohibition and the lack of curated, georeferenced germplasm collections, especially from the putative centers of origin and introduction." See also Ren et al., "Large-Scale Whole-Genome Resequencing."

## 2: *MARIJUANA* AND *GANJA* GO TRAVELING

1. See Bilby, "The Holy Herb," 83–84; Duvall, *African Roots of Marijuana*, 133–140; and Henman and Pessoa Jr., *Diamba Sarabamba*. For a comprehensive list of African words used in Brazil to refer to cannabis, see Mott, "A Maconha na História," 123.
2. Mott, "A Maconha na História," questions the reliability of all written references to psychoactive *Cannabis* in Brazil before 1830 and presents the edict published that year by the Municipality of Rio as the first indisputable Brazilian document concerning the plant. See also Henmann, "War on Drugs," 285.
3. The ethnographic record mentions only one Indigenous Amazonian people using cannabis, the Tenetehara people of the northeast of Brazil, who adopted the plant from local Afro-Brazilians sometime during the nineteenth century. Like many Indigenous peoples across the Americas, the Tenetehara consumed tobacco, and both women and men smoked large cigars made from tobacco as a favorite pastime. According to anthropologists who lived with them in the 1940s, the Tenetehara smoked cannabis in water pipes and long cigarettes rolled in tree bark before working in their gardens or rowing on the river, claiming that it made them "happy to work." They also smoked cannabis in the evenings while discussing community affairs, making speeches, and telling stories and jokes. The quote is from Wagley and Galvão, *Os índios Tenetehara*, 54.

See also Wagley and Galvão, "The Tenetehara," 144; Wilbert, *Tobacco and Shamanism*, 107; and Hutchinson, "Patterns of Marihuana Use," 174.

4. See Duvall, *African Roots of Marijuana*, 140, who writes about the Mexican Spanish word *marihuana/mariguana*: "The *ma*-syllable corresponds to the Bantu plural marker. The *r* is a liquid consonant that persists in many *riamba* cognates. The *g* and *h* represent similar sounds; both initiate the following vowel rather than serving as true consonants that stop air flow. The *marihuana/mariguana* uncertainty follows the Atlantic-wide pattern in which the root word's pronunciation challenged people who were not native speakers of Central African Bantu languages. The *ua* diphthong corresponds to the rising vowel tone central to other *riamba* cognates. The-*na* at the end of *marihuana* is one of the variations on-*mba*. . . ." See also Pessoa de Castro, "Towards a Comparative Approach," 86.
5. See Academia Farmacéutica, *Farmacopea mexicana*, 41, 48. The article appeared in the Mexico City daily *El Republicano* (April 5, 1846) and is quoted by Schievenini and Pérez, "Pasado y presente," 125. Dictionaries routinely identify "marijuana" as a Spanish Mexican word of unknown origin.
6. See Piper, "Origins of the Word 'Marijuana'"; Duvall, *African Roots of Marijuana*, 140–142, 25, 128; and Campos, *Home Grown*, 74–77, for reviews and critiques of the various possible origins of the word *marihuana*. These possibilities include the Chinese Mandarin phrase *ma ren hua*, meaning "hemp seed flowers," which has never served as a name for "cannabis" in China; the combination of the Aztec words *mallín*, meaning "prisoner," and *hua*, meaning "property," which is phonetically similar, but has no apparent connection to the plant; the association of two Spanish names, Maria and Juan, even though *marihuana* was never spelled with a *j* in Mexico during the nineteenth century; the Spanish *mejorana*, meaning marjoram; the folk etymology that associates various herbs in Spanish American culture with the Virgin Mary; and the combination of two "Portuguese" words, *maran* and *guango*, though neither term can be traced in Portuguese.
7. De Las Casas, *Historia de las Indias*, 332. The log of Columbus's first voyage was lost, but Bartolomé de Las Casas made a copy of it and added his commentary. I have adapted De Las Casas's text to make it faithful to the Spanish original. The first two Spanish verbs used to describe smoking are *chupar* and *sorber*, meaning "to suck" and "to sip," respectively.
8. Braudel, *Structures of Everyday Life*, 261, commenting on tobacco's runaway planetary success: "Between the sixteenth and seventeenth centuries, it conquered the whole world and enjoyed even greater popularity than tea or coffee, which was no mean achievement." Kiernan, *Tobacco*, 24, writes: "The great carriers of new things round the coasts of Asia and Africa were the Portuguese, with footholds scattered along the trade routes."
9. See Clarke, *Hashish!*, 53; and Rosenthal, *The Herb*, 65. Dash, *Fundamentals of Ayurvedic Medicine*, 150, states: "The use of *Caras* or the resin which is extracted from the *Cannabis* plant was perhaps unknown in ancient and medieval India. There is no direct reference recommending *Cannabis* for smoking. Only *bhanga* and *ganja* were used, and it is well known that the former has less narcotic effect than the latter." Duvall, "Cannabis and Tobacco," 3, observes that there is no evidence of what Africans smoked in their pipes before the arrival of *Cannabis* "despite many candidate plants," including *datura*.
10. See Gokhale, "Tobacco in Seventeenth-Century India," 484, 491; Ray, "The Hookah," 1319; and Goodman, *Tobacco in History*, 87.
11. Bowrey, *Geographical Account of Countries*, 78–79. Bowrey appears to have written the

earliest first-person account of psychoactive cannabis use after consuming bhang in Bengal on one occasion in the 1670s. However, his manuscript was not published before 1905. Bowrey wrote: "It soon took its operation upon most of us, but merrily, save upon two of our number, who I suppose feared it might do them harm not being accustomed thereto. . . . Myself and one more sat sweating for the space of 3 hours in exceeding measure" (pp. 80–81).

12. Duvall, *Cannabis*, 47, discusses historical cannabis consumption in India: "*Ganja* was primarily roasted, then chewed before the introduction of smoking, but both *bhang* and *ganja* were occasionally administered via smoke inhalation before 1500." Gately, *Tobacco*, 4: "Human lungs have a giant area of absorbent tissue, every inch of which is serviced by at least a thousand thread-like blood vessels, which carry oxygen, poisons, and inspiration from the heart to the brain. Their osmotic capacity is over fifty times that of the human palate or colon. Smoking is the quickest way into the blood stream short of a hypodermic needle." In 1886, Yule and Burnell, *Hobson-Jobson*, 45, identified nine historical references to *bang/bhang/bangue* between 1606 and 1868, with seven mentioning the consumption of cannabis through drinking or eating and none referencing smoking. According to the *Report of the Indian Hemp Drugs Commission, 1893–1894*—an Indo-British study of cannabis usage in late nineteenth-century India published with testimony from more than a thousand individuals, including doctors, yogis, and fakirs—the rate of edible *bhang* consumption in Calcutta was "somewhat less than the rate of [smoked] *ganja* consumption measure for measure" (see Indian Hemp Drugs Commission, *Report of the Indian Hemp Drugs Commission*, 133). See Solvyns, *Les Hindoûs*, 3:261, on tobacco's widespread popularity in eighteenth-century India.

13. See Solvyns, *Les Hindoûs*, 3:273. According to the *Report of the Indian Hemp Drugs Commission 1893–1894*, those who smoked cannabis in late nineteenth-century India included "fakirs and wandering mendicants, sadhus . . . the lower classes of both Hindus and Muhammadans . . . domestic servants of all kinds. . . . Among the upper classes, this habit is generally regarded as exceptional and indicating a special tendency to dissipation, but not so among these lower classes. Bhang is also used to some extent by these classes, but it is more generally used by the more respectable middle and upper classes." Adding: "As a rule, ganja is smoked by the lower class of people and especially by the lowest of ascetics" (see Indian Hemp Drugs Commission, *Report of the Indian Hemp Drugs Commission*, 194, 438).

14. The authorship of the *Dabestan-e Mazaheb* is uncertain; it was probably composed in 1655. The main text's quote is from Shea, *The Dabistan*, 223. The second quote is from Hamilton, *New Account*, 131. See also Farquhar, "The Fighting Ascetics of India," 446; Lorenzen, "Warrior Ascetics," 64; and Dash, *Fundamentals of Ayurvedic Medicine*, 145.

15. The quotes are from Pinch, *Peasants and Monks*, 25–26. See Gross, *Sadhus of India*, 70–71, on armed ascetics attacking British tax collectors; and Ghosh, *Sannyasi and Fakir Raiders*. Lorenzen, "Warrior Ascetics," 70, uses the term "ascetic" to refer to "all manner of yogis, sannyasis, sadhus, fakirs, and bairagis." Dobe, *Hindu Christian Faqir*, 14: "Thus, contrary to recent views that the indiscriminate colonial use of the term fakir for Hindus and others needs to be exclusively marked as Islamic and reflects western confusion, I point out that, along with other vernacular terms such as pir and yogi (vernacular, *jogi*), its indigenous usage reflects rich and fluid identities, which cannot be assigned to one religion." Dobe refers to "yogi-*faqirs*" (p. 13) and to "the figure of the sadhu-*faqir*" (p. 75).

16. The *Report of the Indian Hemp Drugs Commission 1893–94* details how people in India smoked cannabis in the late nineteenth century: "The process of preparing the drug for smoking, the kind of *chillum* or pipe that is used, and the manner of inhaling the smoke are the same all over India. . . . The *chillum* . . . is laid with a foundation of a small quantity of tobacco. On this is placed the washed ganja which has been chopped up and another thin layer of tobacco. . . . The mixture with tobacco has the effect of making the pipe burn properly and go further, and of diluting the smoke which is inhaled" (see Indian Hemp Drugs Commission, *Report of the Indian Hemp Drugs Commission*, 154–155). Regarding Indian ascetics jeering at Europeans during the nineteenth century, see Green, *Islam and the Army*, 92–111. Pinch, *Peasants and Monks*, 30, writes that "any detailed discussion of the social and political dimensions of north Indian monasticism prior to the nineteenth century is fraught with historiographic pitfalls stemming from the general lack of strong documentary evidence with which to confirm or refute religious tradition." On the presence of sadhus among the Indians transported to Jamaica, see Stewart, "Early Encounters," 44–45; and Aïnouche, "Erased from Collective Memory," 146–147, 162, who also discusses the long, matted hair and vegetarian diet of sadhus.
17. Rubin and Comitas, *Ganja in Jamaica*, 15, write: "[Cannabis] may conceivably have been introduced to the West Indies during the slave trade; no evidence, however, has been uncovered of cannabis use for its therapeutic or psychoactive properties by African slaves, or by freemen, in the Antilles during the pre-emancipation period." See Stewart, "Early Encounters," 4, and Borougerdi, *Commodifying Cannabis*, 31, on the reluctance of formerly enslaved people to seek employment on the Jamaican plantations where they were previously forced to work. Bilby, "The Holy Herb," 91, gives the number of 36,000 indentured Indians entering Jamaica between 1845 and 1917 and estimates that 8,000 indentured Africans arrived on the island after 1841. For authors who state that Indian laborers first introduced cannabis to Jamaica, see, for example, Chamberlin and Chevannes, "Ganja in Jamaica," 147; Misra, "Rastafari," 2; and Lee, *Smoke Signals*, 143. In contrast, Duvall, *Cannabis*, 103–104, writes: "Indentured Sierra Leoneans introduced marijuana to Jamaica by 1862."
18. The magistrate, Richard Hill, is quoted in Duvall, *African Roots of Marijuana*, 149. In 1896, missionary Frederick Amir stated in a Jamaican newspaper that some Indians on the island "mixed ganja and tobacco half and half"—see Moore and Johnson, *"They Do as They Please,"* 385. According to the *Report of the Indian Hemp Drugs Commission 1893–1894*, ganja and *charas* "are as a rule smoked admixed with tobacco from a *chillum*" (p. 177).
19. Regarding African and Indian laborers working side by side in Jamaica, see Schuler, *"Alas, Alas, Kongo,"* 149. Aïnouche, "Erased from Collective Memory," 148, refers to this as "a cross-cultural intermingling." See also Chamberlin and Chevannes, "Ganja in Jamaica," 147. Fernández Olmos and Paravisini-Gebert, *Creole Religions of the Caribbean*, 183: "Rastafarianism is an Afro-Jamaican religious movement that blends the Revivalist nature of Jamaican folk Christianity with the Pan-Africanist perspective promulgated by Marcus Garvey, and Ethiopianist readings of the Old Testament. It is a twentieth-century religious and political phenomenon that originated in Jamaica and has gained international attention as a Pan-African approach to the problems of poverty, alienation, and spirituality."
20. See Bilby, "The Holy Herb," 85, 87, who writes that "'ganja' is not the only, nor neces-

sarily the favored, word for cannabis in Jamaican folk culture" and that the term *kaya* "is probably African-derived, from Kikongo 'kaya,' meaning 'leaf.'" Duvall, *African Roots of Marijuana*, 293, writes, "The numerous *kaya* cognates in Central Africa and elsewhere show that the term, shared among several Bantu languages, was used to name smoked herbs, including cannabis." Bob Marley and the Wailers released the album *Kaya* in 1978.

21. See Yawney, "Lions in Babylon," 183, on the smoking habits of Rastafarians living in Jamaica during the 1970s; she estimates that only 5% "never mixed tobacco with their herbs." According to JP Harpignies (personal communication), who briefly interacted with Bob Marley and the Wailers, as well as with other Jamaican musicians in New York during the 1970s, "all the Rastas I met, including the Wailers, smoked big spliffs of just weed when I was around. That doesn't mean some of them might not have mixed some tobacco into the mix at some point, but I never witnessed that. A number of them seemed to me to be into healthy living (staying fit with soccer and jogging, eating a *healthful* diet, etc.) and so didn't smoke tobacco, but I did see a couple of band members smoke cigarettes at different points."
22. See Courtwright, *Forces of Habit*, 42–43; Duvall, *African Roots of Marijuana*, 153; and Johnson, *Grass Roots*, 17–36.
23. Duvall, *African Roots of Marijuana*, 95, refers to "the period after 1840 when Europeans began regularly noting psychoactive cannabis use worldwide."

### 3: FROM GRASS OF FAKIRS TO ASSASSINS' HERB

1. Bar-On et al., "The Biomass Distribution on Earth," estimate that plants represent 81% of the planet's biomass, followed by bacteria with 12.7%; animals, including humans, represent 0.36% of the total. Trewavas, "The Foundations of Plant Intelligence," 1: "The Earth is a planet dominated by green plants." According to Seneca elder John Mohawk: "A culture that makes the possession and use of certain plants illegal strikes me as truly peculiar. If an alien investigator landed and asked me what the strangest thing we did on this planet was, I might well answer: 'Well, we make plants illegal.'" (quoted in Harpignies, *Visionary Plant Consciousness*, 38).
2. Today, the term *hashish* refers to smokable, compressed *Cannabis* resin. However, as the following chapters will show, most Europeans and North Americans writing about hashish in the mid-nineteenth century used the word to describe both the *Cannabis* plant itself and a greasy edible extract or pastes, jams, and confectioneries made from it. In English, *hashish* acquired its current meaning in the late nineteenth and early twentieth centuries. For a description of contemporary hashish production—where the plant's resin glands are sieved and compressed into a smokable paste—see chapter 13.
3. According to Franz Rosenthal's classic study, *The Herb*, which remains the most comprehensive English-language collection of medieval Arabic sources on psychoactive *Cannabis*, the scientific name in Arabic for the plant and its products was *qinnab*. At the same time, *hashish* was a commonly used nickname with an original meaning of *grass*, *fodder*, *medicinal herbs*, or *weeds* (pp. 21–22). Rosenthal comments: "Most likely, it may be simply '*the* herb' as distinguished from all other (medicinal) herbs. Again, we cannot be sure but it seems most likely that the nickname was intended to be of the endearing, rather than the vituperative, kind" (p. 21). See Rosenthal, *The Herb*, 101, 105, 87–89, 109, 124–125, 116, regarding *khamr* and the "covered mind," the "generation

of joy," the recommended punishment for wine users, and "enjoyment and pleasure" as problematic. In the thirteenth century, Muslim societies included Spain, Morocco, Egypt, Arabia, Syria, Iraq, Iran, Afghanistan, Pakistan, and India.

4. See Rosenthal, *The Herb*, 64–65, 106–108, 118–119, 125–126.
5. Quoted in Rosenthal, *The Herb*, 110.
6. See Rosenthal, *The Herb*, 111, 117–120. In the fourteenth century, Egyptian jurist Az-Zarkashī quoted a legal scholar called Ibn al-Attar as saying: "We have no plant whatever that is unclean per se, except plants that are watered with uncleanliness. . . . The statement that hashish is to be declared unclean is not to be considered acceptable, even if (hashish) were intoxicating, for proof comes only in connection with wine, and something other than wine does not correspond to it in all aspects. It is agreed that it is permissible to consume a small quantity of hashish. If it were unclean, this would not be permissible" (see Rosenthal, *The Herb*, 119–120).
7. See Rosenthal, *The Herb*, 120.
8. See Rosenthal, *The Herb*, 101–102.
9. Ibn-al-Baytar, *Traité des simples*, 119–120. I have translated the 1883 French version of the original Arabic text.
10. See al-Makrizi, "De l'herbe des fakirs," on the "grass of fakirs." See Dobe, *Hindu Christian Faqir*, 14, on the original meaning of the word *faqir*. Medieval Egyptian jurist az-Zarkashī wrote that aspects of hashish use "require comment at this time because so many low-class people are affected by it and because many people hesitate to pronounce themselves on the legal situation concerning it, having been unable to find a discussion of it by the ancients" (quoted in Rosenthal, *The Herb*, 102). Rosenthal, *The Herb*, 131, comments: "Wine, in contrast to hashish, was a luxury item that the poorer sections of the population were unable to afford, a fact repeatedly commented upon. The production of hashish was also much less refined and complicated than the cultivation and the processing of grapes, nor should it be forgotten that hashish was a much less bulky and more easily handled merchandise than wine." He adds: "For use as a drug, the wild variety could be used and was, in fact, recommended for use. But primarily, it was cultivated in 'gardens,' as already Ibn-al-Baytar tells us" (p. 132).
11. For a historical source, see al-Makrizi, "De l'herbe des fakirs," 123. See also Khalifa, "Traditional Patterns," 199; Rosenthal, *The Herb*, 51–54; and Green, *Sufism*, 18–22.
12. Quoted in Rosenthal, *The Herb*, 86–87.
13. See Rosenthal, *The Herb*, 82–85. None of the medieval Arabic texts appear to discuss hashish use by women.
14. The quote is from Ibn al-Baytar, *Traité des simples*, 119.
15. See Rosenthal, *The Herb*, 92.
16. The quotes are from Rosenthal, *Man versus Society*, 749, and Rosenthal, *The Herb*, 38, 92.
17. Rosenthal, *The Herb*, 127, 130, discusses the legal practices regarding hashish use in medieval Muslim societies: "We must admit that our knowledge in this respect is almost non-existent. . . . It would seem that the occasions when the government was determined to take drastic steps against hashish (for reasons never stated in satisfactory detail but at best in generalities such as counteracting moral laxity) were infrequent, and the action not very successful."
18. See Rosenthal, *The Herb*, 135–136, 129.
19. According to Egyptian historian Ibn Iyas, *An Account of the Ottoman Conquest of Egypt*,

94–95, an attempt to ban hashish occurred in 1516, when "the Sultan issued a proclamation forbidding all excesses by the people, and that no Jew or Christian was to sell wine, beer, or hashish, under penalty of being hanged forthwith. But no one paid any attention to this order, and things went on just as before." See also Guba, *Taming Cannabis*, 56, and Kozma, "Cannabis Prohibition," 444. Some authors claim that Pope Gregory IX outlawed hemp in Europe in 1231, as he initiated the Inquisition, and Pope Innocent VIII banned its use in Europe in 1484—see, for example, Booth, *Cannabis*, 92; Stoa, *Craft Weed*, 33; and Lee, *Smoke Signals*, 21. However, neither the papal decree of 1231, "*Parens scientiarum*," nor the papal bull of 1484, "*Summis desiderantes affectibus*," mentions hemp.

20. See Guba, *Taming Cannabis*, 60–69.
21. See Guba, *Taming Cannabis*, 69–73. For the quote about the first modern ban, see Stoa, *Craft Weed*, 35.
22. For a copy of Menou's order, see French Republic, *Pièces diverses*, 547–548. I have translated the original French text.
23. Referring to medieval Arabic texts on hashish, Rosenthal, *The Herb*, 77, comments: "It is indeed constantly stressed that wine causes quarrelsomeness, and hashish a kind of languid placidity. It is noteworthy (although the sources themselves rarely comment upon the fact) that no truly violent actions directed against other persons under the influence of hashish are mentioned in any of our stories."
24. Sonnini, *Voyage dans la haute et basse Égypte*, 3:103–104. Sonnini reported that Egyptians consumed the plant in various ways, including smoking it mixed with tobacco in a coconut-based water pipe. He noted: "This method of smoking is one of the most common pastimes for women in the southern part of Egypt." This sentence appears to be one of the earliest references to women consuming cannabis in Egypt. Though Sonnini wrote his report in the 1770s, the French Revolution of 1789 delayed its publication until 1798. Arveiller, "Le Cannabis en France," 453–457, lists the French-language texts published before 1800 that mention psychoactive cannabis.
25. Niebuhr, *Beschreibung von Arabien*, 57. I thank German translator Chris Heidrich for her translation of Niebuhr's sentences into English.
26. In German, the original sentence reads as follows: "Ihm begegneten vier Soldaten auf der Strasse, und er bekam Lust sie alle vor sich her zu treiben." The 1774 French translation (Niebuhr, *Description de l'Arabie*, 51) renders this as "il rencontra dans la rue quatre soldats qu'il lui prit fantaisie de chasser" ("he met four soldiers in the street and he had the whim to chase them away"). The 1792 English translation of the German original reads: "(he) met with four soldiers in the street, and attacked the whole party" (see Niebuhr, *Travels Through Arabia,* 225). Unfortunately, the English translation has misled several contemporary writers in their descriptions of the incident.
27. On the repeated rebellions and uprisings by Egyptians against the occupying French army, see Cole, *Napoleon's Egypt*. Regarding Kleber's assassination, Guba, *Taming Cannabis*, 116, writes: "Nothing in the official records of [the assassin] al-Halabi's interrogations and trial in Egypt discusses hashish as a cause or even as part of the assassination plot."
28. See Guba, *Taming Cannabis*, 78–79, 81.
29. See de Sacy, "Mémoire sur la dynastie des Assassins." Here, I refer to the extended version of de Sacy's memoir published in 1818, not the shorter version with a similar title published in 1809. For an English translation of de Sacy's 1818 text, see Daftary,

*Assassin Legends*, 129–188. European chroniclers of the Crusades began to write about "the Assassins" during the second half of the twelfth century.

30. De Sacy, "Mémoire sur la dynastie des Assassins," 21–35, summarizes the different theories about the word *assassin* formulated by scholars before 1809. Daftary, *Assassin Legends*, 121, writes that "by the middle of the fourteenth century, the word assassin, instead of signifying the name of a sect in Syria, had acquired a new meaning in Italian, French and other European languages; it had become a common noun describing a professional murderer."
31. See de Sacy, "Mémoire sur la dynastie des Assassins," 36–45.
32. The first and third quotes come from the 1818 version of de Sacy's text (p. 55 and p. 83), and the second quote is from the 1809 version (p. 9).
33. Referring to anti-Nizari Isma'ili texts written in Arabic during the 1100s, Daftary, *The Isma'ilis*, 10, 24, writes: "It is important to note that in all the Muslim sources in which the Nizaris are referred to as *hashishi*s, this term is used only in its abusive, figurative sense of 'low-class rabble' and 'irreligious social outcasts.' . . . the fact remains that neither the Isma'ili texts which have come to light in modern times nor any serious contemporary Muslim source in general attest to the actual use of hashish, with or without gardens of paradise, by the Nizaris. . . . Furthermore, the name [*hashishyya*, another plural form of *hashishi*] is rarely used by the Muslim authors who, in contrast to the Crusaders and other Europeans, prefer to designate the sectarians by religious names such as Batiniyya and Ta'limmiyya, or simply as the Isma'iliyya and Nizariyya, if not using terms of abuse such as *malahida* [atheists, infidels, or religious deviants]." For contemporary scholars who consider that de Sacy resolved the mystery of the word *assassin* once and for all, see Daftary, *Assassin Legends*, 122, and Rosenthal, *The Herb*, 44.
34. Lewis, *The Assassins*, 12, writes about Marco Polo's Assassin legend: "Despite its early appearance and wide currency, this story is almost certainly untrue. The use and effects of hashish were known at the time and were no secret; the use of the drug by the [Nizari Isma'ili] sectaries is attested neither by Ismaili nor by serious Sunni authors. Even the term *hashishi* is local to Syria and is probably a term of popular abuse. In all probability, it was the name that gave rise to the story rather than the reverse. Of various explanations that have been offered, the likeliest is that it was an expression of contempt for the wild beliefs and extravagant behavior of the sectaries—a derisive comment on their conduct rather than a description of their practices. For Western observers in particular, such stories may also have served to provide a rational explanation for behavior that was otherwise totally inexplicable." Iversen, *Science of Marijuana*, 5, writes about the "Assassins" [Nizari Isma'ilis]: "It does not seem likely that they would have been able to carry out terrorist acts or politically motivated assassinations while intoxicated by cannabis, nor is there any significant evidence that the drug inspires violence—on the contrary, it tends to cause somnolence and lethargy when taken in high doses."
35. Maalouf, *Samarkand*, 116, writes: "According to texts that have come down to us from Alamut, Hassan [the original leader of the Nizari Isma'ilis] liked to call his disciples *Assassiyyun*, meaning people who are faithful to the *Assass*, the 'foundation' of the faith. This is the word, misunderstood by foreign travelers, that seemed similar to 'hashish.'"
36. Guba, *Taming Cannabis*, 98, writes that de Sacy's memoir "time and again appeared in the footnotes of French and European publications associated with an array of ac-

ademic disciplines, including medicine, pharmacy, psychiatry, history, linguistics, geography, archeology, and agricultural science. . . . With each citation, Sacy's erroneous myth of the Hachichins gained increased veracity, as early as the 1830s becoming an established detail in the fabric of Western knowledge about the Orient and hashish." See also Casto, "Marijuana and the Assassins," 224.

### 4: SHAKY ENTRY INTO WESTERN CULTURE

1. Some authors claim that French soldiers brought hashish back from Egypt in 1801, but no documentary evidence supports this. De Sacy's memoir was widely read throughout Europe in the decades following its publication—see Borougerdi, *Commodifying Cannabis*, 29–30. Regarding the questionable nature of scientific papers on psychoactive cannabis written before 1839, see Mills, *Cannabis Britannica*, 30–39, and Guba, *Taming Cannabis*, 98–100, 108–12. In 1839, O'Shaughnessy, *On the Preparations of the Indian Hemp*, 2, described the lack of European experience with psychoactive cannabis: "The narcotic effects of Hemp are popularly known in the south of Africa, South America, Turkey, Egypt, Asia Minor, India, and the adjacent territories of the Malays, Burmese, and Siamese. In all these countries, Hemp is used in various forms, by the dissipated and depraved, as the ready agent of pleasing intoxication. In the popular medicine of these nations, we find it extensively employed for a multitude of affections. But in Western Europe, its use either as a stimulant or as a remedy is equally unknown. With the exception of the trial, as a frolic, of the Egyptian 'Hasheesh' by a few youths in Marseilles and of the clinical use of the wine of Hemp by Mahneman, I have been unable to trace any notice of the employment of this drug in Europe."
2. See Arveiller, "Le Cannabis en France," 454; Pisanti and Bifulco, "Modern History of Medical *Cannabis*," 195; and Borougerdi, *Commodifying Cannabis*, 75.
3. See O'Shaughnessy, *On the Preparations of the Indian Hemp*; the quotes are on pages 19, 20, and 24.
4. O'Shaughnessy, *On the Preparations of the Indian Hemp*, 27.
5. See O'Shaughnessy, *On the Preparations of the Indian Hemp*; the quotes are on pages 11–12, 14, 37, and 36.
6. The quotes are from Aubert-Roche, *De la peste*, 212–214.
7. The quote is from Aubert-Roche, *De la peste*, 212. For his reference to the nervous system as the locus of bubonic plague, see p. 216.
8. The quotes are from Moreau, *Du hachisch et de l'aliénation mentale*, 5, 6—though a published English translation of this book exists, I have chosen to translate directly from the original. See also Moreau, *Mémoire sur le traitement des hallucinations*, 18–19, and Moreau, *Du hachisch et de l'aliénation mentale*, 34–35.
9. See Moreau, *Du hachisch et de l'aliénation mentale*, 402–403.
10. The quotes are from Moreau, "Recherches sur les aliénés en Orient," 29.
11. Moreau, *Du hachisch et de l'aliénation mentale*, 34.
12. See Moreau, *Du hachisch et de l'aliénation mentale*, 6–7, for the *dawamesc* recipe. Jones, "Mental Illness and Drugs," 383, estimates that Moreau "was probably using doses of 50 to 100 milligrams of THC or more." In contrast, contemporary consumers who try cannabis edibles for the first time are encouraged to start with 1 to 5 milligrams of THC—see, for example, Joyce, "The THC Dosage Guide." According to Arveiller, "Le

Cannabis en France," 461, Moreau and Aubert-Roche obtained their cannabis paste from a French pharmacist in Cairo.

13. The quotes are from Moreau, *Du hachisch et de l'aliénation mentale,* 179, 92–93.
14. Gautier, "Le hachisch," 115–117. As with the texts by Moreau mentioned above, I have prepared my own translation of this quote and all other quotes from the French, although there are other translations in existence.
15. Moreau, *Du hachisch et de l'aliénation mentale*, 20, writes: "Hashish could not have found a more worthy interpreter than the poetic imagination of Mr. Gautier." *Le Club des Hachichins* was named after a text published by Gautier in 1846, which gave a dramatized and partially fictional account of one of Moreau's cannabis gatherings.
16. See Hadengue et al., *Le Livre du Cannabis*, 622. Gautier, "Charles Baudelaire," 58, writes that Baudelaire tried cannabis "once or twice" and "came only rarely and as a simple observer" to the Club's sessions.
17. Baudelaire, *Du vin et du haschisch*, 35–37.
18. The quotes are from Baudelaire, *Du vin et du haschisch*, 26, 41. Baudelaire, *Les paradis artificiels*, 13–14, references de Sacy's work on the link between assassins and hashish.
19. The quotes are from Baudelaire, *Du vin et du haschisch*, 26, 13, 10, 25, and 38.
20. Baudelaire, *Du vin et du haschisch*, 44.
21. Baudelaire, *Du vin et du haschisch*, 42–43.
22. Baudelaire, *Du vin et du haschisch*, 44.
23. Between 1837 and 1845, French psychiatrist Alexandre Brierre de Boismont wrote several publications warning of the dangers of cannabis use—see "Au Rédacteur," "Expériences Toxicologiques," and *Des hallucinations*. Baudelaire based several arguments of *Du vin et du haschisch* on criticisms initially formulated by Brierre de Boismont, but Baudelaire's writings gained much greater prominence, perhaps due to their poetic flamboyance. Some contemporary writers caution against interpreting Baudelaire's statements about wine and hashish too literally. However, many of his readers took them at face value, and some still do today—see Guba, *Taming Cannabis*, 183–184.
24. See Driscoll, *Reconsidering Drugs*, 48. The lethal dose of alcohol depends on several variables, such as the drinker's weight, whether their stomach is empty or full, and the speed at which they drink. Estimates vary on what constitutes a standard dose. For example, Alcohol.org.nz (the alcohol-related harm prevention organization of the New Zealand government) writes: "The lethal dose of alcohol is 5 to 8g/kg. . . . that is, for a 60kg person, 300g of alcohol can kill, which is equal to 30 standard drinks (about 1 liter of spirits or four bottles of wine)."
25. See Gautier, "Charles Baudelaire," 58–59. The quote is on p. 59.
26. Carter et al., "Medicinal Cannabis," 465, write that "a human would have to consume 20,000 to 40,000 times the amount of cannabis contained in one cigarette, in a short period of time, to achieve lethality. Using this as a basis, it has been estimated that it would require 1,500 pounds of cannabis smoked in 15 minutes to induce a lethal effect." Wilson, *Curious about Cannabis*, 248, writes: "The primary reason for the low lethality of Cannabis is due to the small number of cannabinoid receptors on the brainstem and the fact that THC and other cannabinoid receptor agonists found in Cannabis are only partial receptor agonists. Other drugs like alcohol and heroin can be dangerous in much smaller quantities because they have strong effects on the brainstem, affecting the performance of all other organs like the heart and lungs. Overdosing on alcohol or heroin can stop the heart and lungs, but this is generally not the case for Cannabis."

## 5: INSANITY

1. The quotes are from Baudelaire, *Du vin et du haschisch*, 44; De Saulcy, *Narrative of a Journey*, 135; Johnston, *Chemistry of Common Life*, 2:99; and *Scientific American*, "Hasheesh and Its Smokers," 49.
2. See Green, *Islam and the Army*, 96; and Fischer-Tiné, "Britain's Other Civilising Mission." In the nineteenth century, British doctors viewed masturbation as a cause of insanity or an aggravator of mental illness—see Mills, *Madness, Cannabis, and Colonialism*, 111. See also Jiloha, "Lunatic Asylums," 84; and Mills, *Cannabis Britannica*, 85.
3. British Parliamentary Papers, *Papers Relating to the Consumption of Ganja*, 7–8, 4.
4. Simpson is quoted in Cockburn, *Annual Report on the Insane Asylums*, 16. The final report of the government inquiry of 1873 noted that "the total number of cases of insanity is small in proportion to the population, and not large even in proportion to the number of ganja-smokers"—see British Parliamentary Papers, *Papers Relating to the Consumption of Ganja*, 4. Mills, *Cannabis Britannica*, 86, describes this as "a process in which the police, who needed to come up with a cause of insanity to complete the forms correctly, often could find no evidence of what had disrupted the behavior of the individual they wished to incarcerate. This meant that they would have to make up a cause of insanity, and in such a situation, 'ganjah-smoking' was a convenient way of filling the document and one that was likely to be believed."
5. Green, *Islam and the Army*, 96: "Stoned on bhang or opium, raving at the passing sahibs, lying lazy in the shade of a tree, bare of the merest strip of clothing and, as often as not, plainly diseased and scabby, the 'outlaw' *faqir* represented the ultimate affront to Britain's civilizing mission."
6. Wise, "Practical Remarks on Insanity as it Occurs Among the Inhabitants of Bengal," 508, provided his exchange with an Indian mendicant in the early 1850s: "The religious mendicants are a great curse to India. One day, I asked one of them, a notorious gunjah-eater, what was his occupation? Placing his hand on his stomach, he said, 'Eating and smoking gunjah.' The dreadful cannabis! 'But, what is your trade?' He added, 'To contemplate the great God.' 'And where is your home?' He pointed downwards and answered, 'In the earth.' The effects of gunjah are most pernicious." Hasan et al., "Cannabis Use and Psychosis," 2, discuss the difficulty of providing precise statements about "cannabis-induced psychotic episodes," given the variability of definitions of terms such as "schizophrenia" and "psychosis," the fact that not all cases of persistent psychosis are schizophrenia, and "while cannabis-induced psychotic episodes can be transient they can also be a risk factor for developing schizophrenia."
7. Earleywine, *Understanding Marijuana*, 145–146: "Cannabinoid intoxication can also mimic certain aspects of psychoses like schizophrenia. These psychotic disorders typically include odd thoughts, auditory hallucinations, and inappropriate emotions. . . . Large doses of eaten marijuana or hashish can create comparable symptoms, but this cannabis psychosis is not the same as schizophrenia. It usually lacks formal thought problems and inappropriate emotions. It also dissipates relatively quickly, while schizophrenia remains a chronic mental illness." The quotes are from Taylor, "The Vision of Hasheesh," 406–407.
8. See Hill, "Be Clear," S14; Volkow et al., "Effects of Cannabis Use," E3; and Van der Steur et al., "Factors Moderating the Association," 1.

9. The relationship between cannabis use and schizophrenia is complex and remains a topic of debate and research. Most experts believe a strong correlation exists between the two, but emphasize that correlation does not imply causation. Volkow et al., "Effects of Cannabis Use," E3: "It is important to highlight in this context that most individuals who use cannabis do not develop schizophrenia. Therefore, while cannabis use is neither necessary nor sufficient for the development of schizophrenia, available evidence suggests that cannabis use may initiate the emergence of a lasting psychotic illness in some persons (most likely those individuals with a genetic vulnerability), and this finding warrants serious consideration from the point of view of public health policy." Earleywine, *Understanding Marijuana*, 148, notes: "Cannabis may not cause psychopathology, but psychopaths often smoke cannabis." See also Pasman et al., "GWAS of Lifetime Cannabis Use," 1167; and Nutt, *Drugs Without the Hot Air*, 102.
10. The Abbasiya asylum near Cairo was Egypt's only mental asylum until the foundation of the Khanka hospital in 1912, as a result of Warnock's lobbying—see Nour, "'Claiming the Mad,'" 72–73.
11. Clouston, "The Cairo Asylum," 792–793.
12. From the start, Warnock produced statistics concerning admissions to the Cairo asylum and compared them to those generated by British doctors in India—see Clouston, "The Cairo Asylum," 794.
13. Warnock, "Insanity from Hasheesh," 109–110.
14. Warnock, "Twenty-Eight Years' Lunacy Experience," 595–596, 590. Three men out of 460 patients admitted to the Cairo asylum in 1922 were classified as cases of "hashish insanity," representing 0.65% of total cases.
15. See Warnock, "Twenty-Eight Years' Lunacy Experience," 583, 595.
16. Aamodt, "Staying Off the Grass?," 1. Kendell, "Cannabis Condemned," 147: "The role of the Egyptian government and its senior delegate, Dr El Guindy, was clearly crucial. Without the Egyptian initiative and Dr El Guindy's single-minded determination, Indian hemp would never have been brought under the controls of the 1925 Convention." Mills, *Cannabis Britannica*, 179, refers to "El Guindy's spectacular performance in Geneva" and devotes a chapter to contextualizing it.
17. El Guindy spoke in French, and the quotes in the main text directly translate his words. The official English translation varies on several points from the original, rendering, for example, "drogue" as "product"—compare League of Nations, *Records of the Second Opium Conference*, 1:39, to Société des Nations, *Actes de la deuxième conférence de l'opium*, 1:41. See League of Nations, *Records of the Second Opium Conference*, 1:43–51, for the statements of the delegates who supported the Egyptian proposal. In 1923, the Government of South Africa sent a written suggestion to the League of Nations along the same lines as the Egyptian proposal formulated by El Guindy the following year, but South Africa did not attend the 1924 Conference—see League of Nations, *Records of the Second Opium Conference*, 1:53, 134. Kozma, "Cannabis Prohibition in Egypt," 445: "Hashish was banned in Egypt in a series of decrees between 1868 and 1884. . . . On 29 March 1879, the Egyptian government banned the cultivation, distribution, and importation of hashish in Egypt and ordered the destruction of hashish fields."
18. I have translated El Guindy's statement from the French original—compare League of Nations, *Records of the Second Opium Conference*, 1:132–134, to Société des Nations, *Actes de la deuxième conférence de l'opium*, 1:138.

19. League of Nations, *Records of the Second Opium Conference*, 1:133–134, and Société des Nations, *Actes de la deuxième conférence de l'opium*, 1:138–139.
20. League of Nations, *Records of the Second Opium Conference*, 1 :135, and Société des Nations, *Actes de la deuxième conférence de l'opium*, 1:140.
21. All quotes from League of Nations, *Records of the Second Opium Conference*, 1:135–137. For the tasks entrusted to Sub-Committee F, see League of Nations, *Records of the Second Opium Conference*, 1:222.
22. See League of Nations, *Records of the Second Opium Conference*, 1:137–138.
23. I have translated Perrot's statement from the French original because the official English translation is imprecise—compare League of Nations, *Records of the Second Opium Conference*, 1:261, to Société des Nations, *Actes de la deuxième conférence de l'opium*, 1:271.
24. League of Nations, *Records of the Second Opium Conference*, 1:262.
25. League of Nations, *Records of the Second Opium Conference*, 1:262, 222. Delegates from Switzerland, the Netherlands, and the British Empire also raised objections, especially regarding the terms that imposed domestic policy and legislation on individual countries.
26. League of Nations, *International Opium Convention*, 333, 329.
27. See League of Nations, *International Opium Convention*.
28. According to Mills, *Cannabis Britannica*, 181–182, a British report written in early 1924 suggested that the only way Egypt could end illegal hashish importations was to get the League of Nations to establish international regulations to stop the production and commerce of cannabis. Mills writes that "these recommendations almost exactly anticipated El Guindy's agenda, and even his language, at the Second Conference, which itself came just over six months after the report." See also Bewley-Taylor et al., *Rise and Decline of Cannabis Prohibition*, 11.
29. Indian Hemp Drugs Commission, *Report of the Indian Hemp Drugs Commission*, 263–264, 236. Duvall, *Cannabis*, 184: "The first robust study of a non-European *Cannabis* culture came in the 1894 report of the Indian Hemp Drugs Commission, which studied *bhang*, *charas* and *ganja* in British India. No subsequent study of drug *Cannabis* within a society has been as thorough, although recent works have improved understanding of the plant in scattered societies." See also Mills, *Cannabis Britannica*, 4–5.
30. Kendell, "Cannabis Condemned," 150–151: "Perhaps Sir Malcolm Delevingne and his Indian Civil Service colleagues hoped until a late stage that it would be possible to divert Dr El Guindy's proposals into a procedural cul de sac; perhaps they wanted to avoid offending a country that had until 2 years before been a British Protectorate; perhaps they wanted to avoid laying the British open to the charge . . . that they were indifferent to the welfare of the citizens of a country for which they had been largely responsible for the previous 40 years."
31. I have translated El Guindy's original French words, "*formuler un voeu*" ("form a wish"), which the official English translation omits—compare League of Nations, *Records of the Second Opium Conference*, 1:364, to Société des Nations, *Actes de la deuxième conférence de l'opium*, 1:377.
32. League of Nations, *Records of the Second Opium Conference*, 1:135. Bewley-Taylor et al., *The Rise and Decline of Cannabis Prohibition*, 13: "One of the reasons the U.S. had withdrawn from the 1924–1925 Geneva Conference was the producing countries' refusal to commit to specific measures restricting production of raw opium and coca leaves to

medical and scientific needs. Washington saw this as a major gap in the international system of control. Limitation of the available supplies could not be achieved without control at the source: restricting the cultivation of the plants." Mills, *Cannabis Britannica*, 165–173, devotes several pages to "the Egyptian and American ambush" of the Second Opium Conference.

## 6: THE PROHIBITION EFFECT

1. See Gieringer, "The Forgotten Origins of Cannabis Prohibition," 250, and Bonnie and Whitebread, *The Marijuana Conviction*, 61–62.
2. The two quotes are from Fisher, "Racial Myths of the Cannabis War," 953, 972. Fisher points out that the passage of cannabis laws in Massachusetts, Maine, and Vermont was driven by the New England Watch and Ward Society's puritanical crusade against "public vice." Gieringer, "Forgotten Origins," 261, describes these early anti-cannabis laws as "preventative initiatives by drug control authorities to deter future usage." Several historians have claimed that early American antidrug laws were rooted in xenophobic or racist or anti-minority sentiment—see, for example, Musto, *American Disease*, 294–295, and Bonnie and Whitebread, "Forbidden Fruit," 1021. Contemporary scholars such as Fisher, Gieringer, and Rathge (author of "Cannabis Cures") have reassessed this interpretation by focusing on the earliest American drug laws. Fisher, "Racial Myths," 933, writes: "These were laws about Whites. The lawmakers who erected America's earliest drug bans acted first and foremost to protect the morals of their own racial kin. And because the morals of most importance to White lawmakers were those of their own offspring, they acted fastest and most forcefully when a drug took White youth in its clutches." See Hamowy, *Dealing with Drugs*, 10–11, for a state-by-state timeline of the enactment of laws prohibiting the sale of opiates, cocaine, and cannabis.
3. See National Medical Convention, *Pharmacopoeia of the United States of America*, 50; and Wood and Bache, *Dispensatory of the United States of America*, 339, who wrote in 1854: "The *medicinal resin* or *extract of hemp*, directed by the U.S. Pharmacopoeia, is made by evaporating a tincture of the dried tops. Dr. O'Shaughnessy directs it to be prepared by boiling the tops of the gunjah in alcohol until all the resin is dissolved, and evaporating to dryness by means of a water-bath." The *Dispensatory* was one of the country's nationally recognized publications on medical matters.
4. Regarding the shortcomings of cannabis extracts, see Boire and Feeney, *Medical Marijuana Law*, 16; Marshall and Leeuwenberg, *Weed of Wonder*, 94–98; and Booth, *Cannabis*, 143. According to Guba, *Taming Cannabis*, 150–186, the inability to standardize doses, combined with cases of "hashish poisoning," fueled the demedicalization of cannabis in France during the second half of the 1800s. In 1918, the twentieth edition of *The Dispensatory of the United States of America* stated about cannabis extract: "One of the great hindrances to the wider use of this drug is its extreme variability"—quoted in Mikuriya, *Marijuana*, 343. Mills, *Cannabis Nation*, 12, mentions that cannabis was also used as a colorant in corn plasters during this period. Concerning the nonlethal nature of cannabis, see note 26 of chapter 4.
5. See Gieringer, "Forgotten Origins," 242, 250, 255; Bonnie and Whitebread, *The Marijuana Conviction*, 48–49; Musto, "The 1937 Marijuana Tax Act," 419; and Rathge, "Cannabis Cures," 136, 156.
6. Hamowy, *Dealing with Drugs*, 16–17. The National Prohibition Act, commonly re-

ferred to as the Volstead Act, was enacted to implement the provisions of the Eighteenth Amendment, which prohibited the manufacture, sale, and transportation of alcoholic beverages.

7. See Rathge, "Mapping the Muggleheads," 17. I am indebted to Adam Rathge's work for pointing to the archives of the *New Orleans Times-Picayune* as an essential source of information. Gieringer, "Forgotten Origins," 253: "Marijuana (as it is now usually spelled) is said to have first entered Texas around the time of the Mexican revolution and to have likewise appeared in New Orleans by 1910."
8. See "Questions and Answers," *Times-Picayune* (New Orleans), October 23, 1921. Regarding the spelling of "marihuana," the *Times-Picayune* explained in 1929 that the words *muggles* and *mooters* both referred to "the marihuana cigarette, variously spelled marijuana, marajuana, and mary-warner"—see "Crime Trail Widens as Marihuana Fume Descends Upon City," April 21, 1929. For "mirauana," see "'Muggles' Incites Orleans Youth to Crime," *Times-Picayune* (New Orleans), May 29, 1922. For "mariahuana," see "Police Capture Weed, Wine and Owners in Raid," *Times-Picayune* (New Orleans), August 26, 1922. According to Rathge, "Cannabis Cures," 254–255, "between 1923 and 1935, the *Times-Picayune* alone carried some eight hundred and forty-five articles mentioning marijuana." In the main text, I retain all the original spellings without adding (*sic*) to indicate deviance from present-day standards in the hope that readers will appreciate the original flavor rather than deplore the inconsistency of usage.
9. See "New Drug Habit Rapidly Growing, Health Heads Say," *Times-Picayune* (New Orleans), February 18, 1922.
10. Jackson, "Prohibition in New Orleans," 273, gives the local price of moonshine whiskey in 1922 as "30 cents to 50 cents a glass" and notes that prices for this beverage "remained steady" during the entire period of Prohibition.
11. See "'Mary Warner' Epidemic," *Times-Picayune* (New Orleans), May 18, 1923; and "The Victim," *Times-Picayune* (New Orleans), June 3, 1923.
12. See "Use of Mexican 'Dope' Forbidden by City Council," *Times-Picayune* (New Orleans), May 30, 1923; "Council to Act on Sale in City of 'Mary Warner,'" *Times-Picayune* (New Orleans), May 20, 1923; and "The Victim," *Times-Picayune* (New Orleans), June 3, 1923.
13. For the text of the Louisiana law of 1924, see Belenko, *Drugs and Drug Policy*, 135–136. Regarding large shipments from Mexico, see Rathge, "Cannabis Cures," 217, who refers to numerous articles in the *Times-Picayune*. Rathge also finds that of the 225 documented marijuana arrests in New Orleans between 1923 and 1929, only 11 arrestees were identified as Mexicans (pp. 217–218). He writes: "The available arrest evidence from the *Times-Picayune* suggests that the most common marijuana user in the city was a white male in his early twenties. The average recorded age of all those arrested for marijuana was twenty-three" (p. 220). He adds that 25% of all documented arrests concerned teenagers (p. 221).
14. See "Port Termed 'Hypodermic Needle' Feeding Entire Middle West with Drugs," *Times-Picayune* (New Orleans), March 6, 1926; "Thousands of State's Youth Marijuana Addicts, Survey by Criminologist Shows," *Times-Picayune* (New Orleans), August 12, 1926; and Jackson, "Prohibition in New Orleans," 273.
15. The *Chicago Tribune* article is quoted in Belenko, *Drugs and Drug Policy*, 142. In the journalistic jargon of this period, "Americans" usually referred to native-born Whites—see Fisher, "Racial Myths of the Cannabis War," 973.

16. See Rathge, "Cannabis Cures," 252–253, 304, for the Cincinnati and Kansas City quotes.
17. All quotes from "Crime Trail Widens as Marihuana Fume Descends Upon City," *Times-Picayune* (New Orleans), April 21, 1929. Jackson, "Prohibition in New Orleans," 281, writes: "Gangster-type beatings and shootings connected to the illicit liquor trade became much more prominent by 1930 than they had been in earlier years."
18. All quotes and arguments from "Marihuana Law Challenged; High Court May Rule," *Times-Picayune* (New Orleans), April 15, 1931; "State High Court Hears Debate on Marihuana Law," *Times-Picayune* (New Orleans), May 2, 1931; and "Marihuana Law Validity Upheld; Appeal Rejected," *Times-Picayune* (New Orleans), May 26, 1931.
19. See "State High Court Hears Debate on Marihuana Law," *Times-Picayune* (New Orleans), May 2, 1931; and "Marihuana Law Validity Upheld; Appeal Rejected," *Times-Picayune* (New Orleans), May 26, 1931.
20. See "State High Court Hears Debate on Marihuana Law," *Times-Picayune* (New Orleans), May 2, 1931; and "Marihuana Law Validity Upheld; Appeal Rejected," *Times-Picayune* (New Orleans), May 26, 1931.
21. All quotes from Fossier, "Mariahuana Menace," 247–250. Rathge, "Cannabis Cures," 230, describes Fossier's talk as "one of the most influential speeches given on marijuana" and notes that "as a summary of the existing literature on cannabis and hashish with connections to the contemporary marijuana crisis in New Orleans, Fossier's article contained much of what became the general assessment of marijuana in the 1930s." Fossier's text was published in the *New Orleans Medical and Surgical Journal* six months after its initial delivery as a speech.
22. Many countries passed laws restricting cannabis use before 1920, including Egypt in 1868, Turkey in 1875, Greece in 1890, and France in 1916. According to Indian Hemp Drugs Commission, *Report of the Indian Hemp Drugs Commission*, 270, Turkey prohibited hashish in 1875 when the Turkish Imperial Medical Council decided that "the use of hashish in the preparation of medicines was extremely rare, and that, being a narcotic, its use must of necessity be injurious, and that consequently, the suppression of the cultivation of hashish could not fail to prove highly advantageous." France banned "hashish and its preparations" in a 1916 law on "poisonous substances"—see *Journal officiel de la République française*, "Lois et Décrets no. 0190 du 14/07/1916," 6254. Mexico's law of 1920 was called "Dispositions on the cultivation and commerce of products that degenerate the race"—for a discussion, see Campos, *Home Grown*, 124–125, 200–201, and Schievenini, "Historical Approach," 141. Regarding the Canadian law of 1923, Giffen et al., *Panic and Indifference*, 179, note that cannabis was placed under federal law "before it came to be defined as a social problem in Canada. Why this was so remains a mystery." Before the 1925 Geneva Convention, almost all the colonies in Africa had passed cannabis control laws—see Duvall, "Drug Laws." The United Kingdom added the "Coca Leaves and Indian Hemp Drug Regulations" to its Dangerous Drugs Act in 1925 and implemented it in 1928—see Mills, *Cannabis Nation*, 10. Regarding the American states prohibiting cannabis use in 1931, see Bonnie and Whitebread, *Marijuana Conviction*, 52, 56–63, 97–98; Rathge, "Cannabis Cures," 234; Gieringer, "Forgotten Origins," 287–288; and Fisher, "Racial Myths," 945. By 1931, Louisiana, Texas, Colorado, and Kentucky had appealed to the federal government to help restrict cannabis use.
23. See Bonnie and Whitebread, *Marijuana Conviction*, 62. Tennyson drafted the letter, and L. G. Nutt, the deputy commissioner of prohibition, signed it.

24. See Bonnie and Whitebread, *Marijuana Conviction*, 39. The Colorado Law of 1917 concerned the cultivation and unlawful use of "Cannabis sativa (also known as cannabis indica, Indian hemp, and mariguana)"—see Colorado Laws (1917), ch. 39, p. 120. The Colorado Law of 1927 concerned "Cannabis Indica, or Cannabis Sativa, commonly known as Indian Hemp, Hasheesh, or Marijuana"—see Colorado Laws (1927), ch. 95, p. 309. See also Whiteside, *Menace in the West*.
25. The first quote is from Houghton and Hamilton, "Pharmacological," 20. The second is from Stockberger, "Drug Plants," 19, 3. See also Rathge, "Cannabis Cures," 229.
26. Olson and Gerstein, *Alcohol in America*, 4–6: "Several waves of prohibitionist sentiment swept across the country in the nineteenth and early twentieth centuries, culminating in a surge of political action in the first two decades of [the twentieth] century. By 1916, 23 states had passed (mainly by referendum) prohibitionist laws of various kinds. Finally, in 1920, after years of skilled single-issue politicking led by the Anti-Saloon League, the 18th Amendment extended prohibition to the nation as a whole."
27. See Thornton, "Alcohol Prohibition"—the quote is from p. 3—and Jackson, "Prohibition in New Orleans." In 1917, the Eighteenth Amendment passed with a 68% majority in Congress and an 85% majority in the Senate—see "Prohibition Wins in National House by 282 to 128," *The New York Times*, December 18, 1917, and "Prohibition Wins in Senate, 47 to 8," *The New York Times*, December 19, 1917. Prohibition lost support after the onset of the Great Depression, which started in 1929 and revealed numerous causes of human misery besides the consumption of alcohol. Prohibition was repealed in December 1933.
28. The quotes are from Bancroft, *Native Races*, 632–633. Gieringer, "Forgotten Origins," reviews the early history of the term *marihuana* in American publications, showing that it appeared in several variants, including *mariguan* (1894), *mariguana* (1897), *marihuana* (1901), *marihuma* (1905), *mariahuana* (1905), *marihuano* (1912), and *marahuana* (1914). See also Campos, *Home Grown*, 215–223, for an overview of the early uses of the term in American English publications. The first seven statewide cannabis laws passed between 1911 and 1915 did not use the Mexican Spanish name to refer to the plant. In 1917, Colorado passed the first state law mentioning the Mexican word *mariguana* (see the note above). For the USDA booklet of 1915, see Stockberger, "Drug Plants."
29. In the original Mexican Spanish, *marihuana* is spelled with an *h*, not a *j*. When the word entered American newspapers, it was usually spelled with an *h*. During the 1920s, however, English-speaking Americans increasingly spelled it with a *j*—see Gieringer, "Origins of Cannabis Prohibition," 20. Campos, *Home Grown*, 76, notes that "the spelling of marijuana with a *j* in Mexico has always been extremely rare and, to my knowledge, nonexistent prior to the twentieth century."
30. The quotes are from Jacobson, *Whiteness of a Different Color*, 140, 154–155. On the subject of anti-Mexican sentiment, see also Dreyer, "Sustained Animus Toward Latino Immigrants." Some scholars have argued that racism and anti-Mexican sentiment triggered by the presence of large numbers of Mexican immigrants drove the initial impulse to prohibit cannabis in the United States—see, for example, Musto, *American Disease*, 295; Helmer, *Drugs and Minority Oppression*, 54–79; and Bonnie and Whitebread, "Forbidden Fruit," 1012–1016, 1035–1037. Helmer writes: "Public concern about marijuana grew because Americans wanted to drive the Mexicans back over the border, for reasons which had nothing to do with the nature of the drug or its psychological

effects" (p. 56). This argument fails to consider that initial opposition to cannabis occurred in several states without a noticeable population of Mexican immigrants. Fisher, "Racial Myths," 951, writes that Massachusetts had 57 Mexican-American residents in 1920 (representing 0.0014% of the population) and Maine and Vermont together had only three Mexican-American residents in 1920. Likewise, Rathge, "Mapping the Muggleheads," 26, writes that the "1930 census data shows 717 citizens in New Orleans listed as 'Mexican'—accounting for 0.1 percent of the city's 458,762 residents." What seems certain is that authorities in prohibitionist America drew on longstanding anti-Mexican sentiment to convey their disapproval of cannabis by using its Mexican name and highlighting its supposed Mexicanness. Duvall, *The African Roots of Marijuana*, 141, writes: "It is phonetically incorrect to spell *marihuana/mariguana* with a *j*, which in Spanish represents a different sound from *g* or *h*. The *j* arose in American English discourse that tagged the plant drug Juana to strengthen portrayals of its unsavory Mexicanness in the early 1900s." Lee, *Smoke Signals*, 6, describes the initial use of *marijuana* (spelled with a *j* or *h*) in American English as "a derogatory slur." A pamphlet published by the Foreign Policy Association in 1938 entitled *Marihuana: The New Dangerous Drug* presented *marihuana* as distinct from cannabis—as a drug rather than a plant, indeed as "one of the most dangerous drugs known." The pamphlet also spelled out how to pronounce the Mexican word in English: "mar-i-wá-na"—see Merrill, *Marihuana*, 26, 11, 5, 3.

31. The first quote is from *The Concise Oxford Dictionary of Current English* (5th ed., 1972), 744; the second is from *The Concise Oxford Dictionary* (10th ed., 1999), 871.
32. The French definition is from the 2004 edition of *Le nouveau petit Robert,* and the German definition is from https://de.wikipedia.org/wiki/Marihuana.
33. According to Hawaii State Legislature S. B. 786 (2017), "A Bill for an Act Relating to Medical Marijuana,": "The legislature finds that the term 'marijuana' originated as a slang term to describe the genus of plants that is scientifically known as cannabis. 'Marijuana' has no scientific basis but carries prejudicial implications rooted in racial stereotypes from the early 20th-century era when cannabis use was first criminalized in the United States. The term 'cannabis' carries no such negative connotations and is a more accurate and appropriate term to describe a plant that has been legalized for medicinal use in Hawaii, twenty-seven other states, the District of Columbia, and the United States territories of Guam and Puerto Rico." By passing this law, the state of Hawaii replaced the word "marijuana" with "cannabis" in all its statutes and rules. For a contrary view, see Chen, "Why It Can Be Okay to Call It 'Marijuana,'" who writes: "We should develop *more*, not fewer, words for what we need to say." Hudak, *Marijuana*, 26, writes in 2020: "I have chosen to use 'marijuana' in this book—even in its title—because of the contemporary uses of the term. It is mainstream; it is standard; it is the term Americans use almost universally when discussing cannabis and its products. There may well be people who still use the term as a means of invoking racialized language, but most Americans do not." According to Mikos and Kam, "Has the 'M' Word Been Framed?," 22–23, between 2009 and 2018, internet searches in the United States for *marijuana* were three times more numerous than those for *cannabis.* In the United Kingdom, the opposite was true, and people searched for *cannabis* four times more than they did for *marijuana.* In Canada, *cannabis* overtook *marijuana* as the most frequently used term in 2018.

### 7: TO KILL THE WEED THAT KILLS PEOPLE

1. The League of Nations (LN) documents cited in these notes can be accessed at https://archives.ungeneva.org via the reference number provided after each document's title and page number(s). For the League of Nations report, see "Preliminary Note on the Chief Aspects of the Problem of Indian Hemp and the Laws Relating Thereto in Force in Certain Countries," 11, 12, 4, LN, O.C. 1542.
2. The quotes are from "The Abuse of Cannabis in the United States," 4, 5, 4, 4, LN, O.C.1542(c) Addendum. Bonnie and Whitebread, *Marijuana Conviction*, 146, erroneously attribute this memorandum to Harry Anslinger; Fuller identifies himself as its sole author in "Advisory Committee on Traffic in Opium and Other Dangerous Drugs, Minutes of the Nineteenth Session," 28, LN, C. 33.M.14.
3. "Advisory Committee on Traffic in Opium and Other Dangerous Drugs, Minutes of the Nineteenth Session," 22, LN, C. 33.M.14; and "Note on Indian Hemp as it Affects Egypt," 7, 6, 7, LN, O.C.1542(d).
4. "Note on Indian Hemp as it Affects Egypt," 1, LN, O.C.1542(d).
5. In 1950, the secretary-general of the United Nations commented regarding the amended International Opium Convention's definition of "Indian hemp" that "almost every word is such as to cause administrative difficulties in prosecutions" because "it would be quite impossible to prove that an extracted resin . . . had necessarily been prepared from 'the dried flowering or fruiting tops of the pistillate plant.' The resin could be extracted just as readily from fresh, undried tops by means of a suitable solvent. The tops might be picked before flowering had begun. Indeed, the resin might be extracted from some other part of the plant, though the tops bear most of it. The staminate [= male] plants also produce active resin"—see "Preparatory Documentation on the Single Convention, Note by the Secretary-General," 12, UN, E/CN.7/AC.3/2. All United Nations (UN) documents cited in this chapter can be accessed at https://digitallibrary.un.org via the reference number provided after the document's title and page number(s). The quote in the main text is from Société des Nations, *Actes de la deuxième conférence de l'opium*, 2:327.
6. See "Advisory Committee on Traffic in Opium and Other Dangerous Drugs, Report to the Council on the Work of the Nineteenth Session," 7, LN, CRID104/307/17; and "Cannabis. Questionnaire Prepared for the Use of the Experts who have been Invited by the Advisory Committee to Collaborate with it on this Subject," 1–2, LN, O.C.1542(j). The last question in the main text is from "Sub-Committee on Cannabis Sativa, 1st Session, Report Submitted to the Advisory Committee," 3, LN, O.C.1607.
7. See "Report by Dr. J. Bouquet, Hospital Pharmacist, Tunis, Inspector of Pharmacies, Tunis, Containing Answers to the Questionnaire Submitted to the Experts," 21, 26, LN, O.C.1542(o).
8. Quotes 1, 4, and 5 are from "Traffic in Opium and Other Dangerous Drugs for the Year Ended December 31, 1938. Report by the Government of the United States of America," 51, (available in LN file R5031-12-37018-33903.pdf). Quote 2 is from "Memorandum by Dr. W.L. Treadway, Assistant Surgeon-General (U.S.A.) Dealing with Cannabis, with Special Reference to the Questionnaire Submitted to the Cannabis Experts," 1, LN, O.C.1542(r). Quote 3 is from United States Congress, *Hearings on H.R. 6385*, 51. Quote 6 is from "Traffic in Opium and Other Dangerous Drugs for

the Year Ended December 31, 1937, Report by the Government of the United States of America," 65, (available in LN file R5008–12–32015–31960.pdf).

9. See "Excerpts from a Report Prepared in the Treasury Department of the United States of America Concerning Cannabis and from the Decennial Revision of the United States Pharmacopoeia to Cannabis," 2, LN, O.C.1542(q).

10. See "Sub-Committee on Cannabis Sativa, 1st Session, Report Submitted to the Advisory Committee," 3, LN, O.C.1607. The six experts contacted in 1936 came from the United Kingdom, France, Belgium, and the United States. Rattansi, *Racism*, 29, writes that the social Darwinian expression *survival of the fittest* "sanctioned the belief that the technological advances and refined customs of the white races were proof of their greater 'fitness' and the natural necessity that they rule over darker, inferior races. Social Darwinism nurtured eugenics, a stream of racial thinking that dominated the period from the 1880s to the 1930s in both the USA and Europe." Racism continues to haunt the practice of science, as the journal *Nature* recognized in 2022, writing that "science's history is enmeshed with racism and colonization" and referring to racism as "science's toxic legacy"—see Nobles et al., "Ending Racism," 419.

11. Quotes 1, 2, 3, and 9 are from "Cannabis (Indian Hemp Addiction) and the Problems to which it Gives Rise," 59, 75, LN, O.C/Cannabis/3(1). Quotes 4 and 8 are from "Report by Dr. J. Bouquet, Hospital Pharmacist, Tunis, Inspector of Pharmacies, Tunis, Containing Answers to the Questionnaire Submitted to the Experts," 26–7, LN, O.C.1542(o). Quote 5 is from "Sub-Committee to Consider Questions in Regard to Indian Hemp and Indian Hemp Drugs, Provisional Minutes, Fourth Session, First Meeting," 5, LN, O.C./s/c cannabis sat./Indian/4th Session/P.V. 1. Quotes 6 and 7 are from "Supplement to the Report submitted by Dr. J. Bouquet," 30–1, LN, O.C. 1724 Addendum.

12. In 1845, Moreau, *Du hachisch et de l'aliénation mentale*, 8, wrote about *Cannabis* paste: "Its action is far from the same for all individuals. Depending on the individuals, the same dosage can produce extremely varied effects, at least in intensity." The 1930s pharmacologist quoted in the main text is Dr. Ferdinand de Myttenaere—see "Second Note on Indian Hemp," 3, LN, O.C.1542(m). Bouquet's quote in the main text is from "Sub-Committee to Study Questions in Regard to Indian Hemp and Indian Hemp Drugs, Provisional Minutes, Fourth Session, Third Meeting," 8, LN, O.C./s/c Cann. Sat.Indian/4th Session/P.V. 3. Writing in 1938, Walton, *Marijuana*, 152, attributed the decline in the popularity of cannabis extracts in the late nineteenth century to several factors. These included the fact that "the effects are irregular due to marked variations in individual susceptibility and probably also to variable absorption of the gummy resin." Duvall, *Cannabis*, 91: "Genetic variation among people and plants makes subjective effects potentially individualistic. People experience drug *Cannabis* differently depending on personal health and genetics, as well as environmental conditions. . . . The plant's genetic variability expands the range of possible experiences. Farmers and plant scientists have developed many cultivars that each produce distinctive subjective effects."

13. The quotes are from "Cannabis (Indian Hemp Addiction) and the Problems to which it Gives Rise," 80, LN, O.C/Cannabis/3(1) and "Report by Dr. J. Bouquet, Hospital Pharmacist, Tunis, Inspector of Pharmacies, Tunis, Containing Answers to the Questionnaire Submitted to the Experts," 28–9, 31, LN, O.C.1542(o).

14. Quote 1 is from "Sub-Committee on Cannabis Sativa, 3rd Session, Report by Mr. Fuller,

President of the Sub-Committee," 4, LN, O.C.1707. Quote 2 is from United States Congress, *Hearings on H.R. 6385*, 21.

15. According to Musto, "The 1937 Marihuana Tax Act," 432, the Treasury Department's general counsel, Harman Oliphant, came up with the idea of a transfer tax to impose federal restrictions on cannabis during the summer of 1936.
16. The five quotes are from United States Congress, *Hearings on H.R. 6385*, 18, 21, 34, 23, 21. Fisher, "Racial Myths," 935–936, writes that Anslinger "seemed die-cast to play the villain—and earned the part. His sneering prose brewed racial code words with armchair moralisms vilifying society's castaways." Duvall, *African Roots of Marijuana*, 7, writes: "Despite his real role in cannabis history, Anslinger has been made into a semi-fictional straw man, easy to topple as a stand-in for the idea of prohibition."
17. See United States Congress, *Hearings on H.R. 6385*, 48, 50.
18. See United States Congress, *Hearings on H.R. 6385*, 32, 45, 76, 55.
19. See United States Congress, *Hearings on H.R. 6385*, 83, 80.
20. See United States Congress, *Hearings on H.R. 6385*, 61, 74. Botanist Lyster Dewey told the committee: "Of course, it is all introduced from the type that is distributed from the birds, and the birdseed does come up year after year from self-sown seed. . . . It grows as a weed along roadsides, railways, in wastelands, on overflowed lands along rivers and where seed from bird cages has been thrown out in backyards" (pp. 81, 84).
21. See United States Congress, *Hearings on H.R. 6385*, 91, 90. To back up the claim that cannabis might still have therapeutic value, Woodward referred to the authority of French pharmacist Jules Bouquet: "According to what has been quoted from this report of Dr. Bouquet, there are evidently potentialities in the drug that should not be shut off by adverse legislation. The medical profession and pharmacologists should be left to develop the use of this drug as they see fit" (p. 114). During the same hearings, Anslinger quoted Bouquet to affirm the contrary.
22. The full text of the Marihuana Tax Act is available at https://www.druglibrary.org/schaffer/hemp/taxact/mjtaxact.htm.
23. Mack and Joy, *Marijuana as Medicine?*, 157–158, write: "Although the tax act allowed medical use of marijuana, it created a formidable bureaucracy with which few doctors or pharmaceutical firms were willing to contend. Manufacturers and medical users of the drug were required to comply with burdensome registration procedures and pay a tax of $1 per ounce. By contrast, marijuana for nonmedical use—the act's intended target—was taxed at the prohibitive rate of $100 per ounce. In 1942 marijuana lost its legitimacy as a prescription medication when it was removed from the *United States Pharmacopoeia* (*USP*)." See also Musto, "The 1937 Marihuana Tax Act," 429. Shortly after the law's passage, federal and state enforcement authorities had difficulty distinguishing between psychoactive *Cannabis* and hemp grown for fiber, leading them to commit gross mistakes. In one case in 1938, federal and state agents mowed and burned 32,000 *Cannabis* plants in a field in Montana, only to find that they had destroyed a fiber company's experimental hemp crop—see Johnson, *Grass Roots*, 58. One theory popularized in Jack Herer's book *The Emperor Wears No Clothes* holds that the real motivation behind the passage of the Marihuana Tax Act of 1937 was a conspiracy by the Federal Bureau of Narcotics and the Hearst and DuPont business empires to suppress industrial hemp as a competitor to wood-pulp paper and nylon. This theory lacks merit for several reasons. By omitting the plant's stalks from the definition of "marihuana,"

the law did not outlaw the production of hemp fibers, and farmers could cultivate hemp if they paid a $1 tax—see Robinson, "Hemp," 2. Herer's theory ignores that many other countries had passed anti-cannabis laws when the United States finally drafted the 1937 Marihuana Tax Act, and a majority of American states had done the same before the Federal Bureau of Narcotics came into existence in late 1930. Gieringer, "Forgotten Origins," 285, writes: "Herer has never produced an iota of evidence to substantiate this theory."

24. See "Special Committee to Study Questions in Regard to Indian Hemp and Indian Hemp Drugs, Provisional Minutes, First Meeting," 3–4, LN, O.C./$^{s}$/c.Cann.Sat.Ind./3rdSession/P.V. 1 and "Sub-Committee to Study Questions in Regard to Indian Hemp and Indian Hemp Drugs, Provisional Minutes, Second Meeting," 5–6, LN, O.C./$^{s}$/c.Cann.Sat.Ind./3rdSession/P.V. 2.

25. Regarding the subcommittee's unresolved questions, see "Progress Report on the Sub-Committee's Work," 26, LN, O.C.1724(b). Merrill, *Marihuana*, remarked in 1938 regarding the unidentified active principle of *Cannabis*: "Only when chemists can isolate this compound and discover its relation to certain other substances in the plant can chemical tests to identify cannabis be infallible" (p. 8). In the early nineteenth century, scientists identified the active molecules of several psychoactive plants, including morphine in opium poppies, caffeine in coffee beans, cocaine in coca leaves, and nicotine in tobacco. These substances are all alkaloids, which are molecules that contain nitrogen. However, no alkaloids were found in *Cannabis* resin. In the 1930s, researchers elucidated the structure of a non-alkaloid compound called cannabinol (CBN), which has a weak psychoactive effect and was not considered the plant's "active principle." Later, scientists discovered that CBN is the degradation product of THC in aging harvested cannabis. It was not until 1964 that THC, the primary psychoactive compound in cannabis, was identified. Like all other cannabinoids, THC is not an alkaloid. For a summary of early chemical research on cannabis compounds, see Mechoulam and Hanuš, "Historical Overview." The Sub-Committee on Cannabis released a questionnaire coauthored by Jules Bouquet and French psychiatrist Antoine Porot in December 1939, three months *after* the beginning of World War II. Mills, *Cannabis Nation*, 48–49, comments: "If the Sub-Committee had achieved anything, it was to demonstrate that there were great difficulties in trying to establish clear positions on the various scientific, legal, social, and medical aspects of cannabis consumption and that there was certainly no easy consensus among the experts on matters related to the drug and its consumption."

26. See Weinberg, *World War II*. Rattansi, *Racism*, 34, writes: "In the aftermath of the Holocaust and the ending of World War II in 1945, the role of eugenics and scientific racism in underpinning the ideology of Nazism was impossible to ignore." Regarding the history of the League of Nations, see https://www.ungeneva.org/en/league-of-nations. Wikipedia gives detailed histories of the League of Nations and the United Nations.

27. For the preliminary drafts of the Single Convention, see "Preparatory Documentation on the Single Convention, Note by the Secretary-General," UN, E/CN.7/AC.3/1, and "Preparatory Documentation on the Single Convention, Note by the Secretary-General," UN, E/CN.7/AC.3/3. The first quote is on p. 18 of the latter.

28. World Health Organization, *Expert Committee on Drugs*, 11.

29. The quote is from "Commission on Narcotic Drugs: Report of the Ninth Session," 15, UN, E/2606. Regarding the decision to drop the term "Indian hemp" and request an additional study, see "Commission on Narcotic Drugs: Report of the Eighth Session,"

16, UN, E/2423, and "Preparatory Documentation on the Single Convention, Note by the Secretary-General," 13, UN, E/CN.7/AC.3/2.

30. WHO/APD/56, "The Physical and Mental Effects of Cannabis," 11–12, 24, 32. www.tni.org/files/the-problem-of-cannabis-who-wolff.pdf.

31. The quote is from "Commission on Narcotic Drugs: Report of the Tenth Session," 12, UN, E/2768/Rev.1. Regarding India's defense of the medical use of *Cannabis* in Indian medicine, see "Commission on Narcotic Drugs: Summary Record of the Two Hundred and Sixty-Seventh Meeting," 6, UN, E/CN.7/SR.267 and "Commission on Narcotic Drugs: Summary Record of the Two Hundred and Seventieth Meeting," 6, UN, E/CN.7/SR.270.

32. See Bewley-Taylor and Jelsma, "Fifty Years of the 1961 Convention." For the full text of the Single Convention, see "Single Convention on Narcotic Drugs, 1961," UN, E/CONF.34/24/Add.1—in particular Articles 2.5.b, 22, 28, 36.1, and 49.

33. The quotes are from the Single Convention, Articles 3 and 2. Heroin and desomorphine are both semisynthetic opioids derived from morphine, and ketobemidone is a synthetic opioid produced with chemicals unrelated to the poppy plant. According to the "Commission on Narcotic Drugs: Report of the Tenth Session," 19, UN, E/2768/Rev.1, Schedule IV was conceived as a list to which "would be added at a later stage, after consultation with WHO, all particularly dangerous narcotic drugs and specially all new narcotic drugs of particularly great addiction-producing properties not offset by substantial therapeutic advantages which were not obtainable from less dangerous drugs." The Schedules of the Single Convention partly overlap, as all substances in Schedule IV are also listed in Schedule I.

34. The first two quotes are from "Technical Committee, Sixth Meeting," 107, UN, E/CONF/34/24/Add.1, and "Technical Committee, Tenth Meeting," 116, UN, E/CONF/34/24/Add.1. The UN provides official commentaries to accompany its international drug control conventions, which guide their interpretation. According to United Nations, *Commentary on the Single Convention*, 95: "If it is found that a drug has particularly dangerous properties and lacks therapeutic value, as paragraph 5 [of the Single Convention] requires for inclusion in Schedule IV, its deletion from general medical practice will usually be desirable from the point of view of public health. The recommendation of the World Health Organization and the decision of the Commission on Narcotic Drugs to place a drug in Schedule IV will in fact be motivated by such a desire."

35. See "Ad Hoc Committee on Article 39 of the Third Draft, First Meeting," UN, E/CONF/34/24/Add.1; and "Single Convention on Narcotic Drugs, 1961," Articles 1, 28, UN, E/CONF.34/24/Add.1. No other plant species has had its sexual organs specifically targeted by international law. In the case of the opium poppy, regulations apply to all parts of the plant that contain opium alkaloids, with primary attention given to the latex (opium) extracted from the unripe seed pods. However, international legislation does not single out the flowers or other reproductive organs on their own. Poppy seeds, which contain negligible amounts of alkaloids, remain legal for culinary use in many countries. As for the coca plant, international controls focus exclusively on the leaves, which contain the psychoactive alkaloid cocaine. International law does not impose direct controls on the plant's reproductive structures, such as its flowers or seeds.

36. See Mechoulam and Shvo, "Hashish" (1963) and Gaoni and Mechoulam, "Isolation, Structure, and Partial Synthesis" (1964).

37. In 2019, the WHO reversed its position on the therapeutic potential of *Cannabis* and its resin and recommended their reclassification in the Single Convention. In December 2020, the fifty-three member states of the UN Commission on Narcotic Drugs voted to remove *Cannabis* and its resin from Schedule IV—twenty-seven votes in favor, twenty-five against, and one abstention. Those in favor included India, Jamaica, and the United States. Those against included Egypt, Pakistan, and Turkey—see "Decision 63/17: Deletion of Cannabis and Cannabis Resin from Schedule IV of the Single Convention on Narcotic Drugs of 1961 as Amended by the 1972 Protocol," 5, UN, E/2020/28/Add.1. The WHO recognized that the Single Convention of 1961, which listed cannabis in Schedule IV, may have led to some "impairment of research"—see "Commission on Narcotic Drugs, Sixty-Third Session, Compilation of All Questions and Answers on the WHO recommendations on Cannabis and Cannabis-Related Substances Raised During the Fourth and Fifth Intersessional Meeting of the Commission at its Sixty-Second Session," 28, UN, E/CN.7/2020/CRP.4. The Single Convention on Narcotic Drugs currently retains *Cannabis* and its resin in Schedule I, a category reserved for substances, such as cocaine and morphine, whose properties might give rise to dependence, and which present a severe risk of abuse and are subject to all the Convention's control measures. The UN's reclassification of *Cannabis* does not imply the legalization or depenalization of its use. However, it opens the door to scientific studies of the plant's therapeutic potential, which its Schedule IV listing impeded for fifty-six years. Regarding the imprisonment of cannabis users and producers since 1961, the worldwide numbers do not seem to have been compiled; however, in 2020, Ghosh, "Prisoners with Drug Use Disorders," writes: "Nearly half a million people are incarcerated worldwide for drug possession and an additional 1.7 million for other drug-related offenses."

## 8: SPREADING ON THE WINGS OF MUSIC

1. The first and third quotes are from Bergreen, *Louis Armstrong*, 282–283. The second quote is from Jones and Chilton, *Louis*, 116. See also Hammond, *John Hammond on Record*, 106; and Lee, *Smoke Signals*, 10.
2. See Komp, "Cannabis & 'Muggles.'"
3. See Jones and Chilton, *Louis*, 112–113, 115–116. See Gieringer, "Origins of Cannabis Prohibition," on California's 1913 cannabis law.
4. The quote is from Bergreen, *Louis Armstrong*, 316.
5. The quote is from Bergreen, *Louis Armstrong*, 327.
6. See Singer and Mirhej, "High Notes," 11, 14. Booth, *Cannabis*, 244–245, writes that the Federal Bureau of Narcotics kept files "on every jazz musician who had used marijuana—and their acquaintances. The list was formidable, a real jazz Hall of Fame including Louis Armstrong, Duke Ellington, Lionel Hampton, Dizzy Gillespie, Jimmy Dorsey, Count Basie, and Thelonious Monk. Many musicians were kept under covert surveillance, and orchestras or bands containing marijuana users were constantly monitored."
7. See Klein, "Nigeria and the Drugs War," 55; and Klantschnig, "Origins of Cannabis Prohibition," 232–233.
8. Asuni, "Socio-Psychiatric Problems," 3, writes in 1964 about *Cannabis*: "It grows profusely in this tropical climate with little or no care. Farms of the plant, scattered over

southern Nigeria, have been reported by the police." Asuni calculated that a kilo of cannabis worth $336 in 1962 in Lagos could be sold for eighteen times as much once smuggled into the UK (p. 5). See Klantschnig, "Origins of Cannabis Prohibition," 229–230, on the indifference of the colonial government to cannabis use in urban Nigeria during the 1940s and '50s, and Mills, *Cannabis Nation*, 77, 82, on the lack of interest of British authorities concerning cannabis use during the 1950s.

9. See Hoare, "Soho's Early Jazz Clubs"; and Schofield, *Strange Case of Pot*, 68. Mills, *Cannabis Nation*, 62, writes that the 1948 British Nationality Act "made it possible for those living in the Empire to migrate to the UK without a visa, and by the early 1960s over half a million had acted on this invitation." The quote in the main text is from the February 22, 1936, issue of *Melody Maker*, as cited by Booth, *Cannabis*, 260.
10. See Veal, *Fela*, 54; and Olaniyan, *Arrest the Music!*, 9, 24.
11. John Lennon is quoted in Goodden, *Riding So High*, 45.
12. The quote is from Mills, *Cannabis Nation*, 117. See also Booth, *Cannabis*, 269.
13. Mills, *Cannabis Nation*, 118. According to the Wootton Report, an official inquiry published in 1969: "Several witnesses discounted the significance of immigrant influence on cannabis use and asserted that international movement of young people and new attitudes to experimentation with mood-altering drugs were the main explanation of increased cannabis use by white persons in the United Kingdom since 1945"—see Home Office, *Cannabis*, 9.
14. According to the Wootton Report, referring to the period from 1945 to 1967: "In the early part of the period, most seizures were of green plant tops, found in ships from India and African ports and thought to be destined for petty traffickers in touch with colored seamen and entertainers in London docks and clubs. By 1950, illicit traffic in cannabis had been observed in other parts of the country where there was a colored population. In 1950, however, police raids on certain London jazz clubs produced clear evidence that cannabis was being used by the indigenous population; by 1954 the tendency for the proportion of white to colored offenders to increase was well marked, and in 1964 white persons constituted the majority of cannabis offenders for the first time"—see Home Office, *Cannabis*, 8. By the end of the 1960s, the report found that "hashish now formed some eighty percent of the traffic" (p. 9). Presumably, as the laws became stricter and their enforcement more rigorous, smugglers favored the more concentrated substance—hashish, or compressed cannabis resin—over the bulkier and weaker dried herbal cannabis. The quote about the Jamaican man is from Hallam, "Reaching Out," 10. Ballotta et al., "Cannabis Control," 104, write that "the 1961 Convention suggests to apply the most stringent control system to cannabis, yet leaves some flexibility in their interpretation of the necessity of such control."
15. Richards, *Life*, 226, writes that the judge called him "scum" and "filth" for allowing his premises to be used for cannabis smoking. See Bockris, *Keith Richards*, 133–136, and Norman, *The Stones*, 189–215. The day after Keith Richards's initial sentencing, *The Times*, the bastion newspaper of the British establishment, ran an editorial stating that many people "resent the anarchic quality of the Rolling Stones' performances, dislike their songs, dislike their influence on teenagers and broadly suspect them of decadence."
16. Coon and Harris, *Release Report*, 27, wrote in 1969: "It is disturbing to find how important class still is in the legal system. The chances of a defendant going to prison or borstal are increased if his education is limited and his background working class.... Young people of the working class are at a definite disadvantage when compared to the

situations of their contemporaries of a higher status. They are more easily intimidated by authority and accept the advice of the police more readily. People with a higher education are more capable of speaking for themselves and are better equipped to gain access to a lawyer."

17. The quote is from Richards, *Life*, 227. See Carrier and Klantschnig, "Free the Weed," 3, regarding the "aura of defiant cool" imparted by the use of illegal cannabis.
18. The quote is from Coon, "We Were the Welfare Branch," 189.
19. The quote is from Asuni, "Socio-Psychiatric Problems," 8. According to Lambo, "Medical and Social Problems," the number of cannabis-related prosecutions doubled between 1962 and 1963, and the amount of cannabis seized by the police increased eightfold. Klantschnig, "Origins of Cannabis Prohibition," 231, writes: "It is likely that actual cannabis use and cultivation increased steadily after the war but that the police, the government, and the media only took note of this rise in the first half of the 1960s, when they appear to discover a boom in cannabis smoking."
20. The quotes are from Asuni, "Socio-Psychiatric Problems," 4, 13–14, 7.
21. The quotes are from Lambo, "Medical and Social Problems," 3.
22. See Klein, "Nigeria and the Drugs War," 55.
23. Darnton, "Nigeria's Dissident Superstar," 13.
24. See Carrier and Klantschnig, *Africa and the War on Drugs*, 36; and Klein, "Nigeria and the Drugs War," 55.
25. In July 1977, Darnton, "Nigeria's Dissident Superstar," 3, estimating that Fela had already been arrested some half dozen times for cannabis infractions, reported in *The New York Times*: "Last February, Nigerian Army soldiers attacked his house, burned it, beat him and wounded scores of others, including innocent bystanders. Altogether, some 60 civilians were hospitalized, and Fela was jailed. . . . Fela himself, beaten unconscious, was held under armed guard in a hospital room, and as rumors about his condition swept the city, he awoke to find that he had become, overnight, Nigeria's superstar dissident."
26. Carrier and Klantschnig, "Free the Weed," 3.
27. Mezzrow and Wolfe, *Really the Blues*, 72.
28. See Fachner, "Out of Time?," 107; Fachner, "Ethno-Methodological Approach," 91, 82–83, 90; Pearlson, *Weed Science*, 112; and Webster, "Marijuana and Music," 94.
29. Hammond is quoted in Shapiro, *Waiting for the Man*, 32. See also Hammond, *John Hammond on Record*, 106. Fachner, "Out of Time?," 107, mentions the importance for musicians of having experience with the drug's effects to have the ability to "shape the musical-temporal space of sounds, their sound staging, in listening, composition and improvisation, due to the drug-induced changes in the altered metric context."

## 9: A DRUG FOR THE PURSUIT OF KNOWLEDGE

1. In the following notes, I have translated Michaux's original French texts into English, though in some cases previously published translations exist. Michaux, *Oeuvres complètes*, 2:767, writes: "To the devotees of the simple point of view who may be tempted to judge all my writings as those of a drug addict, I regret to say that I am more the water-drinking type. Never alcohol. No excitants, coffee, tobacco, or tea for years." According to Michaux, *Connaissance par les gouffres*, 177: "The author of the present work has for

five years experimented with most of the hallucinogenic drugs, those destroyers of mind and person, lysergic acid, psilocybin, mescaline about twenty times, hashish several dozen times, alone or combined, in various doses, not especially to enjoy them, above all to surprise them, to surprise mysteries hidden elsewhere." Michaux remained vague about the practical details of his cannabis experiments. In one instance, he mentions "smoking" hashish—see Michaux, *Oeuvres complètes*, 2:719. In another text, he writes about having "eaten too large a quantity" of hashish—see Michaux, *Oeuvres complètes*, 2:908. See also Pic, "Par la voie des nerfs," 143; and Emmanuel, "Dans la cristallerie," 8.

2. The quotes are from Michaux, *Oeuvres complètes*, 2:708–709. Michaux had a similar experience with sound, writing: "I seem to be hearing in an unusual way. A real sound, so faint that I would not ordinarily have heard it, is perceptible through three closed doors. I can even follow its slightest movements as I would a flying swarm of bees. I am endowed with *stereo-hearing*" (*Oeuvres complètes*, 2:711–712).
3. Michaux, *Oeuvres complètes*, 2:716.
4. The quotes are from Michaux, *Oeuvres complètes*, 2:706, 720. Michaux found mescaline considerably more challenging to work with than cannabis, hence the title of his first book, *Misérable miracle*, published in French in 1956 and English seven years later. In the single chapter devoted to cannabis, Michaux explains that he conducted only "superficial" experiments with the substance "for the sake of comparison" with mescaline.
5. The quotes are from Michaux, *Connaissance par les gouffres*, 91, 170–173.
6. The quotes are from Michaux, *Connaissance par les gouffres*, 91, 171.
7. Most references to cannabis in this paragraph relate to examples discussed in previous chapters. For historical instances of people who consumed the plant before singing and playing games, see Duvall, *African Roots of Marijuana*, 105–106.
8. One notable exception is German philosopher Walter Benjamin, who published an article in 1932 in a German newspaper about his experience with cannabis, in which he suggested that users could attain "experiences that approach inspiration, illumination." Benjamin claimed that "under the influence of hashish, we are enraptured prose-beings raised to the highest power"—see Benjamin, *On Hashish*, 117, 53. A book containing his cannabis writings was published posthumously in German in 1972 and subsequently translated into other languages—see Benjamin, *On Hashish*. According to Marincolo, *What Hashish Did to Walter Benjamin*, 136, the German philosopher "never really succeeded in writing the book on hashish or drugs he wanted to write because he had to flee from the Nazi régime." Benjamin's 1932 newspaper article was translated into French and published in 1935 in a literary journal under the title "Haschich à Marseilles," which Michaux may have read but did not mention in his books. The quote is from Baudelaire, *Les paradis artificiels*, 8.
9. The quote is from Michaux, *Connaissance par les gouffres*, 9. Michaux's emphasis on worldly knowledge contradicted the views of another famous author, British writer Aldous Huxley. In *The Doors of Perception*, a book published in 1954—two years before Michaux's first book in French on the subject—Huxley claimed that mescaline could facilitate transcendent experiences and mystical insight. Contrary to Michaux, Huxley disregarded cannabis, which he lumped together with alcohol, opium, and barbiturates, as "religion's chemical surrogates." Huxley agreed with Baudelaire's concept of "artificial paradises" and argued that mescaline stood apart from such dubious drugs because it was an "almost totally innocuous" substance that could impart "an experience of the most enlightening kind." (The quotes are from Huxley, *The Doors of Perception*, 62–67.)

The insights attributed to mescaline by Huxley were primarily metaphysical and/or religious.

10. The quotes are from Ginsberg, *Howl and Other Poems*, 32, 12. In 1958, Burroughs had written only one book, *Junkie: Confessions of an Unredeemed Drug Addict*, published as pulp fiction in 1953 under the pseudonym William Lee. See also Morgan and Peters, *Howl on Trial*; Merrill, *Allen Ginsberg*; and Hamdan, "Jazz Aesthetics."
11. The quotes are from Ginsberg, "Henri Michaux," 444–445. At the time of their first meeting, Michaux was 59, Burroughs 44, and Ginsberg 32.
12. See Miles, *The Beat Hotel*, 138–139; Lane, *French Genealogy*, 7; and Miles, *Call Me Burroughs*, 335. Burroughs and Ginsberg published *The Yage Letters* in 1963, a remarkable book about their experiences with yage/ayahuasca.
13. See Morgan, *I Greet You*, 63–64; Michaux, *Oeuvres complètes*, 2:1287; and Lane, *French Genealogy*, 187–214.
14. The first quote is from Burroughs, "Letter from a Master Addict," 127. In 1961, Burroughs presented "Points of Distinction Between Sedative and Consciousness-Expanding Drugs" to the 69th Annual Convention of the American Psychological Association. The paper was published in a literary magazine in 1964 and reprinted in Solomon, *Marijuana Papers* (the quotes are from the latter, 443, 445–446).
15. A court in Massachusetts convicted Burroughs's book *Naked Lunch* of obscenity in 1965—see Whiting, "Monstrosity on Trial."
16. The quotes are from Ginsberg, "Henri Michaux," 446–448. See also "Ginsberg Makes the World Scene," *The New York Times*, July 11, 1965.
17. Concerning Ginsberg's activities during this period, see Schumacher, *Dharma Lion*.
18. The quote is from Ginsberg, "First Manifesto," 231, an extended version of the original essay published in *The Atlantic Monthly*.
19. The quotes are from Ginsberg, "First Manifesto," 244–245, 231.

## 10: WORLD SUNLIGHT CHAMPION

1. The quote is from Small, *Cannabis*, 1. See Courtwright, *Forces of Habit*, 44, on the globalization of cannabis smoking during the 1970s. See National Commission on Marihuana, *Marihuana*, 7, for the national survey.
2. Regarding the cannflavins and other flavonoids found in cannabis, see Barrett et al., "Isolation from Cannabis Sativa L."; dos Santos and Romão, "Cannabis," 4–7; and Fordjour et al., "*Cannabis*," 14. The plant produces more than twenty flavonoids, most of which have potent antioxidant properties—see Bautista et al., "Flavonoids in *Cannabis sativa*," 5119. Regarding the traditional use of fan leaves in bandages, see Singh and Singh, "Indigenous Plant," 116.
3. See Potter, "Propagation," 58–68; Bernstein et al., "Interplay Between Chemistry and Morphology," 185; and Jin et al., "Secondary Metabolites," 5–6. Terpenes are hydrocarbons, terpenoids are oxygen-containing terpenes, and cannabinoids are terpenoid derivatives—see Hanuš and Hod, "Terpenes/Terpenoids in *Cannabis*," 26.
4. See Pollan, *Botany of Desire*, 121–124. Clarke and Merlin, "*Cannabis* Domestication," 308, write about the seeds found in the commercial marijuana imported into temperate parts of North America and Europe during the 1960s and '70s: "Seeds from Colombian and Thai marijuana grown at more nearly equatorial latitudes rarely produced plants that matured to the floral stage when cultivated outdoors before cold autumn

weather killed them, whereas Mexican and Jamaican varieties often matured earlier, before harsh winter weather set in."

5. See Franz and Novak, "Sources of Essential Oils," 41, on the diverse ecological functions of terpenes. Zhao et al., "Volatile Terpenes and Terpenoids," 1: "In the subfamily Formicinae, a wide variety of monoterpenes are utilized as alarm pheromones, such as citronellal, citronellol, alpha-pinene, beta-pinene, limonene, and camphene."
6. Small, *Cannabis*, 199, writes: "About 30% of flowering plants possess 'glandular trichomes' [= stalked resin glands] producing secondary chemicals, usually at the tip of the structure, often in distinctive head-like containers. . . . The substances manufactured are frequently known to serve the plant as protective agents but are also immensely useful to humans as natural pesticides, food additives, fragrances, and pharmaceuticals." Substances produced by living organisms that are not essential for growth, development, or reproduction are called secondary metabolites. Bernstein et al., "Interplay Between Chemistry and Morphology," 192, comment: "Considering the defensive role secondary metabolites play and the importance of reproductive tissues to the plant, it is not surprising that many secondary metabolites are concentrated in floral tissue. This seems to be the case for most of the cannabinoids in cannabis as well."
7. See Tanney et al., "Cannabis Glandular Trichomes," 4; Franz and Novak, "Sources of Essential Oils," 42; and Potter, "Cannabis Horticulture," 66–67. Like many other plant species, *Cannabis* tailors its terpene production to specific needs.
8. See Booth and Bohlmann, "Terpenes in *Cannabis sativa*," 67. See Komenda and Koppmann, "Monoterpene Emissions," on the terpenes of the European pine species. A handful of plant species other than *Cannabis* have high terpene scores, such as the Central American shrub *Lippia integrifolia*, which contains 152 terpenoids—see Franz and Novak, "Sources of Essential Oils," 48; and Marcial et al., "Intraspecific Variation," 203. Sommano et al., "Cannabis Terpenes," 5, estimate that over 200 volatile terpenes or terpene-like substances have been identified in *Cannabis*.
9. See Liktor-Busa et al., "Analgesic Potential," 1273, and Potter, "Propagation," 17–20, on the chemical building blocks used to produce terpenes and cannabinoids. Russo, "Taming THC," 1352, identifies caryophyllene oxide as the volatile substance detected by drug-sniffing dogs. See also Booth and Bohlmann, "Terpenes in *Cannabis sativa*," 69. Pearlson, *Weed Science*, 221, describes CBD and THC as "completely odorless." The chemical formula of basic terpenes is $C_{10}H_{16}$, while the formula of cannabinoids such as CBD and THC is $C_{21}H_{30}O_{2}$, indicating that cannabinoids are larger and heavier molecules than basic terpenes. Small, *Cannabis*, 205, notes that several cannabinoids occur in other plant species and comments: "However, virtually all specialists on the cannabinoids are of the view that they are more characteristic of *Cannabis* than any other plant, and the major cannabinoids of *C. sativa* occur only in this species."
10. See Oswald et al., "Minor, Nonterpenoid Volatile Compounds," who report that volatile sulfur compounds comprise about 0.05% of the molecular constituents of *Cannabis* aroma but substantially impact the plant's odor. Some of these compounds are chemically identical to those of garlic. Others appear unique to the *Cannabis* plant—see Oswald et al., "Identification of a New Family."
11. The basic chemical building block of terpenes and cannabinoids is isoprene, which degrades into isovaleric acid when exposed to oxygen—see Liktor-Busa et al., "Analgesic Potential," 1273. See also Jones, "Isovaleric Acid," 498; Wolff, "What Makes Cannabis Smell"; and Ara et al., "Foot Odor."

12. Kerr, "Report of the Cultivation," 104.
13. The quotes are from Kerr, "Report of the Cultivation," 112, 114, 120, 104.
14. The quote is from the Indian Hemp Drugs Commission, *Report of the Indian Hemp Drugs Commission*, 92. The Commission defined *ganja* as "the dried flowering tops of cultivated female hemp plants which have become coated with resin in consequence of having been unable to set seeds freely" (p. 59).
15. For the 1915 United States Department of Agriculture (USDA) booklet, see Stockberger, "Drug Plants Under Cultivation," 19. The USDA made an original contribution in 1917 by noting that *Cannabis* selected for fiber tended to produce resins that "lacked the active principles" necessary to meet "the standard requirement for this drug." As a result, it advised farmers to grow seeds only from plants with well-established qualities—see Stockberger, "Production of Drug-Plant Crops," 171. Neither publication cites Kerr's report.
16. A Google Ngram Viewer search of Spanish-language texts published before 1979 shows no mention of the word *sinsemilla*. The *Oxford English Dictionary* states that *sinsemilla* first appeared in written American English in 1975. How and when the term was coined remains unclear. It seems that people in Mexico were growing seedless *Cannabis* flowers during the 1960s, and traveling Americans first imported this potent form of cannabis to the United States toward the end of that decade. According to Krassner, "The Secret History of Sinsemilla," an American named Blake Wheeler learned how to grow unpollinated *Cannabis* plants while living in Mexico. The local farmers who shared their knowledge with him referred to their cannabis as *sin hueso*, meaning "boneless"—a carnivorous compliment for a high-end plant product, suggesting "tenderloin grass" or "filet mignon marijuana." Wheeler claimed that when he first imported seedless Mexican marihuana to California, he renamed it *sinsemilla*, and the term spread from there. Levine, *Deep Cover*, 38, notes that Wheeler tended to tell self-aggrandizing and unverifiable stories. Gonzalez, "Sinsemilla," points out that since the eleventh century, brewers have applied the sinsemilla technique to the hop plant—*Cannabis'* closest botanical relative—because seeded hops yield bitter-tasting beer. Gonzalez writes that the sinsemilla technique "is simple enough that anyone with an eye for detail or a basic understanding of botany could figure it out at any point in history." Considering the sinsemilla technique relevant to more than one crop, Gonzalez comments: "Pollination, in itself, is a key concept that would have been understood by several ancient human cultures involved in early agriculture and plant breeding. The question of who created the first sinsemilla plant is not dissimilar to asking who invented pottery or who discovered fire. The answer is almost certainly that sinsemilla was discovered and rediscovered at many points on Earth and in history until it slowly became a staple of all human culture." In Spanish, when fruits such as grapes and oranges are seedless, they are referred to as *sin semillas* but never as *sinsemilla*. I thank historian Isaac Campos for his helpful feedback on the origin of the word *sinsemilla*.
17. See Johnson, *Grass Roots*, 125; and Clarke and Merlin, *Cannabis*, 227, 300.
18. The quotes are from Mountain Girl, *Primo Plant*, 10, 15, 14, 30, and 69. See Lee, *Smoke Signals*, 175–176, on Garcia's biography. Research confirms Garcia's claims about the nutritional qualities of *Cannabis* seeds—see Werz et al., "Cannflavins," 53, who write: "The nutritional value of hemp seeds can hardly be underestimated, since, virtually unique between food plants, they contain all essential amino acids and fatty acids in sufficient amount and ratio to meet human demand." See also Callaway and Pate,

"Hempseed Oil," 192–193; and Callaway, "Hempseed as a Nutritional Resource," 65. According to Hazekamp et al., "Chemistry of Cannabis," 3:1036, a large female plant can produce over 1 kg of seeds.

19. The quotes are from Mountain Girl, *Primo Plant*, 69–70, 14. Clarke and Merlin, *Cannabis*, 300–301, write that in the 1970s, "understanding how to grow *sinsemilla* allowed many growers to realize that most of the flowers could be cultivated without developing seeds; however, a few branches could be intentionally fertilized with only a tiny amount of select pollen to produce seeds of known parentage. This in turn gave birth to, or at least accelerated, the intentional breeding of potent drug *Cannabis*."
20. Rendon, *Super-Charged*, gives a detailed account of these events. See also Clarke, "Sinsemilla Heritage," and Marshall and Leeuwenberg, *Weed of Wonder*.
21. See Clarke and Merlin, *Cannabis,* 37, 357–358, who estimate that 10% of flowering plant species are wind-pollinated, and only one-third of these produce male and female flowers on separate plants. See also Rendon, *Super-Charged*, 17. Small, *Cannabis*, 430, writes that *Cannabis* pollen "is produced in prodigious quantities and is carried by the wind for very long distances" and estimates that a single flower on a male plant can produce about 350,000 pollen grains, and each plant can have hundreds of such flowers (p. 54). Male plants invest their resources not so much in resin glands as in generating vast quantities of pollen grains—which helps explain why their flowering tissues have many times fewer resin glands than those of females and consequently produce and hold much smaller amounts of cannabinoids.
22. Clarke and Merlin, *Cannabis*, 302: "In the absence of careful selection and breeding, drug *Cannabis* also begins to turn weedy, and as natural selection takes over, it loses its vigor, palatability, and potency."
23. The quotes are from "Marijuana Crops Revive California Town," *The New York Times*, March 11, 1979.
24. The quote is from Kayo, "Some Big Well-Lighted Places," 46.
25. See Kayo, *The Sinsemilla Technique*.
26. According to Pellati et al., "*Cannabis sativa* L.," 10, commenting on the difficulty of gaining medical approval for *Cannabis* extracts, "it must be recognized that, in some cases, an unconscious prejudice seems to hover on *C. sativa*, mainly because of its history of drug of abuse." Hesami et al., "Recent Advances," 16, write that due to the long history of *Cannabis* prohibition, "there is a huge gap in human knowledge of this valuable plant in many fields of science."
27. Llewellyn et al., "Indoor Grown Cannabis Yield," 8, report that increasing light intensity proportionally increases inflorescence biomass, which boosts inflorescence yield but does not change cannabinoid concentrations (potency). In other words, the quantity of cannabinoids produced by a plant may vary with the size of its inflorescences, but their quality does not. See also Naim-Feil et al., "Cannabis Plant." Rodriguez-Morrison et al., "Cannabis Yield," 12, write that increasing inflorescence size and density as light levels rise "is a common response in herbaceous plants, including *Cannabis*." See also Livingston et al., "Cannabis Glandular Trichomes," 43–49. Flores-Sanchez and Verpoorte, "PKS Activities," 1771–1773, report that the female inflorescences of psychoactive varieties produce eight times more THC than the male inflorescences of the same variety, and cannabinoid concentrations increase with the growth and density of new resin glands. Potter, "Propagation," 104, estimates that pollination can reduce the potential THC content of a female inflorescence by more than half. Regarding the

number of studies on cannabis, see NORML, "Analysis: Over 35,000 Scientific Papers Published About Cannabis During the Past Decade," December 23, 2024, https://norml.org/blog/2024/12/23/analysis-over-35000-scientific-papers-published-about-cannabis-during-the-past-decade/.

28. Light intensity can be measured in photosynthetic photon flux density (PPFD) units, which are "micromoles per square meter per second" (μmol $m^{-2}$ $s^{-1}$, or μmol for short). "Peak sun" corresponds to 2,200 μmol. Chandra et al., "Light Dependence," 42, report on the light use capacities of several psychoactive *Cannabis* varieties: "In this study, an increasing trend in photosynthesis was observed with light intensity up to the highest level tested (i.e., 2,000 μmol $m^{-2}$ $s^{-1}$) in all the *C. sativa* varieties." Rodriguez-Morrison et al., "Cannabis Yield," 9, report that *Cannabis* inflorescence yield increases linearly and more than fourfold as the light intensity rises from 120 μmol to the highest level tested, 1,800 μmol. They write: "The lack of a saturating yield response at such high light intensity is an important distinction between cannabis and other crops grown in controlled environments. . . . Effectively, within the range of practical indoor PPFD levels—the more light that is provided, the proportionally higher the increase in yield will be" (p. 10). They also refer to "the exceptionally high capacity that *Cannabis* has for converting photosynthetically active radiation into biomass" (p. 2). Llewellyn et al., "Indoor Grown," refer to the plant's "exceptional tolerance of high light intensity" and note that *Cannabis* tolerates light levels "several fold higher than for many other indoor-grown commodities" (p. 2). Eichhorn Bilodeau et al., "Update on Plant Photobiology," 4, write that the "light saturation point for *Cannabis* has not yet been determined." Poorter et al., "Meta-Analysis," 1084, show that herbaceous species have a much greater capacity than woody species to use increased light intensity, as measured in "whole-plant daily net photosynthesis" and biomass increase. Miner and Bauerle, "Seasonal Responses," 1565, write that "sunflower and maize did not always saturate" at light levels of 2,000 μmol, "near the full intensity of the sun." However, the light use efficiency of both plants diminishes at such high levels—see Collison et al., "Light," 4, and Samidjo, "Growth Pattern," 5. According to Gatot Supangkat Samidjo (personal communication, November 9, 2023), when sunflowers are exposed to full sunlight, their light use efficiency "will level off at 95% of the light intensity received." Hansen, "Growing Under High Light Intensities," and Neocision, "Grow Light Intensity," report that *Cannabis* yields continue to increase to light levels of 2,500 μmol—an upper limit determined by the research facility's power capabilities.

29. Llewellyn et al., "Indoor Grown," 9, write that plant species exposed to intense UV light commonly produce "photoprotective compounds, particularly in epidermal regions, to diminish UV penetration deeper into plant tissues." See Potter, "Propagation," 71–72, for a discussion of the *Cannabis* plant's allocation of cannabinoids to floral tissues.

30. As noted in chapter 1, the process by which THCA sheds carbon dioxide and becomes THC is called decarboxylation. It has been described as a predominantly thermal process, perhaps because humans heat the raw plant material to turn it into a psychoactive substance. Still, Hanuš et al., "Phytocannabinoids," 1362, write that "decarboxylation of cannabinoid acids is a continuous process, generating neutral cannabinoids already in early stages of the plant growth, and next continuing during all the vegetation stage." Filer, "Acidic Cannabinoid Decarboxylation," 267–270, reports concentrations of THC and CBD in the resin glands of flowering plants between 1% and 2% of those of their respective precursor molecules, THCA and CBDA, demon-

strating that decarboxylation occurs in the living plant. As the minimal thermal threshold for decarboxylation—about 115°C/240°F—seems difficult to achieve in a *Cannabis* plant during average summer temperatures, Filer writes that the UV energy of sunlight "might also facilitate the small amount of acidic cannabinoid decarboxylation observed in the plant." Filer notes that the process is understudied in living *Cannabis* plants, even though decarboxylation of acidic organic compounds by sunlight occurs in many other plant species. Additionally, Bernstein et al., "Interplay Between Chemistry and Morphology," 192, report that the top parts of *Cannabis* plants with more direct UV exposure have significantly greater concentrations of most cannabinoids, including twice as much THC in upper flowers than in lower ones, commenting: "This observation supports the suggested role of cannabinoids like THC as UV protectants, correlating with the areas of the plant receiving the most photoradiation." Ryu et al., "Conversion Characteristics," 7, write that *Cannabis* "undergoes decarboxylation in its natural state owing to the effects of environmental factors such as light, and also generates neutral cannabinoids." Fordjour et al., "*Cannabis*," provide the estimate of the number of cannabinoids. See also Small, *Cannabis*, 206.

31. Birenboim et al., "Quantitative and Qualitative Spectroscopic Parameters," provide precise measurements of the number of absorbed photons at a specific wavelength for 13 cannabinoids, including CBDA, THCA, CBD, THC, and CBN. Their study uses 280 nm as the benchmark wavelength, which corresponds to the UV-B range (280–320 nm), the most dangerous radiation naturally reaching Earth. At this wavelength, CBN exhibits the largest absorbance, followed by CBDA, THCA, CBD, and THC—with CBN absorbing roughly twice as many photons as CBDA, eight times more than CBD, and nine times more than THC. See also Vacek et al., "Antioxidant Function," 265–268; Llewellyn et al., "Indoor Grown," 2. Regarding CBD's degradation product HU-331, see Wilson et al., "HU-331 and Oxidized Cannabidiol," 139.
32. Hampson et al., "Cannabidiol," 8271, discuss the capacity of CBD and THC to donate electrons as the basis of their antioxidant properties and refer to the "ability of cannabinoids to be oxidized." Potter, "Propagation," 100, discusses the function of cannabinoids in preventing the oxidation of terpenes. See also Pellati et al., "*Cannabis sativa* L.," on CBD's antioxidant properties.
33. See Stack et al., "Cannabinoids Function in Defense Against Chewing Herbivores," 5–6. See also Park et al., "Contrasting Roles"; and McPartland and Sheikh, "Review of *Cannabis sativa*-based Insecticides."
34. Regarding cannabinoids as heat and drought shields and insect-glue components, see Gülck and Moller, "Phytocannabinoids," 987, and Small, *Cannabis*, 218.
35. Stasilowicz-Krzemien et al., "Determining Antioxidant Activity," 1, refer to the "multidirectional biological activity" of cannabinoids. See also Berman et al., "New ESI-LC/MS Approach," 11.
36. Stack et al., "Cannabinoids Function in Defense Against Chewing Herbivores," 1: "In the decades since the first cannabinoids were identified by scientists, research has focused almost exclusively on the function and capacity of cannabinoids as medicines and intoxicants for humans and other vertebrates. Very little is known about the adaptive value of cannabinoid production, though several hypotheses have been proposed, including protection from ultraviolet radiation, pathogens, and herbivores." See also André et al., "Dietary Antioxidants"; Lowe et al., "Non-Cannabinoid Metabolites," 2; and Hanuš and Hod, "Terpenes/Terpenoids," 25. Hybertson et al., "Oxidative Stress,"

244, note that "oxidative stress may indeed be *associated* with 200 diseases, and even *contributory* to all of them, but not necessarily *causative* in every case."

37. Pellati et al., "*Cannabis sativa* L.," 2: "Thanks to its lack of psychoactivity, CBD is one of the most interesting compounds, with many reported pharmacological effects in various models of pathologies, from inflammatory and neurodegenerative diseases to epilepsy, autoimmune disorders like multiple sclerosis, arthritis, schizophrenia, and cancer." See Atalay et al., "Antioxidative and Anti-Inflammatory Properties," 5, on CBD's aptitude to protect lipids and proteins against oxidative damage. See also Burstein, "Cannabidiol," 1382; and Tomko et al., "Anti-Cancer Potential," 12. With a focus on Parkinson's disease, Taylor et al., "Neuroinflammation and Oxidative Stress," 805, consider the link between oxidation, inflammation, and illness and discuss how the oxidation of proteins in human cells can disrupt intracellular signaling and lead to cellular degeneration and death. Trac et al., "Cannabidiol Oxidation," 1, discuss the capacity of CBD's degradation product, HU-331, "to shrink tumors without causing cardiotoxicity" in mice. See also Wilson et al., "HU-331 and Oxidized Cannabidiol"; and Caprioglio et al., "Cannabinoquinones."

38. See Filer, "Acidic Cannabinoid Decarboxylation," 267. Since not all the THCA in fresh cannabis converts to THC when heated, the "total THC" of a cannabis sample is calculated by adding the amount of THCA multiplied by the correction factor 0.87 to the amount of THC—see Jin et al., "Secondary Metabolites," 5. THCA has potential anti-inflammatory, immunomodulatory, neuroprotective, and antitumor properties—see Moreno-Sanz, "Can You Pass the Acid Test," 124. See also Russo and Marcu, "Cannabis Pharmacology," 81. Banister and Connor, "Chemistry and Pharmacology," 170, write that CBN and CBND "are also psychoactive but are found in cannabis at concentrations likely insufficient to contribute to psychoactivity." Citti et al., "Novel Phytocannabinoid," report on a newly discovered psychoactive cannabinoid—see discussion in chapter 16.

39. See Hampson et al., "Cannabidiol," on CBD and THC as antioxidants and neuroprotectants. Piomelli and Russo, "*Cannabis sativa* Versus *Cannabis indica*," 44, write that one of THC's degradation products, delta-8-THC, "is probably less psychoactive, but is present only in trace amounts" and another, cannabinol (CBN), "is about 25% of the potency of THC." See Pellati et al., "*Cannabis sativa* L.," on the similar properties of THC and CBD; Russo, "Taming," on THC's anti-inflammatory and anti-spasmodic activities and CBD's capacity to attenuate THC's psychoactive effects; and Laprairie et al., "Cannabidiol," on CBD's capacity to reduce the potency and efficacy of THC. According to Piomelli and Russo, "*Cannabis sativa* Versus *Cannabis indica*," 44, although CBD is nonintoxicating, "it certainly has antianxiety, antipsychotic, and even antidepressant effects, so properly they must be considered psychoactive."

40. Pollastro et al., "Cannabis Phenolics," 1182: "*Cannabis sativa* L. has an incredible biosynthetic capacity. The plant not only produces cannabinoids . . . but also a plethora of non-cannabinoids second metabolites which are unique of this plant, such as prenylated flavonoids, stilbenoids derivatives and lignanammides." Zandkarimi et al., "Comparison," 2, refer to the plant's production of cannabinoids and terpenes as "staggering and remarkable." Liktor-Busa et al., "Analgesic Potential," 1273, describe the biochemical diversity of *Cannabis* as "unique and remarkable." Regarding Potter's hypothesis, see Rendon, *Super-Charged*, 39; Potter, "Propagation," 84; and Potter, "Cannabis Horticulture," 82.

41. Potter, "Propagation," 100, writes that the high terpene content of the plant's resin

glands "is possibly due to human influence over many centuries. The highest monoterpene contents are most likely found in plants producing high quantities of resin. These plants would have been selectively planted by growers striving for maximum resin yields. Many would also find the odor attractive and of greater demand for those using fragrant resins as religious sacraments. Focused plant breeding of cannabis for pharmaceutical use continues to make dramatic changes to the glandular trichome contents."

### 11: INDOOR OUTDOOR HYPERACCUMULATOR

1. See Rendon, *Super-Charged*, 18; Frank and Rosenthal, *Marijuana Grower's Guide*, 265; and Clarke and Watson, "*Cannabis* and Natural *Cannabis* Medicines," 12. Clarke, "Sinsemilla Heritage," 4, notes: "Traditional cultivars gave modern growers a strong start, having been favored and selected for potent landrace varieties for hundreds of years."
2. See Small, *Cannabis*, 242; Thornton, "Potency of Illegal Drugs," 733; and Block, "Drug Prohibition." Leggett, "Review of the World Cannabis Situation," 97, notes that several factors beyond selective breeding contributed to the rise in potency during this period, including the sinsemilla technique, more rigorous manicuring of harvested flowers, and a better understanding of ripeness and storage techniques.
3. The DCE/SP launched nationwide operations in 1982, and between then and the end of 1989, the program eradicated 538,758,880 plants and made 40,790 arrests (see DEA, "1989 Domestic Cannabis Eradication/Suppression Program," 7). From 1990 to 1998, it eradicated an additional 2,563,800,000 plants—98.3% of which were classified as "ditchweed," defined as "uncultivated marijuana that grows wild"—and its operations resulted in 113,828 arrests (see United States General Accounting Office, "Drug Control," 47). Taken together, these figures show that over the 17-year period from 1982 to 1998, the DCE/SP eradicated approximately 3.18 billion *Cannabis* plants and made a total of 154,618 arrests. Eradication methods often relied on herbicides such as paraquat, glyphosate, and 2,4-D, which the DEA described as "environmentally safe" and capable of destroying millions of plants "at significant cost savings" (see DEA, "1993 Domestic Cannabis Eradication/Suppression Program," 2). In some cases, large-scale eradication efforts involved physically removing plants rather than spraying them with herbicides. In one notable instance—in West Virginia in September 1991—National Guard troops and state police pulled up and burned more than 200,000 wild-growing plants, some reportedly "15 feet tall or taller," with stalks described by a state trooper as "big around as your arm." Despite the apparent success of the operation, some local authorities expressed doubt that it would prevent the plants from sprouting again the following year (see "Town's 'Crop' Is Marijuana, Wild and Free," *The New York Times*, December 1, 1991). Indeed, removing them in late summer likely caused countless ripe seeds to fall to the ground, aiding the regeneration of this annual species in subsequent seasons. Yet eradication operations, which typically relied on helicopters to detect *Cannabis* from the air, were timed for late summer, when the plants were large enough to be visible from above. See note 34 of chapter 16 for further information on ditchweed eradication. See Johnson, "American Weed," 13, and Rendon, *Super-Charged*, 104, 147, on the move indoors in response to the increased dangers of outdoor cultivation.
4. See Cervantes, *Indoor Marijuana Horticulture*, 230, on the acceleration of breeding through indoor cultivation. The quote is from Potter, "Cultivation and Processing of Cannabis," 35.

5. See Hager, "Inside Cannabis Castle," 4; Dronkers, "History of Cannabis," 43–44; Lee, *Smoke Signals*, 181; Rendon, *Super-Charged*, 111–112; and Jansen, "Economics of Cannabis-Cultivation." Duvall, "Brief Agricultural History," 15, notes that the Netherlands relaxed legal controls on cannabis in 1976, which allowed Amsterdam to become the center of the global *Cannabis* seed industry.
6. See Clarke and Watson, "*Cannabis* and Natural *Cannabis* Medicines," 15; Rosenthal, "Growing Dutch"; and Rosenthal, "Miniature Forest," 84.
7. See Cervantes, *Indoor Marijuana Horticulture*, 200–201.
8. See Clarke and Merlin, *Cannabis*, 307; Burdon and Aimers-Halliday, "Risk Management," 158; and Koeppel, *Banana*, 32.
9. The quote is from Leggett, "Review of the World Cannabis Situation," 30.
10. Regarding light levels in grow rooms, see Mills, "Carbon Footprint," 59, and Summers et al., "Greenhouse Gas Emissions," 646–647. The total amount of light delivered to a specific area over twenty-four hours, known as the daily light integral (DLI), is calculated in moles per square meter per day; at the sunniest locations on Earth, values can reach 70 moles/m$^2$/day; by comparison, they approach 100 moles/m$^2$/day in certain grow rooms—see Poorter et al., "Meta-Analysis," and Hansen, "Growing under High Light Intensities." See Cervantes, *Indoor Marijuana Horticulture*, 157, and Stevens, *How to Grow the Finest Marijuana Indoors*, 24, on the impact and implications of $CO_2$ enrichment on plant growth.
11. The quote is from Cervantes, *Indoor Marijuana Horticulture*, 176, and the italics in the quote are original. For other examples, see Stevens, *How to Grow the Finest Marijuana Indoors*, 70, 73; Frank, *Marijuana Grower's Insider's Guide*, 57, 255; and Rosenthal, *Closet Cultivator*, 112. Ahmed et al., "Arthropod and Mollusk Pests," 10: "The major hemp pests in indoor hemp cultivation are most commonly aphids, mites, thrips, and whiteflies, whereas the major outdoor pests would be beetles, armyworms, borers, maggots, snails, and stink bugs." Clarke and Merlin, *Cannabis*, 51: "In fact, pest populations can become rampant indoors under lights and in glasshouses, and the highest THC clones are no more resistant to pathogens and pest infestations than lower THC clones."
12. Johnson, "American Weed," 13: "By the time California became the first state to relegalize marijuana for medicinal use in 1996, indoor growing was the established norm for America's cannabis cultivators." For studies focusing on the THC content of confiscated herbal cannabis, see King et al., *Overview of Cannabis Potency;* ElSohly et al., "Potency Trends of Delta-9-THC"; and Mehmedic et al., "Potency Trends of Delta-9-THC and Other Cannabinoids," 1209, who report an upward trend in mean THC content of all confiscated cannabis preparations in the United States from 3.4% to 8.8% between 1993 and 2008.
13. The quote is from Clarke and Merlin, *Cannabis*, 324. Small, *Cannabis*, 6, 34, 49, 129: "Numerous domesticated plants and animals have been so drastically altered by selection that they cannot survive without the assistance of humans. Domesticated kinds of *C. sativa* and domesticated horses, however, are frequently very hardy, and when they escape to the wild, they are often capable of living on their own. . . . *Cannabis sativa* growing outside of cultivation is basically a weed, growing mostly in habitats created or modified by humans. . . . There are numerous animal, fungal, bacterial, and viral species that affect domesticated *C. sativa*, although the species is resistant to all biotic attacks. . . . As a crop, *C. sativa* is remarkably resistant to many pests." Institut Technique du Chanvre, *Le Chanvre industriel*, 22: "Hemp is known to be immune to most viral,

bacterial and fungal diseases. This is mainly because the varieties available to farmers are . . . heterogeneous regarding pathogen resistance. Generally speaking, the crop does not experience loss of yield linked to disease and requires no anti-fungal treatment between sowing and harvest."

14. The quote is from McPartland et al., *Hemp Diseases and Pests*, 95. See also McPartland, "Survey of Hemp Diseases," 110, and Potter, "Propagation," 170–171. Potter notes that male plants are significantly more susceptible to attacks by insect predators than females. Small, "Classification of *Cannabis sativa* L.," 1: "Wild plants tend to be comparatively resistant to drought, cold, shade, and wind, and probably also to damaging biotic agents ranging from microorganisms to large grazing animals." See McPartland et al., *Hemp Diseases and Pests*, 152–153, 93, on wild animals and gray mold.
15. The quote is from Small, *Cannabis*, 407.
16. Rendon, *Super-Charged*, 204, writes about the stems of certain indoor plants: "Marijuana plants have been bred to produce such a large volume of buds that, like factory-farmed chickens that have trouble walking because they have such large breasts, these plants cannot support themselves. The plants need to be staked so that as the buds mature, they don't break the stem that supports them." Matzneller et al., "Canopy Management," 208, write about indoor *Cannabis* production: "Modern elite cultivars often have a high inflorescence-to-branch ratio. If left unsupported, stems may break, or plants can fall due to the inflorescence weight during late flowering. Plant supports such as trellising or stakes are common in other high-intensity cropping systems, where plants are unable to support their own weight." See Potter and Duncombe, "Effect of Electrical Lighting Power," 620, and Moher et al., "Light Intensity," 4, on light intensity increasing the proportion of floral material. See also Clarke and Watson, "*Cannabis* and Natural *Cannabis* Medicines," 5.
17. See Ahmad et al., "Phytoremediation Potential," 199.
18. See Raimondi et al., "Phytomanagement of Chromium-Contaminated Soils"; Ahmad et al., "Phytoremediation Potential"; Ćaćić et al., "Evaluation of Heavy Metals Accumulation"; and Girdhar et al., "Comparative Assessment."
19. See chapter 1 of this book on the plant's evolutionary past as a hyperaccumulator and pioneer plant.
20. McPartland and McKernan, "Contaminants of Concern," 467, report that plants with mycorrhizal fungi growing in their roots translocate more heavy metals from roots to shoots, perhaps because such plants are healthier, "and therefore can better tolerate heavy metal stress."
21. See Johnson, *Grass Roots*, 147–153, and Lee, *Smoke Signals*, 202–207, 227–257. In a study published in 2021, Hussain et al., "*Cannabis sativa* Research Trends," 4, divide scientific research on *Cannabis* into three periods (1840 to 1937; 1937 to 1996; and 1996 to present), the last of which "began with the historical Compassionate Use Act of 1996 in California approving medical cannabis." During the third and most recent period, they report that "cannabis research witnessed an unprecedented growth" with the publication of 67,777 articles, compared to 8,888 articles in the second period and only 1,923 in the first.
22. The two quotes are from Russo, "Letter: Cannabinoid Medicines," 16. Hazekamp, "Letter: The Herbal Way," 20, notes that medicinal users who smoked the plant "put cannabis back on the political and pharmaceutical agenda in the first place." Morgan, quoted in Randall, *Cancer Treatment and Marijuana Therapy*, 175: "Marijuana also

allows for careful patient control over the dose. Titration—matching dose to need—can be done by the patient and can be carefully controlled. This helps to assure that patients get relief without throwing themselves into an overdose situation. Synthetic THC, which is swallowed, does not permit this type of patient titration." See also Grotenhermen, "Pharmacokinetics and Pharmacodynamics," 328. Bernstein et al., "Interplay Between Chemistry and Morphology," 188: "Cannabinoid content in organs found at various heights along the cannabis plant was analyzed. Significant variation was observed among differing heights for almost every cannabinoid studied. For most cannabinoids, the content increased with plant height and was greatest at the top of the plant. . . . Interestingly, this gradient was present both in the flowers as well as the inflorescence leaves. The variation was substantial, and THC content in the upper flowers was 12% compared with only 6% found in the lower flowers." They note that this "natural chemical variation observed even within the same plant makes accurate dosing almost impossible" (p. 190).

23. See Scholten, "Guidelines for Cultivating Cannabis," 59, 56, for the first two quotes, and Hazekamp, "Letter: The Herbal Way," 20, for the next two.

24. Gorelick and Bernstein, "Chemical and Physical Elicitation," 440, discuss the "inherent biological variation" of plants. Russo, "Letter: Cannabinoid Medicines," 16, writes about the tension between "crude herbal materials" and "modern medicines."

25. See Sullivan et al., "Determination of Pesticide Residues," 2, regarding the 2009 case in Los Angeles. National Library of Medicine, "Bifenthrin," 91, reports that the concentration of bifenthrin in soil takes more than three years to fall to half its initial value. The quotes are from Voelker and Holmes, "Pesticide Use," 3, 12.

26. See National Library of Medicine, "Bifenthrin." See Sullivan et al. "Determination of Pesticide Residues," 5, on the percentage of residues present in cannabis that transfer directly into the mainstream smoke. Chin et al., "Pharmacodynamics, Pharmacokinetics, and Potential Drug Interactions," 51: "When smoked, about 50% of the cannabis contents (THC) combust to smoke, of which 50% of inhaled smoke is exhaled again while some of the remaining smoke undergoes localized metabolism in the lungs. The bioavailability or the percentage which actually makes its way into the bloodstream, of inhaled THC, is highly variable and ranges between 10% and 25%." Other experts make different estimates on this question; for example, Van der Kooy et al., "Cannabis Smoke Condensate II," report that 32.7% of THC in the plant material transfers to the smoke; and Hazekamp et al., "Evaluation of a Vaporizing Device," 9, note that a person who consumes cannabis with a vaporizer ends up taking in "30–40%" of the THC contained in the herbal material, "which is comparable to the efficiency reached by smoking of cannabis." See also Kalant, "Medical Uses," 83.

27. The lack of research on the effects of inhaling pesticide residues in smoke is mainly due to the tobacco industry's successful efforts to evade regulatory scrutiny, even though tobacco is one of the most chemically treated crops. The industry has long maintained that pesticides are essential to tobacco production and that the crop cannot be grown economically without them. For decades, tobacco companies have lobbied against stricter regulations, funded misleading research, and downplayed health risks—see López Dávila et al., "Pesticides Residues"; McDaniel et al., "Tobacco Industry"; and Proctor and Schiebinger, *Agnotology*. See Dryburgh et al., "Cannabis Contaminants," 2474, on the cannabinoids and contaminants delivered by inhalation techniques like smoking, vaporizing, and dabbing. Research suggests that smoking cannabis through a cotton

filter significantly diminishes the amounts of inhaled pesticides—see Sullivan et al., "Determination of Pesticide Residues."

28. Veit, "Quality Requirements," 813: "As the inflorescences are not cut and further crushed when producing the flowers as medicinal cannabis, complete homogenization is not possible, and therefore a certain variability of the cannabinoid contents must be expected. This ranges from ± 10 to 20% for the cannabinoids THC and CBD to be declared. However, this is only possible if the plants are spread, cultivated, and harvested under strictly controlled conditions and stable clones are used as cultivars and, so far, usually only when cultivation takes place indoors." See Wikipedia, "Medical Cannabis," for the list of countries that have legalized the use of medical cannabis.
29. See Chandra et al., "Cannabis Cultivation," 309; Scholten, "Guidelines for Cultivating Cannabis," 61; and Yadav et al., "Production Technology," 459. The quote is from Bilalis et al., "*Cannabis sativa* L.," 146.
30. The quote is from Potter, "Cultivation and Processing of Cannabis," 31.
31. See Russo, "Current Therapeutic Cannabis Controversies," 11, and Punja, "Emerging Diseases," 2, on the contamination risks of indoor, greenhouse, and outdoor cultivation.
32. See Chandra et al., "Cannabis Cultivation"; Potter, "Cultivation and Processing of Cannabis"; Yadav et al., "Production Technology"; and Wartenberg et al., "Cannabis and the Environment." The quote is from Jin et al., "Cannabis Indoor Growing Conditions," 925.
33. In 2023, Yadav et al., "Production Technology," 460, write: "Today, there are signs of recent growth in indoor cultivation, notably in the United States, Canada, Chile, Uruguay, Colombia, and Ecuador. While precise estimations of growing area and output volumes are impossible because of a lack of empirical data, *Cannabis* cultivation has been recorded in 151 countries between 2010 and 2018, demonstrating the enormous geographical extent of production activities." In their view, "commercial *Cannabis* cultivation is primarily done indoors" (p. 464). Potter et al., "Domestic Cannabis Cultivation," report based on an anonymous online survey conducted in 2021 across 18 countries, that most growers cultivate *Cannabis* indoors; only Australia and Uruguay had a majority of outdoor growers. Regarding studies of pesticide residues in illicit cannabis, see Stempfer et al., "Analysis of Cannabis Seizures"; Cuypers et al., "Use of Pesticides"; and Gagnon et al., "High Levels of Pesticides." The quote is from Seltenrich, "Into the Weeds," 1.
34. See Wikipedia, "Legality of Cannabis," for the list of countries that have legalized the recreational use of cannabis. The first quote is from Ministère de la Santé, "Cultivation of Cannabis at Home," 2, and the second from Constitutional Court of South Africa, "Minister of Justice and Constitutional Development v Prince," 50. See also Justice Canada, Cannabis Act, 13–14, and Parliament of Uruguay, Ley No. 19.172, 5. Thailand currently allows adults to cultivate an unlimited number of plants in outdoor plantations as long as they register the crops with local authorities—see Sabaghi, "First Asian Country to Decriminalize Recreational Cannabis."
35. According to Hahner, "Zweitgutachten erwartet im Schweizer Pilotprojekt 'Weed Care,'" laboratory analysis found 0.1 to 0.2 parts per million (ppm) of synthetic fungicide fluopyram in the pilot project's *Cannabis* plants. Swiss authorities had set the maximum tolerated amount of pesticide residue at the lowest limit of detection, namely 0.01 ppm—see Confédération Helvétique, *Ordonnance sur les essais pilotes*, 13. The European Food Safety Authority, "Modification of the Existing Maximum

Residue Levels," currently tolerates up to 0.8 ppm of fluopyram on various food crops. I thank Pure Production's Head of Innovation & Regulation, Marc Brüngger, for explaining the case details.

### 12: BEYOND THC

1. Clarke and Merlin, *Cannabis*, 52, write that "ancient populations of *Cannabis* may have had an average THC content around one to two percent and that some plants may have reached levels reaching three percent or more." They write about present-day self-sown *Cannabis*: "The vast majority of plants contain approximately equal amounts of CBD and THC and a total cannabinoid content of 2 to 3 percent dry weight" (p. 49). Small, *Cannabis*, 33, writes that wild-growing *Cannabis* plants "are either escapes from domesticated forms or the results of thousands of years of widespread genetic exchange with domesticated plants, making it virtually impossible to determine if unadulterated primeval or ancestral populations still exist." See also McPartland, "Cannabis Systematics," 206. See Wang et al., "Quantitative Determination of Delta9-THC," on THC and CBD as the "major constituents" of the plant's cannabinoid profile. De Backer et al., "Evolution of the Content of THC," 918, write that THCA and CBDA, along with their direct and common precursor CBGA, are "the most common types of acidic cannabinoids." To keep things simple in the main text, I mainly refer to the neutral forms of these cannabinoids instead of their acidic forms.
2. Clarke and Merlin, *Cannabis*, 319, write that "the average cannabinoid ratio is easily altered through selective breeding and within a few generations a population can produce almost all THC or almost all CBD. The chances that a population would naturally evolve to be high in THC and low in CBD are very low." See also Pearlson, *Weed Science*, 85, 146; and Grassa et al., "New *Cannabis* Genome Assembly," 1665. Recent studies suggest that THC-rich plants carry a nonfunctional gene for the CBD enzyme and multiple copies of the gene for the THC enzyme, which causes them to convert all the precursor compounds into THC—see Grassa et al., "New *Cannabis* Genome Assembly"; Weiblen et al., "Gene Duplication"; McKernan et al., "Sequence and Annotation"; Laverty et al., "Physical and Genetic Map"; and Vergara et al., "Gene Copy Number." See Swift et al., "Analysis of Cannabis Seizures," on hybrid *Cannabis* varieties with more than 30% THC and negligible amounts of CBD. Small, *Cannabis*, 210, notes that high-THC plants crossed with low-THC plants result in a first generation of hybrids with intermediate potency levels between those of the parent plants.
3. Clarke and Merlin, *Cannabis*, 52–53: "Modern European selections for extremely low THC content (formerly below 0.3 percent and now below 0.2 percent dry weight) cultivars were developed in reaction to perceived increases in *Cannabis* smoking by Europeans, based on the unreasonable fear that fiber hemp can and will be smoked for mind-altering purposes. . . . In sum, extremely low-THC with high-CBD hemp varieties and extremely high-THC with low-CBD drug varieties were derived by disruptive human selection for either industrial hemp or recreational and/or medicinal drugs from a median THC level, likely around 1 to 3 percent dry weight." See also Clarke and Watson, "*Cannabis* and Natural *Cannabis* Medicines," 12. Chandra et al., "Cannabis Cultivation," 304–305, noting that some taxonomists suggest the existence of two different *Cannabis* species or subspecies based on the different cannabinoid profiles, write: "However, the differences in cannabinoid profile are purely the result of

human intervention. No strict natural relationships between fiber characteristics and cannabinoid content exist, only artificial associations for which exceptions occur.... The fact that CBD is the dominant cannabinoid in Western hemp is purely the result of plant breeding and local politics. Licenses to grow hemp in Europe are only issued if growers use varieties that are capable of producing no more than 0.2% weight for weight of THC." See also de Meijer, "Chemical Phenotypes," 98–99.

4. The quote is from Riboulet-Zemouli et al., "International Scientific Assessment of Cannabis," 5. Regarding aspirin, see Joy et al., *Marijuana and Medicine*, 19. See also Mead, "International Control of Cannabis," 45. Archer et al., "Nabilone," 86, write, referring to the Lilly Research Laboratories cannabinoid research program initiated in 1968 and centering on the pharmacological actions and metabolism of THC: "The Lilly cannabinoid program set as its objectives the discovery of compounds that would possess the beneficial effects of marijuana while minimizing the disturbing side effects of tachycardia and dysphoria [= state of unease] found with THC."
5. American pharmaceutical companies involved in cannabinoid research during the 1970s include Eli Lilly, Pfizer Inc., Sisa Inc., and Unimed—see Archer et al., "Nabilone," and Segal, "Cannabinoids and Analgesia." See Joy et al., *Marijuana and Medicine*, 200, regarding regulatory decisions about cannabinoids not found in the plant.
6. Stephanie Shubat, Director of the United States Adopted Names Program, kindly provided information on the origin of the name dronabinol (personal communication, April 14, 2024). For more on dronabinol, see Iversen, *Science of Marijuana*, 127.
7. Silverberg, quoted in Randall, *Cancer Treatment and Marijuana Therapy*, 129.
8. The New York report is quoted in Randall, *Cancer Treatment and Marijuana Therapy*, 231–232. See Pearlson, *Weed Science*, 113, for an account of the study at Johns Hopkins University. In 2003, Russo, quoted in Russo and McPartland, "Cannabis Is More," 432, stated that he "has had the opportunity to interview an estimated 200 patients who have employed Marinol and clinical cannabis, whether smoked or ingested. In no instance were the effects of the former considered of equal efficacy to cannabis, but rather more productive of dysphoric and sedative adverse effects." See also Calhoun et al., "Abuse Potential," 187.
9. For early studies of active cannabinoids besides THC, see Karniol et al., "Cannabidiol Interferes"; Dalton et al., "Influence of Cannabidiol"; and Zuardi et al., "Action of Cannabidiol." See also Zuardi, "Cannabidiol: From an Inactive Cannabinoid"; Zuardi et al., "Cannabidiol as an Antipsychotic Drug"; Russo and Guy, "Tale of Two Cannabinoids"; and McPartland and Russo, "Non-Phytocannabinoid Constituents." Crocq, "History of Cannabis," 226: "CBD was first isolated from marijuana in 1940, and its structure was reported in 1963. However, since CBD was not psychoactive, it was neglected and eclipsed by THC." See Segal, "Cannabinoids and Analgesia," 114, and Iversen, *Science of Marijuana*, 128, on the failure of pharmaceutical companies to dissociate the desired medical effects from the psychotropic actions of synthetic cannabinoids during this period. Archer et al., "Nabilone," 88, estimate nabilone to be 7.5 times as psychoactive as THC. See also Lemberger, "Nabilone," 564; Russo, "Solution to the Medicinal Cannabis," 170; and Ware, "Synthetic Psychoactive Cannabinoids." Riboulet-Zemouli, "'Cannabis' Ontologies I," 20–21, classifies nabilone among "synthetic cannabinoid analogs," meaning "non-naturally occurring compounds, obtained by full chemical synthesis" that mimic the pharmacological effects of THC without being derived from, nor directly relatable to it.

10. Lichtman and Martin, "Analgesic Properties of THC," 516, write that CP 55,940 is "approximately 4 to 25 times more potent than delta9-THC depending upon the pharmacological measure." Regarding Pfizer's work on synthetic cannabinoids, see Johnson and Melvin, "Discovery of Nonclassical Cannabinoid Analgetics"; Pertwee, "Cannabinoid Pharmacology"; and Thomas et al., "Cannabinoid Designer Drugs." See Flemming et al., "Chemistry and Biological Activity," 34, on CP 55,940's painkilling potency. Also see Weissman et al., "Cannabimimetic Activity," and Herkenham et al., "Characterization and Localization," 564. Nutt, *Drugs Without the Hot Air*, 114, writes regarding the synthetic cannabinoids tested on healthy human volunteers that "they were all so unpleasant and powerful that none were taken forward as a medicine." Banister and Connor, "Chemistry and Pharmacology," 175, 171: "In the 1970s and 1980s, Pfizer undertook systematic deconstruction and elaboration of Delta9-THC, filing patents around libraries of compounds lacking the pyran ring of classical cannabinoids. . . . Classical cannabinoids are those based on the tricyclic benzopyran scaffolds of Delta9-THC, such as nabilone and HU-210. . . . Nonclassical cannabinoids lack the pyran ring of Delta9-THC and are typified by CP 47,497-C8 and 55,940." See Howlett et al., "International Union of Pharmacology," 168, and Pertwee et al., "International Union of Basic and Clinical Pharmacology," 592, for other definitions of nonclassical cannabinoids. Lambert and Fowler, "Endocannabinoid," agree that "CP" stands for Charles Pfizer. Still, Seely et al., "Spice Drugs," 236, point out that the nonclassical cannabinoids developed by Pfizer in the 1970s belong to the chemical category "cyclohexylphenol ('CP')."
11. See Melvin and Johnson, "Structure-Activity Relationships," 40; Iversen, *Science of Marijuana*, 32; Segal, "Cannabinoids and Analgesia," 115; Di Marzo et al., "Endocannabinoid System," 771; Appendino, "Early History," 926; and Pertwee, "Cannabinoid Pharmacology," S166. In 1987, Mechoulam et al., "Stereochemical Requirements," 23, created a synthetic cannabinoid with potent analgesic effects and no psychoactive activity, achieving for the first time "a complete dissociation between the cannabimimetic and the analgetic effects in a cannabinoid." The substance in question is the mirror image compound ("enantiomer") of an experimental synthetic cannabinoid with about one hundred times the potency of THC.
12. Opioid receptors are found on nerve cells, mainly in the central nervous system and the gut. Serotonin receptors are present in neurons and other cells distributed throughout the body. On the roles played by nicotine, LSD, and opium in the discovery of neuronal receptors and the endogenous substances that bind to them, see Changeux, "Discovery of the First Neurotransmitter Receptor"; Pollan, *How to Change Your Mind*, 23–24, 146–147, 292–293; and Pasternak and Pan, "Mu Opioids," 1260.
13. See Devane et al., "Determination and Characterization of a Cannabinoid Receptor," for the initial discovery; Howlett et al., "Stereochemical Effects," 164, on the inability to find a binding site with radiolabeled THC; Appendino, "Early History," 926, on THC sticking to biological membranes, making its binding site challenging to detect; Meyer and Quenzer, *Psychopharmacology*, 472; and Howlett et al., "Spicy Story," 1. Mechoulam and Ben-Shabat, "From *Gan-zi-gun-nu* to Anandamide," 134, refer to "the enigma" posed by the molecular basis of THC action, which lasted well into the 1980s.
14. See Herkenham et al., "Characterization and Localization." Madras, "Cannabinoid and Marijuana Neurobiology," 28: "Intriguingly, the brain distribution of this receptor also corresponded to the known pharmacological actions of THC, with high to moderate expression levels in regions implicated in coordination, learning, memory, thinking,

decision-making, pleasure, and emotional states." See also Pollan, *Botany of Desire*, 153; McPartland et al., "Evolutionary Origins"; and Clarke et al., "Endocannabinoid System," 8, who write in 2021: "The endocannabinoid system [ECS] is phylogenetically ancient, but its presence and function in invertebrate systems have only recently been the subject of investigation. CB1/CB2-type receptors-encoding genes were previously thought to be found only in those invertebrate groups most closely related to the vertebrates, specifically the phylum Chordata. Recent studies, however, have provided evidence supporting a role for the ECS in numerous invertebrate phyla."

15. See Devane et al., "Isolation and Structure" on discovering anandamide. See Devane et al., "Novel Probe," 2067, on the potency of HU-243; they note that differences in binding potency and psychoactive potency "are not unexpected, as binding and in vivo pharmacological data are not directly comparable." Mechoulam and Hanuš, "Historical Overview," 9, describe their grounds for searching for an endogenous cannabinoid: "It was quite unacceptable to most neuroscientists that the brain will waste its resources to synthesize a receptor in order to bind a constituent of a plant. The only reasonable assumption which could be made was that the brain must produce a neuronal mediator, a specific compound (or a family of compounds) which binds to and activates the cannabinoid receptor. The plant cannabinoid, THC, by coincidence happens to bind to the same receptor." Firman et al., "Chemoinformatic Consideration," 22–23, identify 223 compounds as synthetic cannabinoids and find that "a mere ten . . . bear structural relation to THC."
16. See Pertwee, "Pharmacological, Physiological and Clinical Implications"; Rinaldi-Carmona et al., "SR141716A"; Di Marzo and Fontana, "Anandamide"; Howlett et al., "International Union of Pharmacology"; and McPartland and Russo, "Cannabis and Cannabis Extracts," 105, who write that the term "phytocannabinoid" was coined in 1999.
17. See Di Marzo et al., "Endocannabinoids"; Felder and Glass, "Cannabinoid Receptors"; Di Marzo et al., "Endocannabinoid System"; Gorelick and Bernstein, "Chemical and Physical Elicitation," 441; Sam et al., "Rimonabant"; and Papaseit et al., "Cannabinoids."
18. The quote is from Russo, "Synthetic and Natural Cannabinoids," 8. See also Di Marzo et al., "Endocannabinoids"; and Mechoulam and Ben-Shabat, "From *Gan-zi-gun-nu* to Anandamide," 135, who write that THC and anandamide have "almost identical" binding values relative to CB receptors. See Pertwee et al., "International Union of Basic and Clinical Pharmacology," 593, on the similar binding affinities of anandamide and THC.
19. See Pertwee, "Sites and Mechanisms"; Madras, "Cannabinoid and Marijuana Neurobiology," 31; and Nutt, *Drugs Without the Hot Air*, 107–108.
20. The quote is from Di Marzo et al., "Endocannabinoids," 528. See also Meyer and Quenzer, *Psychopharmacology*; Chin et al., "Pharmacodynamics, Pharmacokinetics, and Potential Drug Interactions," 50–52; Pearlson, *Weed Science*, 64, 72–82, 87; Gorelick and Bernstein, "Chemical and Physical Elicitation"; Fride et al., "Critical Role of the Endogenous Cannabinoid System"; and Melamede, "Harm Reduction."
21. Bisogno et al., "Molecular Targets," 846, refer in 2001 to "the endocannabinoid signaling system." In 2003, Grotenhermen, "Pharmacokinetics and Pharmacodynamics of Cannabinoids," 342, writes: "Cannabinoid receptors and their endogenous ligands together constitute the 'endogenous cannabinoid system' or the 'endocannabinoid system'

that is teleologically millions of years old." The following year, Di Marzo et al., "The Endocannabinoid System and Its Therapeutic Exploitations," discuss the endocannabinoid system, without quotation marks, as an established bodily system. See also Lowe et al., "Endocannabinoid System," 1; Russo, "Synthetic and Natural Cannabinoids," 7; and Rendon, *Super-Charged*, 37.

22. See Di Marzo et al., "Leptin-regulated Endocannabinoids." On rimonabant's development and demise, see Sam et al., "Rimonabant"; Papaseit et al., "Cannabinoids," 1291; and Mechoulam et al., "Early Phytocannabinoid Chemistry," 760. See also Gadde and Allison, "Cannabinoid-1 Receptor Antagonist;" and Armentano, "FDA Halts the Anti-Pot Pill."

23. Madras, "Cannabinoid and Marijuana Neurobiology," 25, writes that the endocannabinoid system, compared to other signaling systems, "is among the most dense, widely distributed, and versatile in animals and humans."

24. See Zuardi, "Cannabidiol," 276; Zuardi et al., "Cannabidiol as an Antipsychotic Drug"; Russo and Guy, "A Tale of Two Cannabinoids"; Chin et al., "Pharmacodynamics, Pharmacokinetics, and Potential Drug Interactions," 50; Gingrich et al., "Oral Toxicity of Cannabidiol (CBD)," 7; and Pearlson, *Weed Scienc*e, 86–87.

25. The first quote is from Solowij et al., "Randomized Controlled," 18. The second is from Pearlson, *Weed Science*, 89. See also Gingrich et al., "Oral Toxicity of Cannabidiol (CBD)"; Childs et al., "Dose-Related Effects"; Small, *Cannabis*, 223; and Earlenbaugh, "Why Less Is Often More." See Curran et al., "Keep off the Grass?," 1, on THC and CBD's range of opposing effects.

26. Recent studies have demonstrated a statistically significant increase in the risk of heart attack and stroke among regular cannabis users, including young and otherwise healthy individuals. Jeffers et al., "Association of Cannabis Use," 6, report that this risk rises with frequency of use. Storck et al., "Cardiovascular Risk," 1, estimate that cannabis users face a 29% higher risk of heart attack, a 20% higher risk of stroke, and a 100% higher risk of death from cardiovascular disease compared to nonusers. Kamel et al., "Myocardial Infarction," 3, suggest the risk is highest within an hour of consumption. While such events remain relatively uncommon, they occur often enough to raise legitimate concern. Older adults and people with preexisting heart conditions are particularly vulnerable, as their cardiovascular systems are more fragile and cannabis use can further strain the heart. Experts suggest that cannabis users can reduce their risk of cardiovascular events by consuming less frequently and choosing products with lower THC concentrations—see Bajaj, "Marijuana's Links to Heart Attack and Stroke Are Becoming Clearer," *The New York Times*, June 19, 2025. See also Russo, "Synthetic and Natural Cannabinoids," 7; and O'Sullivan, "Phytocannabinoids," 219.

27. Lindigkeit et al., "Spice," detected JWH-018 and CP 47,497-C8 in Spice samples collected from German shops in late 2008 and early 2009. They noted that these products retailed for 10 euros per gram, compared to 5 to 7 euros per gram for locally available cannabis; they attributed their higher price to the added value of their "non-illegal" status. Auwärter et al., "'Spice' and Other Herbal Blends," 832–834, estimate that the "nonclassical cannabinoid" CP 47,497-C8 and the "synthetic cannabimimetic" JWH-018 have, respectively, four and ten times more potency than THC. See also Weissman et al., "Cannabimimetic Activity." Papaseit et al., "Cannabinoids," write in 2018: "As for the vast majority of NPS [= new psychedelic substances], there are no clinical trials specifically designed to evaluate their human pharmacology" (p. 1290). Pertwee et al., "In-

ternational Union of Basic Clinical Pharmacology," 592, write that substances such as JWH-018, which belong to the aminoalkylindole family of CB receptor agonists, "have structures that differ markedly from those of both classical and nonclassical cannabinoids." Seely et al., "Spice Drugs," 235, write that JWH-018 is "easily synthesized and has high pharmacological activity." Huffman et al., "Design, Synthesis, and Pharmacology," 563, differentiate between nontraditional cannabinoids, such as CP 55,940, which have some structural similarity with THC, and aminoalkylindoles, such as JWH-018, "which bear no obvious structural resemblance to other active cannabinoids." See also Yeager, "How K2 and Other Synthetic Cannabinoids Got Their Start."

28. Wiley et al., "Hijacking of Basic Research," 4, write that "perusal of the scientific literature is the likely method through which illicit manufacturers of synthetic cannabinoids discovered their existence and potential abuse liability." See also Lindigkeit et al., "Spice," 5; and Howlett et al., "Spicy Story," 17–22, who write that "each new chemical entity is essentially a pharmacological unknown with the inherent potential to produce unanticipated effects in users or their descendants" (p. 22). Lapoint and Nelson, "Synthetic Cannabinoids," 28, write regarding the legal circumstances in 2011 in the United States: "The authority for temporary schedule status was enacted by the Controlled Substances Act, specifically the Analogue Act, which essentially allows structurally similar drugs with abuse potential to be removed from sale pending investigation or legal action. Since these synthetic cannabinoids are not analogs of delta9-THC, they are not under the auspices of the Analogue Act." Seely et al., "Spice Drugs," 236, discuss the delay in prohibiting these compounds caused by their structural dissimilarities with THC.

29. See Uchiyama et al., "New Cannabimimetic Indazole Derivatives," who noted: "There is a recent trend that various synthetic cannabinoids are appearing on the illegal drug market shortly after their appearance in academic journals or patents. In fact, AB-FUBINACA was first reported by Pfizer in a patent only in September 2009, and then we found it in an illegal herbal product in 2012, as shown in this study" (p. 99). See also Trecki et al., "Synthetic Cannabinoid-Related Illnesses." Pulver et al., "EMCDDA Frameworks and Practical Guidance," 268, 271, 261, 257, outline how new synthetic cannabinoids are assigned names derived from their four main structural components—core, tail, linker, and linked group—which contain information about their structure. In the case of AMB-FUBINACA, *AMB* stands for the linked group "1-amino-3-methyl-1-oxobutane"; *FUB* stands for the tail element "4-fluorobenzyl"; *INA* stands for the core "1H-indazole"; and *CA* stands for the "carboxamide linker commonly found in synthetic cannabinoids, indicating the carbonyl group and amide moiety."

30. See Adams et al., "'Zombie' Outbreak"; Pearlson, *Weed Science*, 84; and Lobato-Freitas et al., "Overview of Synthetic Cannabinoids," 16, 20. The quote is from Santora, "Drug 85 Times as Potent as Marijuana."

31. See Morrow et al., "Outbreak of Deaths"; Johnson et al., "Controlling New Psychoactive Substances"; and Lobato-Freitas et al., "Overview of Synthetic Cannabinoids," 21.

32. See Finlay et al., "Synthetic Cannabinoid Receptor Agonists"; Morrow et al., "Outbreak of Deaths"; Lobato-Freitas et al., "Overview of Synthetic Cannabinoids," 21, 15; Papaseit et al., "Cannabinoids," 1290–1292; Adams et al., "'Zombie' Outbreak," 236; Kumar et al., "Structure of a Signaling Cannabinoid Receptor"; and Thomsen, "Factors Influencing the Toxicity of AMB-FUBINACA."

33. See Lobato-Freitas et al., "Overview of Synthetic Cannabinoids," 14. Morrow et al., "Outbreak of Deaths," 3, write that 67% of the deaths "were reported to be of Maori descent."
34. See Rosenberg and Schweber, "33 Suspected of Overdosing"; Adams et al., "'Zombie' Outbreak"; Santora, "Drug 85 Times as Potent as Marijuana"; and Lobato-Freitas et al., "Overview of Synthetic Cannabinoids," 2, 5, 16. Banister et al., "Dark Classics," 2167: "The toxicity of Delta9-THC has been explored in many species, and is very low compared to most other recreational and pharmaceutical drugs. . . . No cases of fatal overdose in humans following acute Delta9-THC consumption have been conclusively identified, and extrapolated estimates of lethal human doses have ranged from 4 g to greater than 15 g. Typical human doses of Delta9-THC (1–20 mg) are three orders of magnitude lower than the estimated lethal dose range, suggesting a very large therapeutic index." Castaneto et al., "Synthetic Cannabinoids Pharmacokinetics," 1, write that synthetic cannabinoids induce "adverse physiological effects not observed with cannabis intake."
35. The quotes are from Howlett et al., "Spicy Story," 24, 20. In contrast, Potter, "Propagation," 10, defines cannabimimetic as "pertaining to the pharmacological properties of Delta9-tetrahydrocannabinol." See also Riboulet-Zemouli, "'Cannabis' Ontologies I," 24; Flemming et al., "Chemistry and Biological Activity"; Banister and Connor, "Synthetic Cannabinoid Receptor Agonists"; and Firman et al., "Chemoinformatic Consideration," 17. Shevyrin et al., "On a New Cannabinoid Classification System," 186–188, note that substances such as AMB-FUBINACA belong to the group of "indazole carboxamides," which differ substantially from classical, nonclassical, and hybrid cannabinoids, as well as from aminoalkylindoles such as JWH-018.
36. Nutt, *Drugs Without the Hot Air*, 117–118: "At the onset of the UK spice epidemic, in 2017, I wrote to the UK Minister for Health suggesting that he made resurrecting rimonabant a priority. He replied that making medicines was a job for the pharmaceutical sector, not the Department of Health. But even the hundreds of spice incidents a year are not enough to make developing an antidote cost-effective for a pharmaceutical company without some degree of government support or subsidy. As a result, there has been no progress, and prisoners and the homeless continue to die." See Banister et al., "Pharmacology of Valinate and *tert*-Leucinate Synthetic Cannabinoids," 1246, on rimonabant reversing specific effects of substances similar to AMB-FUBINACA. Gray et al., "Use of Synthetic Cannabinoid Receptor Agonists," write that these chemicals were directly implicated in 60 deaths in England and Wales in 2018 and more than 100 deaths in Europe in 2016–2017. Also see EMCDDA, *Synthetic Cannabinoids*, 2, 5, 28; and EMCDDA, *Fentanils and Synthetic Cannabinoids*, 17. Meyer and Quenzer, *Psychopharmacology*, 498, refer to "the unusually high degree of toxicity of synthetic cannabinoids." The Ancient Greek word *pharmakon*—from which English gets *pharmacy*, *pharmaceutics*, and *pharmacology*—can refer to a remedy or a poison. Paracelsus famously made this point with the adage: "The dose makes the poison." Exceptions include asbestos, benzene, and lead, which are toxic even in minute amounts.

## 13: POTENCIES RISING

1. The concept of "intoxication" comes from the word *toxic*, meaning harmful, which traces back to the Latin *toxicum*, meaning poison. Dictionaries primarily define intoxication as "the condition of having physical or mental control markedly diminished by

the effects of alcohol or drugs," and secondarily as "an abnormal state that is essentially a poisoning." The word also carries a third sense meaning "excitement or exhilaration." When scientists discuss cannabis intoxication, they almost always use the term in its primary sense—"physical or mental impairment caused by the drug's effects." While this usage may seem clinical and value-neutral, the term itself emphasizes impairment and fails to capture the full range of cannabis's effects. Scientific English lacks truly neutral terms for psychoactive experiences and the modification of consciousness.

2. See ElSohly et al., "Potency Trends of Delta9-THC"; Mehmedic et al., "Potency Trends of Delta9-THC"; Poulsen and Sutherland, "Potency of Cannabis"; and King et al., "Overview of Cannabis Potency." The quote is from UNODC, *World Drug Report 2006*, 27.
3. See Ashton, "Pharmacology and Effects"; Arseneault et al., "Causal Association"; Hall, "Is Cannabis Use Psychotogenic?"; and D'Souza et al., "Psychotomimetic Effects."
4. See UNODC, *World Drug Report 2006*, 155, 173–174, 180.
5. The quote is from Hardwick and King, *Home Office Cannabis Potency Study 2008*, 3.
6. See King et al., "Overview of Cannabis Potency," 30, 40; Clarke and Watson, "*Cannabis* and Natural *Cannabis* Medicines," 6, 12; Leggett and Pietschmann, "Global Cannabis Cultivation," 200; and Clarke and Merlin, *Cannabis*, 302. Small, *Cannabis*, 253: "Hashish in illicit markets typically has a THC content of 5% to 25% (levels as high as 45% have been reported)."
7. See King et al., "Overview of Cannabis Potency"; Potter et al., "Potency of Delta9-THC"; and Hardwick and King, *Home Office Cannabis Potency Study 2008*.
8. The first quote is from Di Forti et al., "High-Potency Cannabis," 488. The second and third quotes are from Di Forti et al., "Proportion of Patients," 236. Pow et al., "Cannabis Use and Psychotic Disorders," 7066, define high-potency cannabis in recent UK-based studies as "characterized by 12–18% THC content."
9. See Colizzi and Murray, "Cannabis and Psychosis," 195; Volkow et al., "Effects of Cannabis Use," E3; Hill, "Be Clear About the Real Risks," S14; Pearlson, *Weed Science*, 142; Panlilio et al., "Cannabinoid Abuse and Addiction"; Pasman et al., "GWAS of Lifetime Cannabis Use"; Power et al., "Genetic Predisposition to Schizophrenia"; Volk et al., "Reciprocal Alterations"; and Sohn, "Balancing Act," S17.
10. Pearlson, *Weed Science*, 150.
11. See Englund et al., "Cannabidiol Inhibits THC-Elicited Paranoid Symptoms"; Colizzi et al., "Descriptive Psychopathology of the Acute Effects"; and Freeman et al., "How Does Cannabidiol (CBD) Influence the Acute Effects?"
12. See Englund et al., "Cannabis in the Arm."
13. See Englund et al., "Cannabis in the Arm"; Leung et al., "Do Cannabis Users Reduce Their THC Dosages?"; Grotenhermen, "Pharmacokinetics and Pharmacodynamics," 332; Wood et al., "Canada's THC Unit," 4; Freeman and Lorenzetti, "'Standard THC Units,'" 7; and Hammond, "Communicating THC Levels," 4. Spindle et al., "Acute Effects," find that vaporizing is slightly more efficient than smoking. In contrast, Newmeyer et al., "Free and Glucuronide Whole Blood Cannabinoids' Pharmacokinetics," 1579, conclude that the two inhalation methods "provide comparable cannabinoid delivery."
14. See Englund et al., "Cannabis in the Arm"; Kitdumrongthum and Trachootham, "Individuality of Response"; and Freeman and Lorenzetti, "'Standard THC Units.'" Earleywine, *Understanding Marijuana*, 123, writes that 11-hydroxy-THC "only differs from

THC by a few atoms, but it may be three times as potent because it reaches the brain more readily." Iversen, *Science of Marijuana*, 45: "A further complication of the oral route is that one of the metabolites formed in the liver is 11-hydroxy-THC. This is a psychoactive metabolite with potency approximately the same as that of THC. The amount of 11-hydroxy-THC formed after smoking is relatively small (plasma levels are less than one-third of those for THC), but when cannabis is taken by the oral route—in which all the blood from the intestine must first pass through the liver—the amount of 11-hydroxy-THC in plasma is approximately equal to that of THC, and it probably contributes at least as importantly as THC to the overall effect of the drug."

15. The examples of volunteers' experiences are taken from D'Souza et al., "Psychotomimetic Effects," 1562, and Colizzi et al., "Acute Effects of Intravenous Delta-9-Tetrahydrocannabinol," 10. Englund et al., "Cannabis in the Arm," 4911, situate the low dose of intravenous THC for an individual weighing 70 kilos (154 pounds) between 0.79 mg and 1.84 mg and the high dose between 2.59 mg and 4.85 mg. See also Englund et al., "Cannabidiol Inhibits THC-Elicited Paranoid Symptoms."
16. See Englund et al., "Can We Make Cannabis Safer?"; Solowij et al., "Randomized Controlled Trial"; Pennypacker et al., "Potency and Therapeutic THC and CBD Ratios"; and Zamarripa et al., "Orally Administered Delta-9-Tetrahydrocannabinol." Lawn et al., "Acute Effects of Cannabis," 1291, write that "doses of vaporized CBD, which are near to naturally occurring levels, do not mitigate the acute harms of THC. Harm reduction messages should avoid stating that CBD found in recreational cannabis protects against the acute psychotomimetic or verbal memory-impairing effects of cannabis."
17. The quote is from Englund et al., "Does Cannabidiol Make Cannabis Safe?," 874.
18. See Wood et al., "Canada's THC Unit," 3; Kitdumrongthum and Trachootham, "Individuality of Response"; and Bosnyak et al., "Novel EEG-Based Objective Test."
19. See Kalinowski and Humphreys, "Governmental Standard Drink Definitions"; and Mongan and Long, "Standard Drink Measures." Hammond, "Communicating THC Levels," 3, writes that a standard drug dose "should be set sufficiently low to allow consumers to effectively titrate to minimize adverse effects while attaining the desired therapeutic or psychoactive effect" and describes the amount of alcohol in a standard drink as "lower than the level that would induce intoxication or impairment for most consumers." Sarne, "THC for Age-Related Cognitive Decline?," 3628, estimates that doses of 1–20 mg THC correspond to "the range of self-titrated doses smoked by users of recreational marijuana," suggesting that the dose considered "standard" by different individuals can vary twenty-fold.
20. See Hammond, "Communicating THC Levels;" Leos-Toro et al., "Cannabis Labelling"; and Jugl et al., "Much Ado."
21. Bassir Nia et al., "Sex Differences in the Acute Effects of Intravenous Delta-9-Tetrahydrocannabinol (THC)," 1621, report that female participants had higher scores for intoxication than male participants relative to a THC dose adjusted for body weight, but "no other sex differences were observed in acute subjective, psychotomimetic, cognitive, or physiological effects of THC." Aghaei et al., "Sex Differences in the Acute Effects of Oral THC," reached a similar conclusion. Sholler et al., "Sex Differences in the Acute Effects of Oral and Vaporized Cannabis," found that all else being equal, women reported higher rates of anxiety and nervousness, heart racing, and restlessness compared to men, leading the researchers to suggest that public health messaging "recommend lower starting doses for females and warnings about acute anx-

iogenic reactions." See also Struik et al., "Modulating Role," and Castelli et al., "Male and Female Rats." Broyd et al., "Acute and Chronic Effects of Cannabinoids on Human Cognition," 561, write in 2016: "Potential sex differences are insufficiently addressed in studies of predominantly male cannabis users (approximately 75% of studies reviewed), although many argue that this overrepresentation matches the sex distribution of cannabis users in the general population. Where sex effects were tested (approximately 12% of studies), very few were found."

22. Calakos et al., "Mechanisms Underlying Sex Differences," 4: "Interestingly, CB1 receptor availability has been shown to increase with age in women, but not men, in basal ganglia, limbic regions, lateral temporal cortex, and hippocampus."

23. See Sarne, "THC for Age-Related Cognitive Decline?" The quote regarding the advised procedure for cannabis initiation is from MacCallum and Russo, "Practical Considerations," 13, who add, "For cannabis inhalation, patients should start with one inhalation and wait 15 minutes. Then, they may increase by one inhalation every 15–30 minutes until desired symptom control has been achieved."

24. EMCDDA, *EU Drug Market*, 5: "The average potency of herbal cannabis and cannabis resin increased by about 60% and nearly 200% respectively between 2011 and 2021, while prices remained largely stable." Regarding changes in production methods, see Chouvy and Macfarlane, "Agricultural Innovations," 87.

25. See Smart et al., "Variation in Cannabis Potency," 6–7. For the quotes by the California distributor, see Roberts, "Cannabis Industry's Greatest Lie." See also Hammond et al., "Trends in the Use of Cannabis Products." Maccoun, "Commentary on Niesink et al. (2015)," 1951, describes events in Holland that contradicted the Iron Law of Prohibition, in which periods of lenient law enforcement saw more significant increases in cannabis potency, while periods with stricter enforcement coincided with its leveling off. Smith, "How Iron-Clad Is the Iron Law of Prohibition?," 5, quotes psychiatrist Keith Humphreys, who qualifies the driving force behind the rise in cannabis potency as "pure and simple profit."

26. Szejko et al., "Medicinal Use of Different *Cannabis* Strains," 136, surveyed German consumers of medicinal cannabis, showing that a significant majority preferred varieties that were "THC dominant with high THC concentrations." See also Donnan et al., "Characteristics That Influence Purchase Choice."

27. Pardal and Wadsworth, "Strictly Regulated," 1006, write that "as the biological THC ceiling is approximately 35%, it limits high-potency products such as solid concentrates, which can be greater than 80%." Grassa et al., "New Cannabis Genome Assembly," 1676: "It is highly abnormal for a plant to allocate 20–30% of its flowering biomass to one or two specialized metabolites, as do modern *Cannabis* cultivars." Resin, "Growing for Terpenes," 37, writes that improvements in the cultivation of *Cannabis* have made "phytocannabinoids and terpene percentages rise to unbelievable levels. It's not uncommon for some strains to test as high as 35–38% total cannabinoids, with some flower testing as high as 5.5–6% total terpene content." Knodt, "44% THC—Does Berlin Have the Strongest Weed in the World?" refers to a German case, most probably based on incorrect measurements.

28. See Mongan and Long, *Standard Drink Measures*, 14, on the tendency to overpour high-potency spirits.

29. Hammond, "Communicating THC Levels," 2, considers the absence of visual cues concerning the strength of cannabis products to be problematic in the case of oil

concentrates and edibles, "given that very small differences in consumption can translate into substantial differences in dose. In the case of edibles, a single product can include as many as 10 or 20 'servings' of THC, whereas oils are often measured in fractions of a milliliter. Thus, the margin of error for dose titration is very narrow, particularly if consumers are uncertain of a product's THC content." See also Kitdumrongthum and Trachootham, "Individuality of Response." Freeman and Lorenzetti, "'Standard' THC Units," 8, discuss infrequent users who experience severe reactions to doses considered standard by regular users.

30. The quotes are from Russo et al., "Cannabinoid Hyperemesis Syndrome," 336, and Perisetti et al., "Cannabis Hyperemesis Syndrome," 571. See also DeVuono and Parker, "Cannabinoid Hyperemesis Syndrome."
31. See Chan et al., "User Characteristics"; Loflin and Earleywine, "New Method"; Miller et al., "Exploring Butane Hash Oil Use," 2; Black, "Concentrated Cannabis"; and Saunders and Cannon, "Dab Myths Debunked." The quote is from Rosenthal, *Beyond Buds*, 97.
32. The quotes are from Black, "Generation Dab," 55, 58. Rosenthal, *Beyond Buds*, 105: "THC lowers blood pressure, so dab sitting down. Overdosing on THC, while not toxic, can be uncomfortable, causing anxiety and even nausea. Don't dab unless you're an experienced marijuana user. Even then, dabs aren't a contest. Do small dabs and enjoy yourself. Huge dabs increase your tolerance, can result in acute intoxication, and can be part of a constellation of behaviors that could ultimately lead to harm."
33. To estimate the total daily amount of THC consumed by some frequent dabbers, I consider an average dose containing 19 mg of THC (as reported in Pearlson, *Weed Science*, 163) used 16 times a day (as reported by some participants in a study by Russo et al., "Cannabinoid Hyperemesis Syndrome Survey and Genomic Investigation")—which gives a daily consumption of 304 mg of THC. By contrast, the basic dose for an infrequent cannabis user is estimated at 2.5 mg of THC—see Kitdumrongthum and Trachootham, "Individuality of Response," 5, and Russo, "Current Therapeutic Cannabis Controversies," 4. The lethal dose of alcohol depends on several variables, such as the drinker's weight, whether their stomach is empty or full, and the speed at which they drink. Estimates on what constitutes a standard dose also vary. Pearlson, *Weed Science*, 208, writes, discussing the lethal dose of different drugs: "This can be gauged by comparing the amount needed to produce psychoactive effects with the fatal dose, to give a safety ratio. For alcohol, this number is 10. If two units (i.e., two beers, two glasses of wine) cause intoxication, then 20 units will kill you. . . . So, when it comes to drug-specific mortality, cannabis is clearly safer than both alcohol and opioids because although huge doses may push you into an extremely unpleasant delirious state, there are virtually no fatalities. Perhaps none at all."
34. See Bidwell et al., "Naturalistic Administration of Cannabis Flower," 794.
35. Meyer and Quenzer, *Psychopharmacology*, 488, suggest that the absence of withdrawal symptoms in many cannabis users may be due "to the long elimination half-life of THC, which causes the cannabinoid receptors to remain partially occupied for a significant time even after termination of drug treatment." Pearlson, *Weed Science*, 109, notes that "many people exhibit cannabis tolerance without other symptoms of dependence such as drug withdrawal." See Volkow et al., "Adverse Health Effects," on withdrawal symptoms. Banister et al., "Dark Classics," 2167–2168, write that "withdrawal signs during abstinence of Delta9-THC are very mild (irritability, insomnia, appetite disrup-

tion, craving) compared to those experienced during withdrawal from opioids, alcohol, or benzodiazepines." Treatments for Cannabis Use Disorder usually include therapy, counseling, and exercise—see Patel and Marwaha, *Cannabis Use Disorder*, and Brellenthin and Koltyn, "Exercise as an Adjunctive Treatment."

36. Vuilleumier et al., "Cannabinoids in the Treatment of Cannabis Use Disorder," 7, report that a month-long treatment based on a daily dose of 400 mg of CBD reduces a patient's cannabis use and the same regimen with a dose of 800 mg reduces withdrawal symptoms, leading them to consider CBD as "a cannabinoid of substantial interest with regard to the treatment of Cannabis Use Disorder." See also Hurd et al., "Cannabidiol for the Reduction of Cue-Induced Craving"; Hindocha et al., "Cannabidiol Reverses Attentional Bias"; Paulus et al., "Cannabidiol in the Context of Substance Use Disorder Treatment"; and Thomsen et al., "Cannabinoids for the Treatment of Cannabis Use Disorder."

37. Meyer and Quenzer, *Psychopharmacology*, 489: "Educational performance and IQ Survey studies indicate that the amount of cannabis use by young people is inversely related to educational performance. That is, greater use is associated with poorer grades, more negative attitudes about school, and increased absenteeism. . . . In contrast, a recent longitudinal study found that low to moderate use in early adolescence has relatively minor effects on academic performance that appear to dissipate after a period of abstinence. Yet another study found that daily tobacco use assessed at the age of 15 had a stronger impact on academic test scores one year later than weekly cannabis use. These findings must be viewed cautiously, however, because of the possibility that heavier use of cannabis (e.g., daily use) would have produced greater impairments than seen in moderate users." Lubman et al., "Cannabis and Adolescent Brain Development," 7: "Cannabinoid receptors and endogenous cannabinoids emerge early in the developing brain and are involved in a wide range of processes during embryonic and prenatal neural development, including proliferation and differentiation of progenitor cells, glial cell formation, neuronal migration, and axonal pathfinding. In addition to being critically involved in neural development during prenatal and early postnatal life, there is growing evidence that the endocannabinoid system continues to undergo functional development during adolescence and may also play a role in regulating neurogenesis into adulthood. The $CB_1$ receptor is expressed at high levels throughout the cerebral cortex, hippocampus, basal ganglia, and cerebellum, with many of these regions undergoing substantial remodeling during adolescence." Rettew et al., "Blunted," 77: "Early cannabis exposure alters brain reward pathways, thus facilitating the subsequent use of other drugs. If cannabis use starts in childhood, one in six starters will become addicted. Daily users of cannabis are at the highest risk for developing cannabis dependence." Gobbi et al., "Cannabis Use in Adolescence," 431: "The brain indeed remains in a state of active, experience-guided development from the prenatal period through childhood and adolescence until the age of approximately 21 years. During this period, it is intrinsically more vulnerable to the adverse long-term effects of environmental insults, such as exposure to THC." Ritchay et al., "Resting State Functional Connectivity," 7: "The adolescent brain is particularly sensitive to the effects of THC, and earlier exposure to substances is associated with poorer outcomes in cognition than typically seen in later-onset cannabis users. Thus, encouraging youth to minimize, eliminate, or delay their substance use onset until after the age of 18—via personalized feedback, psychoeducation, and exercise—may reduce some of the difficulties seen in early-onset users." Pearlson, *Weed Science*, 154, writes

that "adolescent cannabis use may not have a detectable impact on cognition unless it occurs at very high levels and/or over many years." Scott et al., "Association of Cannabis with Cognitive Functioning," 585: "Associations between cannabis use and cognitive functioning in cross-sectional studies of adolescents and young adults are small and may be of questionable clinical importance for most individuals. Furthermore, abstinence of longer than 72 hours diminishes cognitive deficits associated with cannabis use. Although other outcomes (e.g., psychosis) were not examined in the included studies, results indicate that previous studies of cannabis in youth may have overstated the magnitude and persistence of cognitive deficits associated with the use." See also Meyer et al., "Role of the Endocannabinoid System."

38. See Calabrese and Rubio-Casillas, "Biphasic Effects of THC," 6. Zamberletti and Rubino, "Dos(e)Age: Role of Dose and Age in the Long-Term Effect of Cannabinoids on Cognition," 11, write that "the most striking factor in predicting the outcome of cannabinoid use seems to be age. Indeed, what is detrimental during brain development (prenatal and adolescent exposure) has been reported to be beneficial in old age. . . . In contrast, it appears difficult to clearly establish a possible role of dosage in the effects of cannabinoids on cognition, particularly when administration occurs during the adolescent period. Indeed, cannabinoids administered during adolescence can have long-term effects on specific cognitive tasks; however, the appearance and persistence of such effects vary greatly depending on the paradigm of administration and behavioral task used, making it impossible to draw any definitive conclusion." See also Bilkei-Gorzo et al., "Chronic Low Dose"; Sarne et al., "Reversal of Age-Related Cognitive Impairments"; Calabrese and Rubio-Casillas, "Biphasic Effects"; Nitzan et al., "Ultra-Low Dose of Delta9-Tetrahydrocannabinol"; Shapira et al., "Hippocampal Differential Expression"; and Pocuca et al., "Effects of Cannabis Use."
39. See Caron, "Psychosis, Addiction, Chronic Vomiting."
40. Mechoulam and Ben-Shabat, "From *Gan-zi-gun-nu* to Anandamide and 2-Arachidonoylglycerol," 136, first proposed an "entourage effect" in which the combined activity of all the compounds found in the plant's resin glands works in a "synergism" to maximize the pharmacological effect of cannabis. See also Schwabe et al., "Uncomfortably High"; Freeman and Winstock, "Profile of High-Potency Cannabis"; Ferber et al., "The 'Entourage Effect'"; Plumb et al., "The Nose Knows"; Finlay et al., "Terpenoids from Cannabis"; and Cogan, "The 'Entourage Effect' or 'Hodge-Podge Hashish.'" Santiago et al., "Absence of Entourage," 166: "With so many bioactive components in cannabis, the systematic, granular elucidation of possible entourage effects poses a substantial combinatorial puzzle and scientific challenge." Recent research suggests that measuring the entourage effect may be possible; see Raz et al., "Selected Cannabis Terpenes," 1. See also Spindle et al., "Vaporized D-Limonene."
41. See Raber et al., "Understanding Dabs," on the risks of concentrate contamination by pesticides and residual solvents. See also Holland, *Weed*, 207
42. Escondido, "Strongest Concentrates on Earth," 60, reviews concentrates presented at the 2018 Cannabis Cup and writes that those "with THC levels dipping below the 90 percent threshold . . . tend to have better flavor and aroma, which contribute to the overall experience." See also Purdy, "Master of Hash," 52, who quotes hashish expert Frenchy Cannoli as saying: "Concentrates produced with butane and propane extraction versus hashish could be compared to the difference between producing wine

and producing liquor like Cognac or distilling full-proof alcohol. These products all serve a purpose, just not the same one."

43. See Louw and Lambrechts, "Grape-Based Brandies," and Bertrand, "Armagnac, Brandy, and Cognac." According to Campbell, "Grain Whisky Distillation," 182, the purity of neutral spirits has a practical limit of 97.2% alcohol by volume when using distillation, although it is possible to reach 100% under vacuum.
44. Kitdumrongthum and Trachootham, "Individuality of Response," 3: "THC is the key and most potent psychoactive ingredient in cannabis, and that which causes highness upon intake." See also Papaseit et al., "Cannabinoids," 1287.
45. See Caron, "Psychosis, Addiction, Chronic Vomiting"; Thomas et al., "Cannabis Vaping"; and Miller et al., "Marijuana and Suicide."

## 14: NEW IDEAS

1. See Tart, *On Being Stoned*, 285, 30. The questionnaire asked participants to rate the "degree of stoned-ness" required to experience the described effects on a scale of five levels: low, moderate, strong, very strong, and maximum (p. 33).
2. See Cardeña, "Festschrift for a Consciousness Hummingbird," 222; Pearlson, *Weed Science*, 110; and Earleywine, *Understanding Marijuana*, 101. The quote is from Tart, *On Being Stoned*, 286.
3. See Tart, *On Being Stoned*, 76, 85, 140, 134, 151, 183, 160, 154, 92–95, 89–91.
4. See Tart, *On Being Stoned*, 137–138, 187, 153–156, 167–179.
5. Tart, *On Being Stoned*, 172.
6. See Tart, *On Being Stoned*, 174, 295. Tart considered cannabis's capacity to enhance creativity a "high research priority" and suggested that future studies include carefully prepared professionals whose work depended on creativity (p. 172).
7. See Global Commission on Drug Policy, *War on Drugs*. For a detailed overview of the scientific research on cannabis and cognition conducted during the last three decades of the twentieth century and focusing almost exclusively on impairment, see Solowij, *Cannabis and Cognitive Functioning*.
8. See Weckowicz et al., "Effect of Marijuana." For studies on cannabis-induced cognitive impairment from the 1970s, see Carlin et al., "Social Facilitation"; Darley et al., "Influence of Marihuana"; and Klonoff et al., "Neurological Effects"—which based their conclusions on research using doses of 15 mg of smoked THC, 20 mg of oral THC, and 13.6 mg of smoked THC, respectively. In 1972, Canada's Commission of Inquiry into the Non-Medical Use of Drugs, *Cannabis*, 40–41, wrote: "In Commission laboratory experiments, we found that when smoking techniques were used which maximized the delivery and absorption of THC, cigarettes containing 5 to 10 mg THC, completely smoked, produced effects which were beyond the range of cannabis experiences of some regular users who had been 'turning on' several times a week for a number of years. As will be discussed in more detail later, acute adverse anxiety reactions occurred in a few individuals at these doses. . . . when cannabis users were allowed to select their own marijuana doses daily for a period of several weeks in the laboratory, on the majority of occasions, subjects consumed 2 to 4 mg of THC to get 'high.'" Solowij, *Cannabis and Cognitive Functioning*, 20, writes: "In general, only a small amount of smoked cannabis (e.g., 2 to 3 mg of available THC) is required to produce a brief pleasurable high for

the occasional user, and a single joint may be sufficient for two or three individuals. A heavy smoker may consume five or more joints per day, while heavy users in Jamaica, for example, may consume up to 420 mg THC per day." Health Canada, *Information for Health Care Professionals,* 48, writes in 2018 that an "inhaled dose of 3 to 6 mg THC" would allow an average adult "to reach the threshold for psychotropic effects"; and an "inhaled dose of 10 to 20 mg THC" would be sufficient "to produce marked intoxication"; in addition, multiplying the smoked dose by 2.5 "can yield an approximately equivalent oral dose of delta-9-THC" (p. 49).

9. See Reber et al., *Penguin Dictionary of Psychology*, 814, on convergent and divergent thinking.
10. The quotes are from Weckowicz et al., "Effect of Marijuana," 396, 393.
11. See Block et al., "Acute Effects of Marijuana"; Pope and Yurgelun-Todd, "Residual Cognitive Effects"; and Curran et al., "Cognitive and Subjective Dose-Response Effects." For other publications from this period focusing on cognitive impairments, see, for example, Bowman and Pihl, "Cannabis: Psychological Effects"; Braff et al., "Impaired Speed"; Lundqvist, *Cognitive Dysfunctions*; Fletcher et al., "Cognitive Correlates"; Solowij, "Do Cognitive Impairments Recover?"; and Block, "Does Heavy Marijuana Use Impair Human Cognition?" I found no studies on cannabis and creativity published in the 1980s or '90s. In 2001, Bourassa and Vaugeois, "Effects of Marijuana Use," revisited the impact of cannabis on creativity by having novices and moderate users smoke cannabis containing 10 mg of THC, finding no positive effects in novices and impairment in moderate users. This study was followed eight years later by Jones et al., "Cannabis and Ecstasy/MDMA," 323, who reported that abstinent cannabis users "had significantly more 'rare-creative' responses than controls" on several psychological tests.
12. See Hart et al., "Effects of Acute Smoked Marijuana," and Kelleher et al., "Effects of Cannabis on Information-Processing Speed."
13. In 2006, Nordstrom and Hart, "Assessing Cognitive Functioning," 2798, proposed that participants' varying histories of cannabis use were "the most likely factor responsible for the disparate findings" across different studies.
14. Green et al., "Being Stoned," review the motivations of cannabis users. For studies that made passing mentions of cannabis's positive effects on cognition, see Block et al., "Acute Effects of Marijuana," 907; Block and Ghoneim, "Effects of Chronic Marijuana Use," 222–226; and Curran et al., "Cognitive and Subjective Dose-Response Effects," 68. According to Curran et al., "Keep Off the Grass?," 10, writing in 2016: "It should be noted that public health messages to users are distorted because funding for research is often targeted to studying the harmful effects of cannabis rather than the benefits. Despite studies aiming to document negative effects, occasionally positive effects are noted, such as enhanced divergent thinking following either oral THC or smoked cannabis. Future research should evaluate perceived benefits to give a more balanced understanding; people clearly do not use cannabis only for its harms."
15. See Parloff, "How Pot Became Legal," and Curran et al., "Keep Off the Grass?"
16. Schafer et al., "Investigating the Interaction," 296.
17. See Schafer et al., "Investigating the Interaction," 297. Regarding definitions of creativity, De Dreu et al., "Human Creativity," 206, write: "Science defines creativity as the production of ideas, problem solutions, and products that are both novel (original) and feasible and potentially useful. Something is novel when it is rarely if ever encountered,

either by the inventors themselves or by outside observers, or when it is different from what was done before." See also Mayseless et al., "Generating Original Ideas," 232; and Brown and Kim, "The Neural Basis of Creative Production," 103–104.

18. The quote is from Tart, *On Being Stoned*, 18.
19. See Warnick et al., "Head in the Clouds?," 1, 6, 23.
20. Macher and Earleywine, "Enhancing Neuropsychological Performance," show the positive impact of motivational statements on cannabis users' cognitive performance. See also Runco and Chand, "Cognition and Creativity," 246, who argued in 1995 that "motivation and knowledge are critical for creative thinking."
21. See Warnick et al., "Head in the Clouds?," 4, 25, 28, 7, 31, 3.
22. See Warnick et al., "Head in the Clouds?," 6, 37–38.
23. Oleson et al., "Cannabinoid Modulation," 1, write that dopamine function is "tightly regulated by the brain's endocannabinoid system and greatly influenced by exogenous cannabinoids." Zabelina et al., "Dopamine and the Creative Mind," 13, find that "dopamine levels in multiple brain areas affect human creativity, with an interaction between frontal and striatal dopaminergic pathways." Fernández-Ruiz et al., "Cannabinoid-Dopamine Interaction," 72, 83, write that "the effects of cannabinoids on dopamine transmission and dopamine-related behaviors are generally indirect and exerted through the modulation of GABA and glutamate inputs received by dopaminergic neurons," while noting that "dopaminergic neurons do not contain CB1 receptors, with some exceptions, but these receptors are located on neurons present in regions innervated by dopaminergic neurons." Pearlson, *Weed Science*, 81 writes that "the dopamine release provoked by THC is significantly less than that provoked by amphetamine or cocaine in reward-relevant brain regions." See also Oleson and Cheer, "Brain on Cannabinoids," and Lee, "Tetrahydrocannabinol and Dopamine D1 Receptor."
24. See Beaty et al., "Creative Cognition," and Shofty et al., "Default Network." Brown and Kim, "Neural Basis of Creative Production," 103: "Creativity has garnered a great deal of attention in the psychology and neuroscience literatures in recent years, in part because of its real-world application to understanding technology development, scientific discovery, and artistic creation, among many other domains."
25. See Beaty et al., "Creative Cognition," 3, on "candidate ideas" generated by the default network. See also De Dreu et al., "Human Creativity," 224; Mayseless et al., "Generating Original Ideas"; Marron et al., "Chain Free Association"; and Gross and Schooler, "Standing Out."
26. According to Gross et al., "Why Creatives Don't Find the Oddball Odd," 12, research suggests "a central role for dopamine in driving motivated, engaged, and exploratory behaviors." See also Faust-Socher et al., "Enhanced Creative Thinking," and Gross and Schooler, "Standing Out."
27. See De Dreu et al., "Human Creativity," 215. Gross and Schooler, "Standing Out," 2–3, write, referring to activity in the salience network: "Following cannabis consumption, information that is typically ignored may be assigned an unusually high level of salience, thereby becoming accessible to working memory resources; this may predispose individuals to novel ideas and associations." See also Smallwood et al., "Cooperation Between the Default Mode Network."
28. See Oleson et al., "Cannabinoid Receptor Activation," 1441, 1451. In 1971, Tart, *On Being Stoned*, 92, observed that the perception of time slowing down is "one of the most characteristic effects of marijuana intoxication."

29. Prieto-Arenas et al., "Gender Differences," 5: "Acute cannabinoid use induces changes in brain neurochemistry, such as an increased dopamine release, reduced glutamatergic transmission, the release of endogenous opioids and the inhibition of acetylcholine secretion. These brain biochemical changes are responsible for the acute effects of cannabis, which can be divided into physiological, psychological or behavioral and cognitive effects." Fernández-Ruiz et al., "Cannabinoid-Dopamine Interaction," 78: "From a pharmacological point of view, cannabinoid agonists (depending on doses and duration of treatment) produce euphoria, stimulate brain reward, are anxiolytic, and decrease motivation and arousal while increasing emotionality, effects that were observed in humans and laboratory animals . . . [and] were paralleled by alterations in mesocorticolimbic dopaminergic neurons." See also De Dreu et al., "Human Creativity," and Pearlson, *Weed Science*, 144, 185.
30. The long-term effects of cannabis use and THC on the human brain are the subject of ongoing research and debate—see Ishrat et al., "Association Between Cannabis Use and Brain Structure"; and Gowin et al., "Brain Function Outcomes." The quote is from Gross and Schooler, "Standing Out," 598. Shofty et al., "Default Network," 1848, write that "despite many studies demonstrating correlation between the default network, other large-scale cortical networks, and creativity, no causal relation has been shown." They add that "several technical limitations associated with transcranial stimulation limit the ability to establish such causality or identify relationships at the individual participant level." Short of performing "awake brain surgery," neuroscientists studying creativity come up against a tangible limit: the human skull.

## 15: INTOXICATION AND AROMATIC PLEASURE

1. See Peterman, *Marijuana Use and Highway Safety.*
2. See Lapham, "Limits of Tolerance," 26; Pearlson et al., "Cannabis and Driving," 3; and Sewell et al., "Effect of Cannabis," 6.
3. See Huestis, "Human Cannabinoid Pharmacokinetics"; Parker, *Cannabinoids and the Brain*, 11; Earleywine, *Understanding Marijuana*, 132–136; Desrosiers et al., "Smoked Cannabis' Psychomotor and Neurocognitive Effects"; Huestis and Smith, "Cannabinoid Pharmacokinetics," 300; Desrosiers et al., "Phase I and II Cannabinoid Disposition"; Milman et al., "Plasma Cannabinoid Concentrations"; Schwope et al., "Identification of Recent Cannabis Use"; Sewell et al., "Effect of Cannabis," 4–6; Di Ciano et al., "Cannabis and Driving in Older Adults," 9; and Pearlson et al., "Cannabis and Driving," 4. THC and CBD share the same chemical formula ($C_{21}H_{30}O_2$) and are much bigger molecules than alcohol ($C_2H_5OH$), weighing almost 7 times more in atomic mass units (314 vs. 46).
4. See EMCDDA, "Cannabis and Driving," 7; Vandrey et al., "Pharmacokinetic Profile," 83; Arkell et al., "Failings of *Per se* Limits," 106; Wong et al., "Establishing Legal Limits," 3; and International Council on Alcohol, Drugs & Traffic Safety, "Cannabis & Driving," 3.
5. See Milman et al., "Plasma Cannabinoid Concentrations," 2; Earleywine, *Understanding Marijuana*, 134; Pearlson et al., "Cannabis and Driving," 4; Huestis, "Human Cannabinoid Pharmacokinetics," 11–14; Parker, *Cannabinoids and the Brain*, 13; Karschner et al., "Do Delta9-Tetrahydrocannabinol Concentrations Indicate Recent Use?"; and McCartney et al., "How Long Does a Single Oral Dose of Cannabidiol Persist?"

6. National Institute of Standards and Technology, "NIST Researchers to Test New Approach," suggests that a single breath measurement does not reliably indicate recent cannabis use—but two measures taken within an hour of each other may show a significant drop in THC levels, which appears to be specific to recent cannabis use and could serve as a marker for it. Research on this question is ongoing.
7. See Miller, "Driving Under the Influence"; Arkell et al., "Effect of Cannabidiol"; Logan et al., *Evaluation of Data from Drivers*," 45–46; and Desrosiers et al., "Smoked Cannabis' Psychomotor and Neurocognitive Effects," 255.
8. The quote is from Preuss et al., "Cannabis Use and Car Crashes," 1. See also White and Burns, "Risk of Being Culpable," 12.
9. See Hudak et al., *Uruguay's Cannabis Law*, 4; Arkell et al., "Failings of *Per Se* Limits," 106; and Brands et al., "Cannabis, Impaired Driving, and Road Safety," 5.
10. See Preuss et al., "Cannabis Use and Car Crashes," 8; Di Ciano et al., "Cannabis and Driving in Older Adults," 2; Pearlson et al., "Cannabis and Driving," 10; Arkell et al., "Failings of *Per se* Limits," 107; Logan et al., *Evaluation of Data from Drivers*, 2–3; Behzad et al., "Association of Driving with Blood THC," 2; Fitzgerald et al., "Driving Under the Influence," 724; Marcotte et al., "Driving Performance," 207; and McCartney et al., "Are Blood and Oral Fluid Delta9-Tetrahydrocannabinol (THC) and Metabolite Concentrations Related to Impairment?," 8.
11. See Desrosiers et al., "Phase I and II Cannabinoid Disposition," 631–634; Logan et al., *Evaluation of Data from Drivers Arrested*," 25–26; Pearlson et al., "Cannabis and Driving," 10; Arkell et al., "Failings of *Per se* Limits," 102–104; and Wong et al., "Establishing Legal Limits," 3–4.
12. See Bergamaschi et al., "Impact of Prolonged Cannabinoid Excretion," 519; Preuss et al., "Cannabis Use and Car Crashes," 9; and Arkell et al., "Failings of *Per se* Limits," 51. Karschner et al., "Do Delta9-Tetrahydrocannabinol Concentrations Indicate Recent Use?," 5, report that women took on average six days more than men to eliminate their last THC metabolites.
13. See Arkell et al., "Failings of *Per se* Limits," 102–106, who found that the 5-nanogram threshold misidentified half the participants as false positives after 30 minutes and nearly half as false negatives three hours later. By comparison, at the 1-nanogram threshold, half the participants were false positives after 30 minutes, and more than a quarter were false negatives three hours later.
14. See Vandrey et al., "Pharmacokinetic Profile," and EMCDDA, "Cannabis and Driving," 7.
15. See Pearlson et al., "Cannabis and Driving," 10; and Logan et al., *Evaluation of Data from Drivers Arrested*, 2, 20. Moscowitz and Florentino, *Effects of Low Doses of Alcohol*, 15: "Virtually all subjects tested in the studies reviewed here exhibited impairment on some critical driving measure by the time they reached 0.08 g/dl."
16. The quotes are from Arkell et al., "Failings of *Per se* Limits," 102; Logan et al., *Evaluation of Data from Drivers Arrested*," 3; and Pollard et al., *Technology to Prevent Alcohol- and Drug-Impaired Crashes*, 14.
17. The quote is from Auwärter et al., "Stellungnahme der Grenzwertkommission zur Frage einer Änderung des Grenzwertes," 35.
18. The quote is from Bundesministerium für Digitales und Verkehr, "Unabhängige Expertengruppe legt Ergebnis zu THC-Grenzwert."
19. The quotes are from Backmund et al., "Empfehlungen der Interdisziplinären Expertengruppe," 1, 6–8.

20. The quotes are from Bundesministerium für Digitales und Verkehr, "Unabhängige Expertengruppe legt Ergebnis."
21. See Science Media Center Germany, "Neuer Grenzwert für THC."
22. Auwärter and Graw's quotes are from Science Media Center Germany, "Neuer Grenzwert für THC."
23. The lyric is from Neil Young's song "My, My, Hey, Hey (Out of the Blue)" on the album *Rust Never Sleeps* (1979).
24. See Bidwell et al., "Association of Naturalistic Administration of Cannabis Flower," 794–795—who report that concentrate users in their study had an average blood THC baseline of 5.5 nanograms per milliliter and a peak of 176 nanograms per milliliter, which was more than double the respective averages of the herbal cannabis smokers. However, the self-reports estimating subjective drug effects had similar scores, suggesting saturation above a certain amount of THC in the system.
25. See Fitzgerald et al., "Driving Under the Influence," who present their 2023 study as "the largest randomized double-blinded-placebo-controlled trial to date that examines the relationship between THC concentrations in various biofluids and performance in a driving simulator" (p. 725).
26. See Shepherd, "Neuroenology," 3. Baldy, *University Wine Course*, 44, writes that wines "are always tasted blind—that is, identified by code only" in scientific teaching.
27. According to the influential twentieth-century wine scientist Emile Peynaud, *Taste of Wine*, 173–174: "The winetaster instinctively detects the aroma of alcohol and is wary of high levels of alcohol.... Put another way, alcohol should be masked, covered by the elements of richness. To make a good wine, sufficient alcohol is necessary, but the other elements must match the level of alcohol." He adds that "10 to 15% of a wine's volume is alcohol"—though certain sweet wines can go up to 20% (p. 33–34). Malfeito-Ferreira, "Fine Wine Flavour Perception," 339: "Increased ethanol [= alcohol] content is known to reduce aroma complexity, increase the hotness sensation in the body, and therefore it should not be related with increased complexity, harmony, balance or flavor richness." Parker, *Parker's Wine Buyer's Guide*, 1084, refers to wine's "pleasure-giving qualities." An Internet search in 2024 under "10 top wines of the world" led to a list of vintages ranging from 13 to 15% alcohol by volume; see https://top100.winespectator.com/2024/wine/wine-no-10/.
28. See Malfeito-Ferreira, "Fine Wine Flavour Perception," 333–342; Malfeito-Ferreira, "Fine Wine Recognition," 3; Parr, "Demystifying Wine Tasting," 230; Shepherd, *Neuroenology*, 162; and Morrot et al., "Color of Odors," 2. Neuroscientist Sam Lyons, quoted by Liu, "Science of Tasting," 2: "The most important aspect of your flavor experience is actually smell—retronasal smell, that is. Retronasal smell occurs when the flavor molecules in the wine are pushed through the back of your mouth and up to your olfactory receptors, which then send this information to the smell parts of your brain.... Despite often being confused with taste, retronasal smell is the primary and most important contributor to flavor."
29. See Charters and Pettigrew, "Dimensions of Wine Quality," 1004; and Peynaud, *Taste of Wine*, 220.
30. Peynaud, *Taste of Wine*, 252: "If I had to define the art of drinking, I would say that it conforms to two rules: moderation and good taste, which can be summed up in two simple formulas: 'Drink little, but drink well,' or else 'Drink little so that you can continue to drink for a long time.'" For other examples, see Shepherd, "Neuroenology,"

5, and Asimov, "In Defense of Wine," D2. Neuroscience supports the view that alcohol is not the primary cause of wine enjoyment. Magnetic resonance imaging (MRI) scans of people drinking wine reveal increased brain activity when alcohol levels are low and unchanged activity when high—see Frost et al., "What Can the Brain Teach Us?"

31. Charlet et al., "Dopamine System," 464–465, note that alcohol-induced dopamine release does not seem to mediate the experience of pleasure, "which is controlled by opioid and also potentially endocannabinoid and gamma-aminobutyric acid (GABA)-benzodiazepine neurotransmitter systems. One hedonic hotspot for opioidergic enhancement was located in the nucleus accumbens. Animal and human studies showed that alcohol stimulates endorphins, which act on mu-opiate receptors in the nucleus accumbens and also stimulate dopamine release in the same brain area via indirect effects on GABAergic neurons."
32. The quotes are from Plumb et al., "The Nose Knows," 74, 80.
33. The quotes are from Plumb et al., "The Nose Knows," 79, who note that a small minority of participants, primarily men, enjoyed high-THC samples. Referring to a California-based competition for organic, outdoor-grown medicinal cannabis, Lastreto and Chaitanya, "Why Choose Sungrown Cannabis?," 7, write: "Once the Emerald Cup began to test the flower entries for potency, it became clear that the winner was never the cultivar with the highest THC. Often, 1st Place came in at 17% or 18% THC. What set the Top 20 winners apart was always the terpenes, the fragrance of the flower—the enchanting, arresting aromas which define the numerous varieties of cannabis while creating the subtle, more complicated highs." See also Roberts, "Emerald Cup." Judges at the Emerald Cup receive no information about the samples and evaluate them according to their appearance, aroma, taste, and overall effect.

## 16: THE FUTURE

1. Harari, *Homo Deus*, 64.
2. See Payne et al., "Cannabis and Male Fertility"; Gervasi et al., "Anandamide Capacitates Bull Spermatozoa"; and Nazaryan, "Your Weed Habit."
3. In 2023, Grinspoon, *Seeing Through the Smoke*, 157, writes: "It has not been conclusively demonstrated that cannabis is (or isn't) safe during pregnancy or breastfeeding. As such, the prudent thing to do is to presume that cannabis use, especially regular heavy cannabis use, is unsafe during pregnancy and breastfeeding until we uncover reasonable evidence that it is safe. Given what's at stake, the burden of proof is on cannabis in this case." In 2025, Lo et al., "Prenatal Cannabis Use," published a meta-analysis of fifty-one studies and reported that prenatal cannabis use was associated with a 52% increased risk of preterm delivery and a 75% increased risk of low birth weight. They rated the reliability of the evidence as "moderate," an improvement from their review published one year earlier, which rated its quality as being of "very low or low certainty." For commonly used intoxicants such as cannabis, alcohol, and tobacco, the risks are dose-dependent, and medical guidance recommends complete abstinence from all three to minimize the risk of adverse birth outcomes. See also Pearlson, *Weed Science*, 206–207; Kozak et al., "Reimagining Research with Pregnant Women"; Wymore et al., "Persistence of Delta-9-Tetrahydrocannabinol"; and Holdsworth et al., "Human Milk Cannabinoid Concentrations."

4. For the recommendations of abstinence, see Nazaryan, "Your Weed Habit," and Grinspoon, *Seeing Through the Smoke*, 157, quoted above.
5. UNODC, *World Drug Report 2022*, 16, estimates that 209 million people used cannabis in 2021.
6. A *Forbes* report in 2025 projected the global legal cannabis market to reach $82.3 billion by 2027—see Robert Hoban, "The Global Trade of Cannabis: A Growing Commodity Market," February 16, 2025, https://www.forbes.com/sites/roberthoban/2025/02/16/the-global-trade-of-cannabis-a-growing-commodity-market/. According to Grand View Research, "Legal Marijuana Market Size, Share & Trends Analysis by Application (Medical, Adult Use), by Product Types (Flower, Oils and Tinctures), by Region, and Segment Forecasts, 2024–2030," available at www.grandviewresearch.com, the size of the global legal cannabis market was US$26 billion in 2024. Its revenue forecast for 2030 is US$102.2 billion, with a compound annual growth rate of 25.7%.
7. Zohaib Ahmed, "Nano THC Drink: The Next Frontier in Cannabis," April 9, 2024, https://nanohemptechlabs.com/nano-thc-drink-the-next-frontier-in-cannabis-beverages/.
8. See Bruni et al., "Cannabinoid Delivery Systems," 15; and Hobbs et al., "Evaluation of Pharmacokinetics."
9. Bidwell et al., "Naturalistic Study," found that individuals who used cannabis at least once a week reported an average smoked THC dose of 51 mg. Prince et al., "Quantifying Cannabis," found that daily users smoked joints containing an average of 0.58 grams of herbal cannabis per use occasion. Assuming a THC potency of 20%, this yields a dose of about 116 mg THC, although many consumers do not finish an entire joint in one sitting. The rough estimate cited in the main text is based on these data. The first quote is from Tilray Brands, "Good Supply Cannabis Brand Launches Canada's Strongest Infused Pre-Rolls," news release, February 22, 2023, https://tilray.gcs-web.com/news-releases/news-release-details/good-supply-cannabis-brand-launches-canadas-strongest-infused. Good Supply is one of Tilray Brands' cannabis brands. The second quote is from Tilray Brands, "Good Supply Introduces New Monsters Resin Vapes: Full Spectrum Highs," news release, February 20, 2025, https://ir.tilray.com/news-releases/news-release-details/good-supply-introduces-new-monsters-resin-vapes-full-spectrum/. This company reported a net revenue of $272.8 million for cannabis products in 2024.
10. See Baker et al., "Race for All-Powerful Pot," on sales to high-dose users; and Twohey et al., "As America's Marijuana Use Grows." See Grinspoon, *Seeing Through the Smoke*, for a discussion on the challenges of determining how many people suffer from cannabis use disorder; and Pearlson, *Weed Science*, 288, on the unknown long-term consequences of prolonged, daily intake of high doses of THC.
11. The first quote is from Tilray Brands, "Tilray Brands Enhances Global Cannabis Supply Chain," news release, February 10, 2025, https://ir.tilray.com/news-releases/news-release-details/tilray-brands-enhances-global-cannabis-supply-chain. The second quote is from Will Yakowicz, "Inside Curaleaf Billionaire Boris Jordan's Hunger to Become the 'Frito-Lay' of Cannabis," *Forbes*, November 27, 2020, https://www.forbes.com/sites/willyakowicz/2020/11/27/inside-curaleaf-billionaire-boris-jordans-hunger-to-become-the-frito-lay-of-cannabis/.
12. Holland, *Weed*, 211.
13. See Marshall and Leeuwenberg, *Weed of Wonder*, 250–253.

14. See Stoa, *Craft Weed.*
15. See TG Branfalt, "California Allows Regional Designation for Cannabis Marketing," *Ganjapreneur*, October 14, 2024, https://ganjapreneur.com/california-allows-regional-designation-for-cannabis-marketing/; and Stoa, *Craft Weed*, 15.
16. See Chouvy, "Why the Concept of Terroir Matters," and Celik et al., "Effects of Terroir."
17. See Holland, *Weed*, 192; Lastreto and Chaitanya, "Why Choose Sungrown Cannabis?"; and Zandkarimi et al., "Comparison of the Cannabinoid and Terpene Profiles."
18. See Chouvy, "Why the Concept of Terroir Matters"; Stoa, *Craft Weed*, 128; and Kathleen Hearons, "Cannabis Terroir: Weed's Induction into High Society Is a Win for Everyone," *Head*, November 5, 2023, https://headmagazine.com/cannabis-terroir-weeds-induction-into-high-society-is-a-win-for-everyone/.
19. See Chouvy, "Why the Concept of Terroir Matters"; and Stoa, *Craft Weed*, 193.
20. See Pardal et al., *Alternatives to Profit-Maximising Commercial Models*, 18–50. Some authors refer to recreational cannabis use as "adult-use."
21. See Pardal et al., *Alternatives to Profit-Maximising Commercial Models*, 18, 44–45.
22. See Manthey et al., "Germany's Cannabis Act," and Jörg Lips et al., "Cannabis Law," October 4, 2024, available at www.cms.law.
23. The German law is known as the "Gesetz zum kontrollierten Umgang mit Cannabis und zur Änderung weiterer Vorschriften (Cannabisgesetz—CanG)," https://www.recht.bund.de/bgbl/1/2024/109/VO.html. The law defines "within sight" based on the understanding that "visibility cannot be reliably assumed beyond 100 meters from the entrance area of the specified facilities" (p. 7). According to Wolfgang Ranft, "Mega-Geldstrafe fürs Kiffen auf dem Oktoberfest," *Bild*, July 16, 2024, https://www.bild.de/politik/inland/5000-euro-strafe-fuer-kiffen-auf-oktoberfest-669524694c0a7158c79bf2b8, Bavarian authorities impose a fine of €1,500 for a first cannabis offense at Oktoberfest and a fine of €5,000 for each subsequent offense committed on the festival grounds—(€1 is more or less equivalent to US$1).
24. For a discussion of the law's inconsistencies, see Jurawelt, a website that serves as a platform for German legal professionals; for example, they write: "The threat of up to three years of imprisonment for producing edibles contradicts the legislator's stated goal of decriminalizing cannabis use as a socially accepted practice" (see "Sind THC Edibles legal?—Rechtliches und Wissenswertes," https://jurawelt.com/sind-thc-edibles-legal-rechtliches-und-wissenswertes/). Regarding the average yield of *Cannabis* plants, few scientific publications consider the question on a per-plant basis; instead, most estimates are established relative to cultivation surface or available light energy. Still, according to estimates by Chuck, "How Much Weed Does One Plant Yield? A Dosaline Case Study," *Futureharvest,* March 19, 2025, https://blog/futureharvest.com/blogs/news/how-much-weed-does-one-plant-yield/, a beginner can expect to harvest on average 42 to 56 grams from a well-maintained indoor plant, while an experienced grower can achieve up to 453 grams. Outdoor plants yield on average 482 grams but can produce up to 992 grams or more under ideal conditions.
25. The quote is from Manthey et al., "Germany's Cannabis Act," 6.
26. See UNODC, *World Drug Report 2022*, 13, 32; ERA Economics, *California Cannabis Market Outlook, 2024 Report*, 13, https://www.cannabisbusinesstimes.com/home/document/15738914/california-cannabis-market-outlook-2024-report; and PsychePen, "Cannabis in Germany in 2024: Why 60% Still Use Illegal Sources?," *Cannadelics*,

https://cannadelics.com/2024/09/17/cannabis-in-germany-2024-why-60-still-use-illegal-sources/.

27. Harari, *Homo Deus*, 396, writes in 2016: "Instead of narrowing our horizons by forecasting a single definitive scenario, this book aims to broaden our horizons and make us aware of a much wider spectrum of options. As I have repeatedly emphasized, nobody really knows what the job market, the family or the ecology will look like in 2050, or what religions, economic systems or political structures will dominate the world." Taleb, *The Black Swan: The Impact of the Highly Improbable*, 136, reflects on "how much of the future lies beyond our abilities" and proposes that there are "structural, built-in limits to the enterprise of predicting."

28. By 2025, nine countries had legalized recreational cannabis to varying extents: Uruguay, Georgia, South Africa, Canada, Mexico, Malta, Thailand, Luxembourg, and Germany (listed in chronological order of effective date). Additionally, legalization had taken place in twenty-four states, three territories, and the District of Columbia in the United States, as well as in one Australian state.

29. Harari's quote appears as an epigraph at the beginning of this chapter. Gidley, *The Future*, 117, suggests that the transition "from the obsolete industrial era mindset to an emerging solar age" represents a path to a more sustainable economic future. For more on the understudied functions of cannabinoids, see Stack et al., "Cannabinoids Function in Defense Against Chewing Herbivores," 1.

30. See Citti et al., "Novel Phytocannabinoid." They write: "Up to now, almost 150 phytocannabinoids have been detected in the cannabis plant, though most of them have neither been isolated nor characterized" (p. 7). In the living plant, THCP is primarily found in its acidic form, THCPA, which is not psychoactive and must be decarboxylated, similar to THCA, to become active. In the human body, THCP binds to CB1 receptors with over thirty times the affinity of THC, although this does not necessarily translate to equivalent potency. To date, the effects of this substance in humans remain poorly understood. See also Elizabeth Johnson, "How Do You Get THCP?," *Blimburn*, 2025, https://blimburnseeds.com/blog/tips-and-tricks/how-do-you-get-thcp/.

31. Gidley, *The Future*, 2, writes that "the single, predictable, fixed future that the trend modelling proposes does not actually exist. Instead, what is out there is a multitude of possible futures." She argues that outcomes ultimately depend on the actions people take between now and then, which remain unknown (pp. 3, 54–56). The quote is from Zohaib Ahmed, "Nano THC Drink: The Next Frontier in Cannabis Beverages," April 9, 2024, https://nanohemptechlabs.com/nano-thc-drink-the-next-frontier-in-cannabis-beverages/.

32. See Gülck and Moller, "Phytocannabinoids," 992, on the fatty-acid derived precursors of the plant's cannabinoids.

33. Abel, *Marihuana*, vii, writes in 1982: "Marihuana is undoubtedly a herb that has been many things to many people"—and lists as examples cordage, canvas, cloth, lamp oil, painkiller, aphrodisiac, and drug to ease sorrow. Marshall and Leeuwenberg, *Weed of Wonder*, 253, write in 2021: "That is the thing about the weed of wonder; it was, is, and will always be many things to many people. Whatever the future brings for our enduring relationship with cannabis, it will never be dull."

34. Small, *Cannabis*, 6, compares rewilded *Cannabis* and horses, as quoted in note 13 of chapter 11. From 1982 to the present, the Drug Enforcement Administration (DEA)'s nationwide Domestic Cannabis Eradication/Suppression Program (DCE/SP) has

destroyed more than 4.7 billion wild *Cannabis* plants across the United States—see *Vote Hemp*, "Billions of Wild Drug-Free Hemp Plants Eradicated by DEA in Effort to Confiscate Cultivated Marijuana Since 1984," December 26, 2006, https://www.votehemp.com/press_releases/billions-of-wild-drug-free-hemp-plants-eradicated-by-dea-in-effort-to-confiscate-cultivated-marijuana-since-1984/. This figure includes the 3.18 billion plants eradicated between 1982 and 1998 (see note 3 of chapter 11 for details) and reflects the program's continued activity in the decades since. Despite these sweeping and sustained efforts, feral *Cannabis* populations continue to thrive across many parts of the central U.S., underscoring the plant's resilience (see Aina et al., "Genetic Diversity," 7, 9). Since 2007, the DEA has ceased reporting ditchweed seizures, and annual eradication totals have dropped sharply as a result. In 2024, for instance, the DCE/SP reported the eradication of 5.28 million "illegally cultivated marijuana plants"—see Paul Armentano, "So Where Did All the Ditchweed Go?," *NORML*, August 5, 2008, https://norml.org/blog/2008/08/05/so-where-did-all-the-ditchweed-go/; and DEA, "Domestic Cannabis Eradication/Suppression Program," 2025, https://www.dea.gov/operations/eradication-program. That figure is nearly one hundred times smaller than the total reported in 1994, when the program claimed to have eradicated 508 million plants, including 504 million wild ones—see United States General Accounting Office, "Drug Control," 47, for the 1994 figures. Mehmedic et al., "Potency Trends of Delta9-THC," 1215, report an average of 0.4% THC in domestic ditchweed samples seized in the U.S. between 1993 and 2008. According to Small, *Cannabis*, 9, "almost all wild-growing plants in North America cannot produce intoxication." Johnson, "American Weed," 10, notes: "Uncultivated cannabis produces very little THC, the major psychoactive compound in marijuana. The hemp produced on Midwestern farms in the 1940s had only trace amounts of it, insufficient for a buzz. Still, it is possible that the escaped progeny of those hemp plants expressed natural genetic variation and began producing small amounts of THC."

35. See Quammen, "Planet of Weeds: Tallying the Losses of Earth's Animals and Plants"; World Wildlife Fund, *Living Planet Report 2024*; Keck et al., "Global Human Impact"; McKeon et al., "Human Land Use"; and Trinity College Dublin, "The Weeds Shall Inherit the Earth," July 4, 2023, https://www.tcd.ie/news_events/articles/2023/the-weeds-shall-inherit-the-earth/.
36. Quammen, "Planet of Weeds," 68.
37. The quotes are from Quammen, "Planet of Weeds," 68.
38. See Mizumoto and Bourguignon, "Evolution of Body Size"; Evangelista et al., "Fossil Calibrations"; and Burke et al., "Phylogenomics and Plastome Evolution," 7.
39. Kimmerer, *Braiding Sweetgrass*, 128, 334.

## CONCLUSION: KNOWLEDGE FOR PEOPLE, FREEDOM FOR PLANTS

1. See Marder, *Philosopher's Plant*, 110; Gagliano, "Breaking the Silence," 90; Raguso and Kessler, "Speaking in Chemical Tongues," 28; Chamovitz, *What a Plant Knows*; and Frongia et al., "Sound Perception."
2. Marder, "What Is Plant-Thinking?," discusses the "non-conscious intentionality" of plants and describes a plant's turning and striving toward the sun as an iconic illustration of its "act of intending" (p. 128). See also Marder, *Philosopher's Plant*, 118; Taiz et al., "Plants Neither Possess nor Require Consciousness"; and Mallatt et al., "Debunking

a Myth." Minorsky, "'Plant Neurobiology' Revolution," gives an informed overview of current disagreements among scientists on how to conceptualize plant behavior.

3. See Small and Naraine, "Expansion of Female Sex." Small, *Cannabis*, 55–56: "Because *C. sativa* produces enormous amounts of pollen, the stigmas of female flowers normally are pollinated, and this stops them from growing longer than about 3 mm. By contrast, illicit and medicinal marijuana are usually produced in the absence of pollen. The style-stigma parts of such virgin pistils expand notably in length (averaging over 8 mm) so that the female flowers have extremely prominent stigmas. Frequently, the stigmas also increase greatly in diameter and occasionally develop more than the normal two branches." The definition of the verb "tend" is from *The Concise Oxford Dictionary* (10th ed., 1999), 1475.
4. See Small and Naraine, "Expansion of Female Sex Organs." Marder, "What Is Plant-Thinking?," 130: "The reproductive intentionality of the plant is of course to 'reproduce its kind' not for itself but for the species it belongs to." See also Cisternas-Fuentes and Koski, "Effective Population Size"; and Hughes et al., "Ecological Consequences."
5. The quote is from Laursen, "Cultivation of Weed," S4. See also Clarke and Merlin, "Cannabis Domestication," 305.
6. Clarke and Merlin, *Cannabis*, 131: "Indoor, artificial light, and glasshouse marijuana crops are now most commonly grown from vegetatively reproduced hybrid cuttings, and seeds are rarely used except to grow replacement cutting stock. This limits crop improvement through selective breeding as seeds are rarely used or produced, and as a result, sexual reproduction stops, and evolution ceases or certainly slows down dramatically." Clarke and Merlin, "Cannabis Domestication," 318: "The past 50 years have seen the genetic diversity of the *Cannabis* genome dwindle away. Indeed, the vast majority of landraces may already be extinct, and we therefore must be careful to preserve and multiply what remains." Aina et al., "Genetic Diversity," 2: "Escaped, naturalized, and regionally adapted feral cannabis populations that exist across the US represent a potential source of untapped genetic diversity. These populations potentially harbor a wide range of genetic variation lost in cultivated crops, providing opportunities to introduce new traits for improved crop performance." See also Torkamaneh and Jones, "Cannabis, the Multibillion Dollar Plant."
7. Clarke and Merlin, *Cannabis*, 371: "No other drug-producing organism can be so readily cultivated in such a wide variety of natural as well as artificial environments—not the opium poppy, coca bush, tobacco plant, nor even psychoactive fungi."
8. Kessler et al., "Changing Pollinators," discuss a species of wild tobacco (*Nicotiana attenuata*) that attracts and rewards pollinators by emitting a bouquet of floral volatiles and offering sugar-rich nectar. It also varies the timing of these emissions and offerings to favor specific pollinators and to exclude others. There are more than seventy species of tobacco, all of which appear to produce nicotine. For more on tobacco and nicotine, see, for example, Narby and Chanchari, *Plant Teachers*, 19–36.
9. See chapter 10; and Rendon, *Super-Charged*, 85.
10. See Stack et al., "Cannabinoids Function in Defense Against Chewing Herbivores," who could not include THCA or THC in their study due to federal regulations. Consequently, the repellent properties of these two compounds, as they affect the living plant, remain undetermined. See Silver, "The Endocannabinoid System of Animals," on the absence of cannabinoid receptors in insects. See also Rami et al., "Overview of Plant Secondary Metabolites," and chapter 10.

11. See Clarke and Merlin, *Cannabis*, 36, 51, 302.
12. See Hines et al., "High-Potency Cannabis Use," 1045, on dose-dependent health problems. As pointed out in 2004 by Dolphin, "Measuring Marijuana's Potency," 2:97: "Some people prefer pot with moderate potency because it's easier to control just how high one gets. Once a basic level of THC potency is reached, the flavor and the experience of the high are often more important than how many tokes it takes to get high." UNODC, *World Drug Report 2006*, 155: "A global blind spot has developed around cannabis, and in this murk, the plant itself has been transformed into something far more potent than in the past. Suddenly, the mental health impact of cannabis use has been thrown into sharp relief, and the drug with which the world has felt so familiar seems strange once again." Twohey et al., "As America's Marijuana Use Grows," A1: "A $33 billion industry has taken root, turning out an ever-expanding range of cannabis products so intoxicating they bear little resemblance to the marijuana available a generation ago. Tens of millions of Americans use the drug for medical or recreational purposes—most of them without problems. But with more people consuming more potent cannabis more often, a growing number, mostly chronic users, are enduring serious health consequences."
13. Small, *Cannabis*, 354: "Compared to most field crops, a relatively small outdoor area is needed to satisfy the needs of a large consumer population." UNODC, *World Drug Report 2006*, 155, 194: "Moreover, cannabis is both easy to grow and highly productive, yielding a large quantity of ready-to-use drug per plant. . . . Today, there are a variety of cultivation styles in evidence in the world. In many developing countries, people simply drop seeds and return months later to collect whatever develops, a practice that is virtually cost-free and thus very difficult to deter." Wartenberg et al., "Cannabis and the Environment," 100, estimate that 460 hectares of cannabis produce enough to furnish the California market, representing only a fraction of the state's 9.8 million hectares of agricultural lands. Gieringer, "Economics of Cannabis Legalization," 2, first compared cannabis to other leaf crops in terms of price. At present, a kilo of cannabis flowers costs an average of about $2,000 on the world market, with significant variation by region, quality, and market segment. In comparison, a kilo of black tea costs an average of $6.

# Index

*Note*: Page references following "n" refer to notes.

# ABOUT THE AUTHOR

Robert Siegenthaler

**Jeremy Narby**, PhD, studied history at the University of Kent and earned a doctorate in anthropology from Stanford University. Since 1990, he has coordinated projects supporting Amazonian communities through the Swiss NGO Nouvelle Planète, focusing on land rights, bilingual education, and sustainable forestry. He is the author of *The Cosmic Serpent: DNA and the Origins of Knowledge* (1998) and *Intelligence in Nature: An Inquiry into Knowledge* (2005), and coeditor with Francis Huxley of *Shamans Through Time: 500 Years on the Path to Knowledge* (2001). His most recent work, coauthored with Rafael Chanchari Pizuri, is *Plant Teachers: Ayahuasca, Tobacco, and the Pursuit of Knowledge* (2021).